ORGANOMETALLIC MODELING OF THE HYDRODESULFURIZATION AND HYDRODENITROGENATION REACTIONS

Catalysis by Metal Complexes

Volume 24

Editors:

B. R. James, *The University of British Columbia, Vancouver, Canada*
P. W. N. M. van Leeuwen, *University of Amsterdam, The Netherlands*

Advisory Board:

The titles published in this series are listed at the end of this volume.

ORGANOMETALLIC MODELING OF THE HYDRODESULFURIZATION AND HYDRODENITROGENATION REACTIONS

by

Roberto A. Sánchez-Delgado

Instituto Venezolano de Investigaciones Científicas (IVIC),
Caracas, Venezuela

SPRINGER-SCIENCE+BUSINESS MEDIA, B.V.

A C.I.P. Catalogue record for this book is available from the Library of Congress.

ISBN 978-94-017-4205-4   ISBN 978-0-306-47619-8 (eBook)
DOI 10.1007/978-0-306-47619-8

Printed on acid-free paper

To Vicky, Eugenia, and Alexander

CONTENTS

PREFACE   xi

CHAPTER 1.
HYDRODESULFURIZATION AND HYDRODENITROGENATION

1.1. Introduction   1

1.2. The Hydrodesulfurization Reaction   3
      1.2.1. SULFUR COMPOUNDS IN PETROLEUM AND FUELS   3
      1.2.2. HDS CATALYSTS   5
      1.2.3. REACTION MECHANISMS   7
      1.2.4. THE ACTIVE SITES   8
          1.2.4.1. The Co-Mo-S model   9
          1.2.4.2. Active sites and the concept of vacancies   12
      1.2.5. THE ACTIVATION OF HYDROGEN   14
      1.2.6. CHEMISORPTION OF ORGANOSULFUR COMPOUNDS   16
      1.2.7. HYDROGENATION AND HYDROGENOLYSIS
             OF ADSORBATES   18
          1.2.7.1. HDS of thiophenes   18
          1.2.7.2 HDS of benzothiophenes and dibenzothiophenes
          and the deep desulfurization issue   20

1.3. The Hydrodenitrogenation Reaction   25
      1.3.1. CHEMISORPTION OF ORGANONITROGEN
             COMPOUNDS   26
      1.3.2. MAIN REACTION PATHWAYS AND MECHANISMS
             IN HDN   28

References   32

CHAPTER 2.
COORDINATION AND ACTIVATION OF THIOPHENES
IN METAL COMPLEXES

2.1. Introduction   35

2.2. η^1S-Bonded metal thiophene complexes   37

2.3. η^2(C=C)-Bonded metal thiophene complexes   43

2.4. η^3(S,C=C)-Bonded metal thiophene complexes   46

2.5. η^4-Bonded metal thiophene complexes   47

2.6. η^5-Bonded metal thiophene complexes   50

**2.7. η^6-Bonded metal benzothiophene and
        dibenzothiophene complexes**   57

2.8. Conclusions and further comments   59

References   59

CHAPTER 3.
HYDROGENATION REACTIONS

3.1 Introduction   62

3.2. Homogeneous Hydrogenation of Aromatic Hydrocarbons   63
    3.2.1. THE CATALYSTS   63
    3.2.2. REACTION MECHANISMS   65
        3.2.2.1. Allyl metal complexes   66
        3.2.2.2. Cyclopentadienyl and arene metal complexes   69
        3.2.2.3. Other metal complexes   71
        3.2.2.4. Tethered complexes on supported metals   72
    3.2.3. CONCLUSIONS AND PERSPECTIVES   73

3.3 Homogeneous Hydrogenation of Heteroaromatic Hydrocarbons 75
    3.3.1. SULFUR CONTAINING HETEROAROMATICS   75
        3.3.1.1. Thiophene hydrogenation   75
        3.3.1.2. Benzothiophene hydrogenation   78
        3.3.1.3. Benzothiophene hydrogenation as a pretreatment for HDS 83
    3.3.2. NITROGEN CONTAINING HETEROAROMATICS   84
        3.3.2.1. Hydrogenation of pyridines   85
        3.3.2.2. Hydrogenation of polynuclear N-heterocycles   85

References   92

CHAPTER 4
**RING OPENING, HYDROGENOLYSIS AND DESULFURIZATION
OF THIOPHENES BY METAL COMPLEXES**

4.1. Introduction   95

4.2. Stoichiometric ring opening, hydrogenolysis,
  and desulfurization of thiophenes   96
    4.2.1. REACTIONS OF THIOPHENES
        ON MONONUCLEAR COMPLEXES   97
      4.2.1.1. Complexes with cyclopentadienyl and related ligands 97
      4.2.1.2. Complexes with phosphine ligands   101
    4.2.2. REACTIONS OF THIOPHENES
        ON DINUCLEAR AND POLYNUCLEAR COMPLEXES   114
      4.2.2.1. Homopolynuclear complexes   114
      4.2.2.2. Heteropolynuclear complexes   119
      4.2.2.3. Curtis' Co-Mo-S clusters   123

4.3. Homogeneous Catalytic Hydrogenolysis
  and Hydrodesulfurization of Thiophenes   127

4.4. Further remarks on the relations between
  homogeneous and heterogeneous hydrogenolysis
  and HDS of thiophenes   134

References   135

CHAPTER 5.
ACTIVATION OF HYDROGEN ON METAL COMPLEXES
WITH SULFIDE LIGANDS AND RELATED
COORDINATION CHEMISTRY OF H$_2$S

5.1 Introduction   138
5.2 Hydrogen activation on dinuclear complexes
  containing sulfido or disulfido ligands   140
5.3 Hydrogen activation on mononuclear Ti=S
  and Ti(η^2S$_2$) complexes   143

5.4. Reactions of H$_2$S with metal complexes   145

5.5 Conclusions and remarks   151

References   151

CHAPTER 6.
MODELING HYDRODENITROGENATION

6.1 Introduction   153

6.2 Binding Modes of N-Heterocycles in
  Transition Metal Complexes   154

6.2.1 COMPLEXES WITH PYRROLE, INDOLE
            AND CARBAZOLE LIGANDS   154
    6.2.2. COMPLEXES WITH PYRIDINE, QUINOLINE,
            AND RELATED LIGANDS   162

6.3 Reactions of N-Heterocycles in Transition Metal Complexes 168
    6.3.1. REACTIONS OF COORDINATED PYRROLES,
            INDOLES, AND RELATED COMPONDS   168
    6.3.2. REACTIONS OF COORDINATED PYRIDINES,
            QUINOLINES, AND RELATED COMPOUNDS   171
    6.3.3. HYDROGENOLYSIS OF N-HETEROCYCLES
            CATALYSED BY TRANSITION METAL COMPLEXES   173

6.4 C-N Bond Activation of Amines, Imines,
    and Amides by Transition Metal Complexes   174

6.5. Final Remarks   177

References   178

EPILOGUE   181
            Vacancies   184
            The nature and the structure of the active sites   185
            Hydrogen activation   186
            Chemisorption modes and reactivity   188
            Sulfur extrusion   190
            A general mechanism for HDS on Co-Mo-S sites   191
            Hydrodenitrogenation   194
            Concluding remarks   194

ABBREVIATIONS   196

LIST OF FIGURES   197

LIST OF TABLES   201

INDEX   202

PREFACE

The fields of hydrodesulfurization (HDS) and hydrodenitrogenation (HDN) continue to attract the attention of researchers in the various disciplines connected to these fascinating problems that represent two of the key outstanding chemical challenges for the petroleum refining industry in view of their very strong environmental and commercial implications. One area that has flourished impressively over the last 15 years is the organometallic chemistry of thiophenes and other related sulfur-containing molecules. This has become a powerful method for modeling numerous surface species and reactions implicated in HDS schemes, and nowadays it represents an attractive complement to the standard procedures of surface chemistry and heterogeneous catalysis, for understanding the complex reaction mechanisms involved in this process. Similar developments have begun to appear in connection with HDN mechanisms, although in a much more modest scale and depth.

Some years ago when, encouraged by Prof. B. R. James, this book was planned, several excellent reviews and monographs treating different aspects of HDS were already available including some on the subject of organometallic models. However, it seemed appropriate to try to summarize the most striking features of this chemistry in an updated and systematic way, and inasmuch as possible in connection with the common knowledge and beliefs of the mechanisms of heterogeneous HDS catalysis. Hopefully, this attempt to build some conceptual bridges between these two traditionally separated areas of chemistry has met with some success. The timeliness of this idea has probably increased during the period of writing and publishing of this manuscript, as evidenced by the increasing number of papers, patents and scientific meetings devoted to the subject of HDS and HDS modeling, and by the frequency with which one now sees organometallic and surface chemists referring to the each other's work in a natural way and as part of a common way of thinking toward the solution of an important environmental and industrial problem. It also seemed appropriate to try to summarize our current knowledge on the subject of organometallic modeling of the HDN reaction, since the literature available, though scarce and generally scattered, clearly indicates that this is a very important and yet a very open field for future research.

The completion of this project was considerably facilitated by being the recipient of a John Simon Guggenheim Fellowship (December 1998-January 2000) during which most of the preparation of the book was accomplished. I am most grateful to the Chemistry Department of the University of California at Berkeley for providing an excellent environment for initiating this work during the period December 1998-March 1999. Particularly to Prof. R. A. Andersen who kindly made the arrangements for my stay there, and dedicated many hours to discuss chemistry with me; to Prof. R. G. Bergman for his generous hospitality, and to many other colleagues and enthusiastic graduate students and postdocs who took some of their time to abundantly exchange information with me. I also want to express my deep gratitude to Prof. G. Parkin and his research group of the Chemistry Department of Columbia University in New York for hosting me in their labs and offering me their time and their friendship during a most productive and pleasant summer of 1999. Many fruitful and enjoyable discussions and group meetings with Prof. J. R. Norton and his students at Columbia are also gratefully acknowledged.

A number of colleagues were kind enough to provide me with their complete bibliography on HDS-related chemistry, including some unpublished work: R. J. Angelici, C. Bianchini, M. D. Curtis, R. H. Fish, J. J. García, S. Harris, W. D. Jones, M. Rakowski DuBois, L. Rincón, and Y. Aray. The innumerable lively discussions in Caracas and in Florence over more than 10 years of a very enjoyable and productive collaboration with Claudio Bianchini, Andrea Meli and the rest of the team of the ISECC-CNR deserve a very special mention, as well as the funding received from CONICIT (Venezuela) and CNR (Italy) for this purpose. R. Galiasso and D. Páez of PDVSA-INTEVEP provided a great deal of information and insight from the industrial point of view, and F. Ruette and Y. Aray at IVIC in the theoretical aspects. D. D. Whitehurst, H. Topsøe, I. Mochida, and M. Vrinat were of great help to my understanding of the current status and challenges of HDS catalysis.

The Chemistry Center of the Venezuelan Institute for Scientific Research (IVIC) is also thanked for allowing my frequent absences related to writing this book, and mostly for providing the right environment and good facilities for our own work on HDS modeling over many years. CONICIT, the Venezuelan Research Council has generously funded our research on this area, and also PDVSA-INTEVEP provided some financial assistance. Of course the many students and coworkers who have accompanied me over the years are responsible for many of the results and ideas presented here, chronologically: R.-L. Márquez-Silva, E. González, N. Valencia, V. Herrera, A. Andriollo, L. Rincón, A. Fuentes, F. López-Linares and M. Rosales. To my current and recent students, my apologies for the times when I was not available.

Finally, my deepest thanks go to my wife Vicky and my daughter Eugenia for their unlimited patience and tolerance over the many stolen days and hours of family life.

ROBERTO A. SANCHEZ-DELGADO

CHAPTER 1
HYDRODESULFURIZATION AND HYDRODENITROGENATION

1.1. Introduction

Hydrotreating is the collective name given to a series of hydrogenation reactions to which crude petroleum and various refinery streams are subjected, in order to saturate a variety of unsaturated hydrocarbons present and to remove unwanted heteroatoms such as sulfur (hydrodesulfurization, HDS), nitrogen (hydrodenitrogenation, HDN), oxygen (hydrodeoxygenation, HDO), and metals (hydrodemetalation, HDM). Such processes are routinely practiced in the refining industry as a means of converting heavy streams into lighter ones and of markedly improving the quality of the various resulting products. Also, hydrotreating is a key pretreatment of the feeds for other refinery processes, such as fluid catalytic cracking (FCC) or catalytic reforming, involved in gasoline production. These refining steps use expensive catalysts that are readily poisoned by such heteroatoms. Other alternative hydrocarbon feedstocks, such as shale oils, liquefied coals and bituminous sands are also treated in this way due to the high contents of this type of contaminants present in them.

Refinery feedstocks in general are extremely complicated chemical mixtures in which each heteroatom is present in the form of literally hundreds of different compounds. Therefore, classical hydrotreating technologies have been mainly designed on the basis of (1) the average composition and the average properties of the feeds; (2) empirical information about the characteristics and performance of the specific catalysts used in each case; (3) the different process variables and (4) the product specifications. These extensively used technologies enabled oil refineries throughout the world for many years to systematically reach a ceiling of *ca.* 90% sulfur removal and 70% nitrogen removal, corresponding to a final content of about 0.1 wt. % S and 0.5 wt. % N at typical operating conditions (25-50 atm, 330-350 °C). New severe environmental regulations call for much lower levels of heteroatom-derived pollutants, and this has stimulated an intensification of research and development efforts toward new or improved catalytic processes. A major shift is thus observed from the classical to a "deep" hydrotreating mode aimed at removing no less than 95% S and N.

This worldwide environmental pressure on fuel manufacturing has awaken a renewed interest in HDS and HDN catalysis research since the currently available technologies, which had been operating satisfactorily for several decades, are no longer adequate for the "deep refining" necessary to meet the new strict standards. Therefore a major and continued effort in this direction is urgently needed and the most varied approaches incorporating different disciplines are to be encouraged. Considering the current volumes of petroleum being processed throughout the world at this moment (several tens of millions of barrels per day), this is no doubt the largest volume

industrial application of any transition-metal catalyzed reaction. This is reflected also by
the fact that hydrotreating catalysts account for *ca.* 10% of the total world market for
catalysts. This shift represents a very complex technical challenge requiring major
breakthroughs in catalyst development that will need to go well beyond minor
improvements or modifications. With the advent of novel sophisticated analytical tools,
the key problems of hydrotreating have been better identified. The knowledge related to
feed and product composition, as well as some intimate properties of the most typical
catalyst formulations and many details of the reaction networks and reaction mechanisms
operating in these processes has advanced impressively. As a consequence, catalyst
design has become a more rational and more advanced science –and art- even if the
appropriate answers for the new challenges have not yet reached the stage of industrial
application.

A good number of reviews and monographs on the general subject of
hydrotreating, with particular emphasis in HDS are available, and the interested reader is
referred to those publications for detailed information on catalysts, reaction networks,
conditions, mechanisms, and engineering aspects [1-18]. Particularly instructive and
recent are a review by Startsev [14] on the various aspects of HDS, a comprehensive
monograph by Topsøe *et al.* [15] on the broader area of hydrotreating, and an extensive
article by Whitehurst *et al.* [16]. Also of interest are the reviews by Shafi and Hutchings
[19] and by and Landau [20] devoted mainly to the "deep hydrotreating" problems and
possible solutions.

In this Chapter, the most relevant aspects of conventional HDS and HDN
catalysis, as well as the major outstanding problems that need to be solved in the near
future, will be briefly reviewed. This will hopefully put the subject in context for
future reference in the following parts of the book, which will be devoted primarily to a
description of the organometallic chemistry of sulfur-containing molecules, particularly
of the thiophenes. The synthesis, structures and bonding, as well as the reactivity of
such complexes are included, with a strong emphasis on their pertinence –and their
possible relations- to HDS catalysis. Additionally, the somewhat less developed field of
organometallic modeling of the HDN reaction, based on the study of the structures and
reactions of metal complexes of adequate N-containing ligands, predominantly aliphatic
and aromatic amines, will also be discussed. Similar studies for the modeling of HDO
and HDM are comparatively scarce, and therefore they will not be dealt with in this
discussion.

It is hoped that summarizing this wealth of information –which has largely
become accessible within the last 15 years- in a critical way, and inasmuch as possible
in connection with the known aspects of HDS and HDN catalysis, will not
unnecessarily repeat the contents of other excellent reviews available on the
organometallic aspects of the HDS and HDN reactions [17, 21-35]. It is important to
point out at this stage that the organometallic modeling approach is to be taken as an
additional tool –complementary to the many state-of-the-art techniques that constitute
today's arsenal of surface chemistry and heterogeneous catalysis research- to stimulate
alternative ways of thinking which might help in the general understanding of the
problem and in the generation of new ideas for greatly needed developments in this
important field.

1.2. The Hydrodesulfurization Reaction

Hydrodesulfurization (HDS) is the reaction through which sulfur is removed from petroleum feedstocks in refineries by their interaction with hydrogen over solid catalysts under rather severe conditions of temperature and pressure, according to the generic transformation represented by Eq. 1.1:

$$[\text{R-S}] \ + \ \text{H}_2 \ \xrightarrow{\text{cat}} \ [\text{R-H}] \ + \ \text{H}_2\text{S} \tag{1.1}$$

This is the most important and most thoroughly studied reaction of the complex overall "hydrotreating" process; HDS in particular is mandatory as a pretreatment of the feeds for subsequent refining steps such as catalytic reforming which is carried out over expensive noble metal catalysts readily poisoned by sulfur-containing compounds. Furthermore, current and future environmental legislation imposes severe restrictions on the amounts of sulfur and other pollutants allowed in transportation fuels, which are by far the most important petroleum products and whose demand continues to grow at an alarming rate; sulfur compounds in fuels generate sulfur oxides upon combustion which are released into the atmosphere and are responsible for the generation of acid rain. Thus, for instance, it is expected that a maximum of 0.05 wt. % S will be allowed in diesel, whereas the levels in gasoline will probably have to be as low as 0.01 wt. % in Europe and 0.0003 wt. % in California not much later than the year 2000; in fact, the tendency toward "zero sulfur" fuels is rapidly becoming dominant.

1.2.1. SULFUR COMPOUNDS IN PETROLEUM AND FUELS

Petroleum is a very complex mixture consisting predominantly of hydrocarbons but containing also varying amounts of heteroatoms (depending on the origin of the crude), mainly sulfur, nitrogen, oxygen and metals (primarily vanadium and nickel). The hydrocarbons are mostly composed of paraffins, naphthenes (saturated 5- and 6-member ring structures) and mononuclear and polynuclear aromatics, usually with short alkyl substituents; olefins are not naturally present in petroleum but they are produced in considerable concentrations (up to 50 wt. %) during some of the primary refining steps, *e.g.* during fluid catalytic cracking (FCC).

The most abundant heteroatom is invariably sulfur, appearing in concentrations from below 0.1 wt. % in North African or Indonesian light crudes to over 5 wt. % in Venezuelan heavy crudes (Boscan) or Canadian tar sands. A wide variety of sulfur containing compounds are present in petroleum and refinery fractions, ranging from thiols to thiophenes; the most important classes of organosulfur compounds of interest for our purposes are represented in Fig. 1.1.

Upon hydrotreating under standard conditions, a good proportion of these species can be removed with relative ease, such as the thiols, sulfides, and disulfides, so they do not represent a major problem with today's technology, but others are more refractory, like the thiophenes, benzothiophenes, and specially the dibenzothiophenes, as a result of their aromatic character.

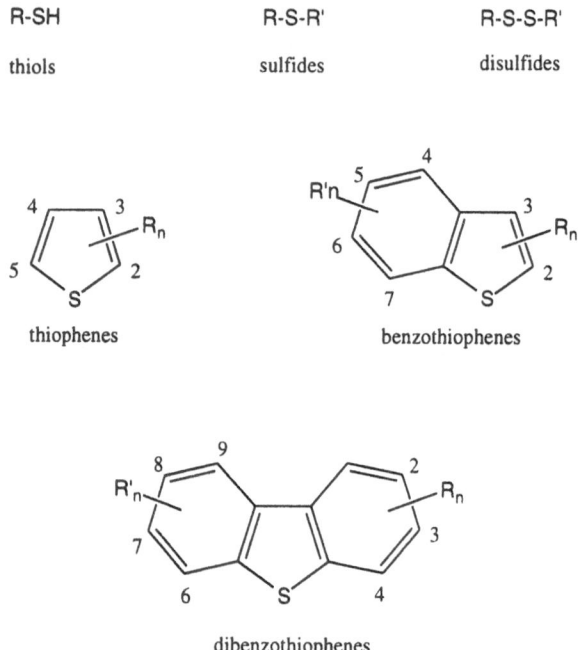

R-SH   R-S-R'   R-S-S-R'

thiols   sulfides   disulfides

thiophenes   benzothiophenes

dibenzothiophenes

Fig. 1.1. Some major types of organosulfur constituents of petroleum and distillates
(R$_n$ and R'$_n$ are short chain alkylgroups).

Of particular current interest is the fact that the most abundant residual sulfur compounds in diesel fuels are found in the form of alkyl-substituted benzothiophenes and dibenzothiophenes, 4-methyldibenzothiophene and 4,6-dimethyldibenzothiophene being the prototypes of the most difficult compounds to degrade. The removal of these very stable molecules through what has come to be known as "deep desulfurization" in order to meet the severe new standards for sulfur contents in fuels is certainly a most difficult task and a satisfactory solution for this problem has not yet been found [14-16, 19, 20].

Consequently, the organometallic chemistry which has evolved in connection with HDS over the last 15 years has not surprisingly concentrated essentially on thiophenes, benzothiophenes and dibenzothiophenes as ligands; the description and discussion of the binding and reactivity of such organosulfur compounds on the metal centers of discrete molecular complexes, with particular emphasis on hydrogenation, hydrogenolysis and C-S bond breaking reactions, form the main body of this book, and whenever possible, interesting connections between molecular and surface chemistry related to HDS will be pointed out throughout the text.

Similar considerations for metal complexes of other types of petroleum constituents (*e.g.* thiols, polynuclear aromatics) have been addressed by some researchers and such metal-containing derivatives have been used successfully in modeling and in mechanistic studies [24, 36]; therefore some illustrative and instructive examples of such work have also been included at the appropriate sections. On the other hand, the organometallic chemistry of alkyl substituted dibenzothiophenes, which is of most current interest, has only very recently begun to emerge in the literature, as described in subsequent Chapters, but it remains largely an open and very attractive area for further research.

1.2.2. HDS CATALYSTS

There are many possible formulations for HDS catalysts, and the selection in any one refinery is usually adapted to both the feed to be processed and to the specifications of the corresponding products. However, the most commonly used catalysts are those referred to as "promoted Mo or W catalysts" which are composed of two metals, predominantly Co-Mo, supported on alumina; other common combinations include Ni-Mo, Ni-W, and less frequently, Co-W. The Co-Mo catalysts are excellent for HDS and less active for HDN and hydrogenation reactions which are better performed over Ni-Mo or the more expensive Ni-W (see Section 1.3 below).

Sometimes silica-alumina, silica, or other common supports are employed instead of alumina; not infrequently the catalysts are modified by addition of phosphorus which is known to increase the selectivity for desulfurization $vs.$ hydrogenation. The cobalt (or nickel) content is commonly 1-5 wt. % whereas molybdenum is present in 8-15 wt. % (and tungsten in 12-25 wt. %). In order to reach maximum activities, the catalysts are calcined at temperatures of 400-600 °C to ensure complete decomposition of the metal salts employed in their preparation, and subsequently sulfided prior to use or alternatively, in the early stages of catalysis through exposure to either H_2S mixed into the hydrogen stream, or to the sulfur-rich feed itself. The operating conditions for industrial HDS reactions range between 300-450° C and 10-250 atm H_2 depending on the particular feed being treated.

Many other metals have been shown to be active in HDS catalysis, and a number of papers have been published on the study of periodic trends in activities for transition metal sulfides [15, 37-43]. Both pure metal sulfides and supported metal sulfides have been considered and experimental studies indicate that the HDS activities for the desulfurization of dibenzothiophene [37] or of thiophene [38, 39] are related to the position of the metal in the periodic table, as exemplified in Fig. 1.2 (a), 1.2 (b), and 1.2 (c). Although minor differences can be observed from one study to another, all of them agree in that second and third row metals display a characteristic volcano-type dependence of the activity on the periodic position, and they are considerably more active than their first row counterparts. Maximum activities were invariably found around Ru, Os, Rh, Ir, and this will be important when considering organometallic chemistry related to HDS, since a good proportion of that work has been concerned with Ru, Rh, and Ir complexes, which are therefore reasonable models in this sense; however, Pt and Ni complexes have also been recently shown to promote the very mild stoichiometric activation and desulfurization of substituted dibenzothiophenes (See Chapter 4).

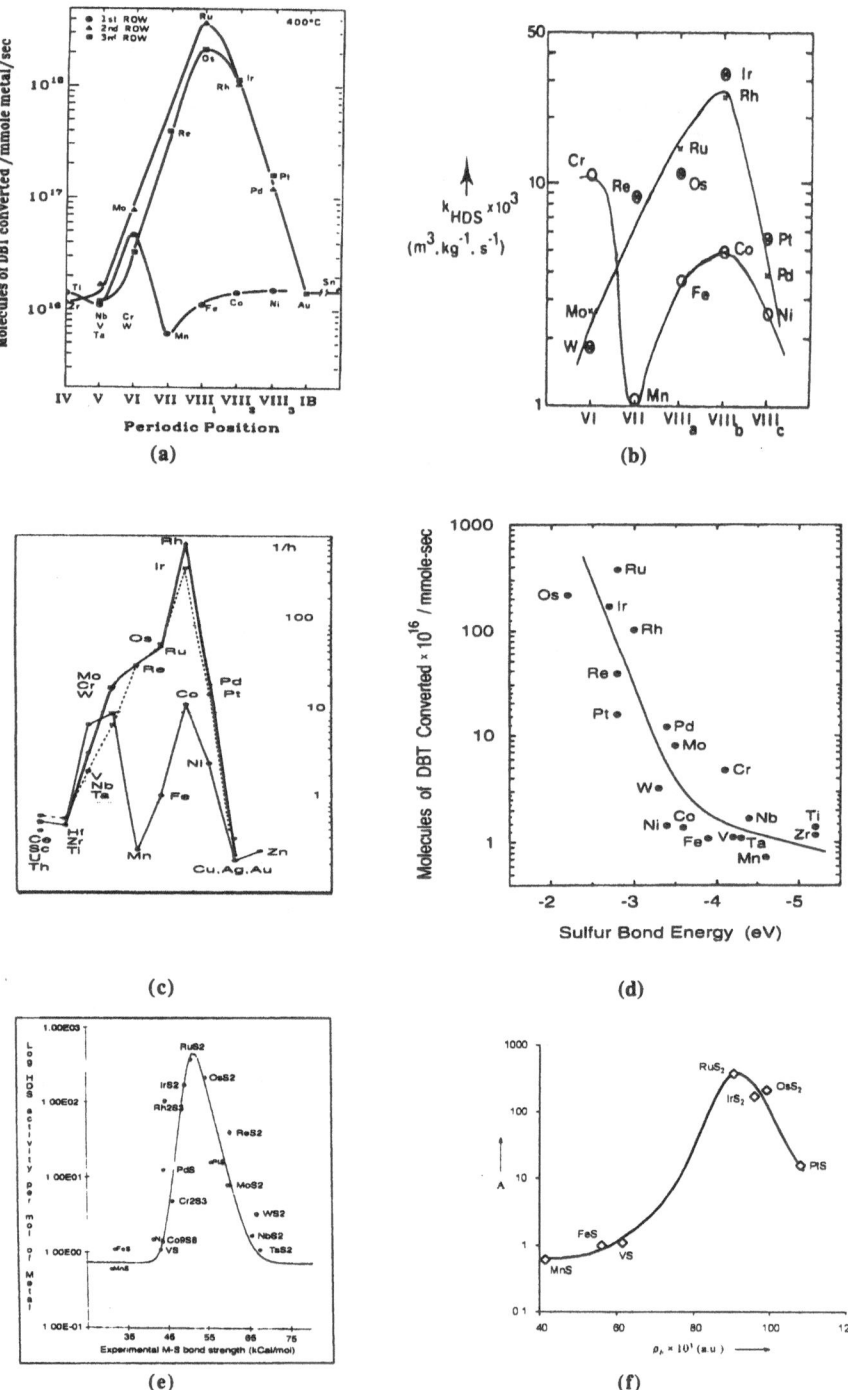

Fig. 1.2. Trends in HDS activity for dibenzothiophene (a, d-f) and thiophene (b,c). (reproduced from Refs. (a) 37, (b) 38, (c) 39, (e) 41, (f) 42).

These periodic trends have been explained in several ways by different authors, always in terms of the strength of the metal sulfur bond. Two different models for assessing the M-S bond strength have been considered: Topsøe et al. use a model which calculates heats of formation on a sulfur atom basis instead of a metal atom basis, obtaining the M-S bond energy from the bulk modulus of the metal and the degree of filling of the metal-sulfide d band (calculated by ab initio methods). According to this model, the trend in HDS activities closely corresponds with trends in M-S bond energies, no volcano-type behavior is encountered but rather the largest activities correspond with those metal sulfides with the least stable surface sulfides (Fig. 1.2 (d)) [15]. In the alternative model developed by Toulhoat et al. [40, 41], the M-S bond strength is estimated as the bulk cohesive energy (experimentally determined or calculated by ab initio methods) per formula unit, divided by the number of metal-sulfur bonds; this leads to a typical volcano plot for the correlation between HDS activity and metal-sulfur bond strength (Fig. 1.2 (e)). This suggests the idea that maximum HDS activity is related to an optimum metal-sulfur bond strength. Recent results obtained by Aray et al. [42, 43] by use of the topological theory of Bader also disclose a volcano type relation (Fig. 1.2 (f)), indicating an optimum value of M-S bond strength for maximum activity, in parallel with experimental results. This is also in accord with Sabatier's principle which states that solids whose heats of formation and hence, whose metal sulfur bond strengths are either too large or too small are practically inactive, while those with intermediate M-S bond strengths are active. In any case, both models are linked to the idea that since catalysis takes place on coordinatively unsaturated sites the HDS activity is directly related to the number of "vacancies" present on the surface, which is in turn determined by the M-S bond energy. This concept of "sulfur vacancies" is further discussed in Sections 1.2.4.2 (p. 12) and 4.2.2.3 (p. 123).

1.2.3. REACTION MECHANISMS

A complicated set of exothermic reactions takes place upon hydrotreating petroleum or petroleum fractions, the main classes of interest here including hydrogenation of aromatics, hydrodesulfurization and hydrodenitrogenation. A large amount of work has been devoted to the study of these reaction pathways, their kinetics, product distributions, intermediates, and reaction mechanisms, both on real feeds and -mainly- on model compounds or mixtures of model compounds. Thermodynamics of the various possible reactions under standard catalytic conditions are known to control to a large extent the relative rates at which they take place. It is also known that the various reactants can compete for the active sites, some of them thereby acting as poisons for the degradation of others [44, 45]. The best understood reaction within the framework of hydrotreating is no doubt hydrodesulfurization, particularly of thiophene which has been extensively used as a model compound. The knowledge of similar reaction schemes for the higher homologues, mainly the benzothiophenes and dibenzothiophenes has also advanced considerably in recent years, as has the understanding of the mechanisms of hydrodenitrogenation (HDN) discussed in Section 1.3.2, and hydrodemetalation (HDM), which falls outside the scope of this book.

Even for the simplest substrates, HDS networks are very complex, since they involve a considerable number of elementary steps -predominantly adsorption/desorption equilibria and hydrogenation and hydrogenolysis reactions- whose kinetics are strongly dependent on the reaction conditions as well as on the structure of the particular substrate being considered and/or on the substituents present in such molecules. Of course the composition of the catalyst, and the nature and concentration of active sites are additional key features to be considered for each set of reactions. To further complicate matters, studies by different authors on the same substrate (*e.g.* thiophene) over similar catalysts and approximately the same reaction conditions have at times produced different -if not contradictory- results. This extensive literature is available in specialized monographs and reviews, and it continues to be the subject of an interesting and lively debate [1-20]. There is, however, a general agreement in that, whatever detailed mechanism is invoked, some common fundamental steps need to be taken into account, namely:

-The generation and the nature of the active sites
-The dissociative adsorption of hydrogen on the surface of the catalyst
-The chemisorption of the organosulfur compound on the catalytic sites
-The reactions of the adsorbates: hydrogenation of unsaturated bonds and hydrogenolysis of C-S bonds

In the following sections, a brief account of the state of knowledge for each one of these important points will be presented.

1.2.4. THE ACTIVE SITES

The characterization of HDS catalysts has been the subject of a large number of papers, and virtually all the surface techniques and analytical tools available today, as well as powerful theoretical methods, have been extensively employed in order to tackle this exceedingly complicated problem [see *e.g.* ref. 15]. The mass of information thus obtained has been interpreted in terms of several different models that have been evolving over the years into a rather sophisticated and well founded picture; however, in spite of all the data available and of over seven decades of industrial practice, the exact nature and the structure of the catalytically active HDS sites of standard catalyst formulations continue to be the subject of controversy and frequent speculation. A great deal of the published work in this area has been devoted to the study of unpromoted catalysts in both calcined and sulfided forms, and this has resulted in the clarification of several important aspects; nevertheless, for the sake of brevity, our description will concentrate essentially on the promoted Co-Mo catalysts in their sulfided forms, which are the ones most frequently used for practical purposes. Many excellent reviews widely cover the various theories and models which have been put forward for HDS active sites (see *e.g.* refs. 14, 15, and references therein) and thus there is no need to repeat that information at length here.

1.2.4.1. The Co-Mo-S model

The most widely accepted model for the structure of promoted HDS catalysts is nowadays the one advanced by Topsøe in terms of a *Co-Mo-S phase*. This proposal was initially based on very elegant work involving *in situ* Mössbauer emission spectroscopy, EXAFS and infrared spectroscopy studies of working catalysts; moreover, further evidence previously and subsequently disclosed by several research groups is essentially in line with this way of thinking [15]. As can be seen from the schematic representation shown in Fig. 1.3, the catalyst is thought to be composed of single-sheet MoS$_2$-like crystallites supported on the alumina surface and "decorated" by the promoter atoms (Co, Ni) at the (10$\bar{1}$0) edge planes. This Co-Mo-S structure is not a unique species but instead it should be regarded as a family with varying Co concentrations ranging from zero to full edge coverage (which has been estimated to take place when Co/(Co + Mo) = 0.3). Other cobalt phases such as Co$_9$S$_8$ and Co in the alumina lattice are also known to be present in the catalysts, but the combined activity/spectroscopy results clearly indicate that most of the catalytic activity is associated with the promoter atoms in the Co-Mo-S phase, and this seems to be a general phenomenon for other related catalysts (Ni-Mo-S, Co-W-S, Ni-W-S).

Other surface techniques that have been applied to this system (*e.g.* XANES, XPS, and high-resolution transmission electron microscopy) also provided further support for the overall picture. As for the precise geometry of the active sites, the debate continues.

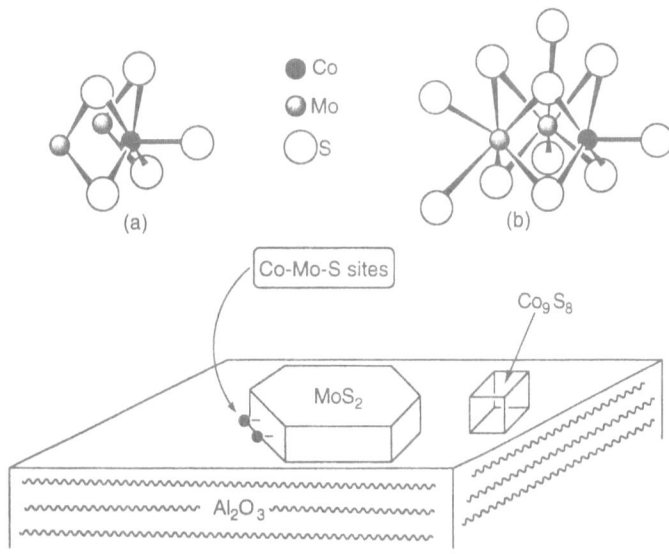

Fig. 1.3. Schematic representation of Topsøe's Co-Mo-S model, showing perspective views of: (a) the coordination around the Co site and (b) a Mo$_2$CoS$_{11}$ unit at the edge Co-Mo-S site.

From EXAFS measurements it appears that each Co is pentacoordinated in an approximately square-base pyramidal arrangement of sulfur ligands (as in structure (a) in Fig 1.3) with two neighboring Mo atoms, while each Mo is in turn hexacoordinated by a trigonal prismatic array of sulfur atoms with three neighboring Mo atoms and one Co atom (partially shown in (b) in Fig 1.3).

The apical sulfur on the Co atoms is known to be loosely bound, so that it is easily removed under catalytic conditions in order to generate the necessary vacant sites for chemisorption of reactants. XPS and EPR data indicate that the active site contains Co(II) and Mo(IV), although low concentrations of Mo(V) have been observed, and evidence for Co(III) being present in important concentrations has been presented. The number of active Co-Mo-S sites also varies with the Co loading. In an ideal structure the Co:Mo ratio should be 0.5. To complicate matters even more, a number of authors believe that more than one type of active sites exist, each type being associated with a different kind of reactivity. For instance, Topsøe claims that "edge sites" are responsible for hydrogenation activity, while "corner sites" would be related to direct sulfur extrusion. Furthermore, the activity also seems to depend on the reagents and conditions used in catalyst preparation, and this has been referred to as "Type I" sites found on monolayer MoS_2 slabs and "Type II" sites which are more active but only present in multilayer slabs. In any case, the Mössbauer emission studies indicated that only about 10% of all the Co-Mo-S sites present actually display activity for sulfur extrusion. By careful control of synthetic parameters and a judicious choice of additives, highly active catalysts (100% Type II) can now be prepared, which are considerably more active than present-day commercial catalysts but still not active enough to meet future sulfur specifications for fuels [14-18].

In a different and very original turn, Whitehurst et al. [16] have argued that in the most active catalysts presently known, the active phase can be as small as seven Mo atoms forming a flat hexagonal crystal with three identical bonding positions at the three $10\bar{1}0$ faces. Each Mo in these units has four orbitals extending away from the $10\bar{1}0$ face, forming essentially a square of sulfide groups on which the Co can bind. As explained above, this face is the one normally believed to be where the Co-Mo-S phase is formed, and molecular models show that under such a situation for very small crystallites there are three sites but only one unique geometry is reasonable for them, thus giving rise to "the two site dilemma", i.e. can a single geometric configuration perform two different functions?

The answer may be affirmative if several conceivable situations are accepted (see Fig. 1.4): (1) the number of Co-S bonds may be less than four, possibly three, which can be easily achieved by breaking M-S-M bridges; (2) the number of Co orbitals actually involved in binding the thiophenic substrate may vary; (3) the orientation of the various Co orbitals available for binding the substrate, with respect to adjacent -SH groups, need not be identical; (4) the preferred mode of binding of each type of substrate to the cobalt center may be determinant; (5) the oxidation states of Co may change without accompanying major geometrical changes, giving rise to different reactivities;

(6) some of the available Co orbitals could be occupied, at least temporarily, by other molecules or fragments (H_2S, -SH, -H, aromatics, a second thiophenic molecule).

A representation of Whitehurst's proposal is shown in Fig. 1.4, where it can be noted that (i) one Co-S-Mo bridge has been broken; (ii) hydrogen has been heterolytically activated into M-H and -SH on Co (see Section 1.2.5, p. 14); and (iii) thiophene has been adsorbed in an S-bonded manner, lying adjacent to the metal hydride and the -SH group. Startsev has also claimed [14, 18] that oxidative addition of hydrogen takes place on Co, thereby oxidizing it to Co(III) as the active form, with the hydride occupying the sixth position of an octahedron, below the square base of sulfurs, in agreement with Whitehurt's proposal.

Another very recent exciting development is the report of the use of scanning tunneling microscopy (STM) to obtain the first atomic-scale images of single layers of MoS_2 [46] and Co-Mo-S [47] nanoclusters grown on Au (111). MoS_2 clusters prepared in this way are of triangular shape. In principle, the morphology of a MoS_2 slab should be determined by two types of edge terminations, a ($\overline{1}$010) S-edge and a (10$\overline{1}$0) Mo-edge for a hypothetical hexagonal cluster, where the edges are simple terminations of the bulk MoS_2 structure. However, STM results, interpreted together with DFT calculations [48, 49] lead to the conclusion that the structure observed is one in which the edge termination is a Mo-edge with one S atom per Mo in a reconstructed geometry. Treatment with hydrogen leads to the removal of some of the sulfur atoms at the edges, forming vacancies, which could also be observed experimentally. Although these nanoclusters have been obtained under very particular conditions, this model provides new insights into the structure of the active sites in molybdenum sulfide. In the case of Co-Mo-S nanoclusters, the presence of the promoter atom causes the shape to change from triangular to hexagonally truncated, possibly driven by a preference for Co to be located at the S-edge of MoS_2 resulting in a stabilization of the S-edge. It seems also that the presence of Co atoms at the S-edges induces an important electronic perturbation on the neighboring S-atoms that may well be related to the increased activity of the Co-Mo-S phase. On the basis of this work, a new structural model was proposed for the Co-Mo-S phase, in which the Co atom is substituted into Mo positions at the S-edge as depicted in Fig. 1.5(a).

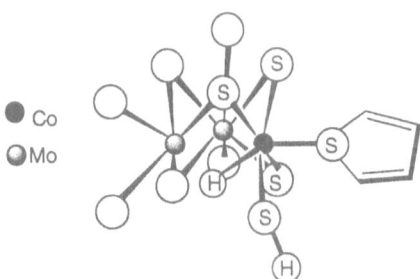

Fig. 1.4. A representation of Whitehurst's model for thiophene, -H and -SH
bonded to a unique Co-Mo-S site (adapted from ref. 16).

Fig. 1.5. A representation of the Co-Mo-S model site from STM measurements (Adapted from ref. 47).

In this case, a tetrahedral environment for Co is produced as shown in Fig. 1.5(b) if the outermost S atoms are assumed to be bridged-bonded monomer sulfur atoms located in the plane of the Mo atoms. These results are also in agreement with DFT calculations [48, 50] and are reasonably consistent with EXAFS measurements which indicated average coordination numbers of $N_{Co-S} = 5 \pm 1$ and N_{Co-Mo} around 2. This model implies intrinsically coordinatively unsaturated sites where the incoming reactant can adsorb, which is an interesting explanation for the enhanced activity of promoted catalysts. These novel proposals, however, must be taken with some caution, as these nanoclusters have been prepared, sulfided and observed under very special conditions which may not necessarily lead to the same structures that are dominant in real systems. Nevertheless, they represent an important breakthrough in that they are the first atomic-scale observations of metal sulfides, and in that they provide new alternative ways of envisaging the morphology and the properties of HDS active sites.

Although very few complexes capable of closely mimicking the structure of HDS active sites are known, in the following Chapters a wealth of results obtained from coordination and organometallic chemistry will be presented, some of which are very clearly in line with a number of the possibilities discussed above, thus adding to this very lively discussion.

1.2.4.2. Active sites and the concept of vacancies.

The nature and the geometry of the Co-Mo-S sites have been discussed in the preceding section, but a key element in the generation of the active sites during actual catalysis has been the concept of "vacancies" (or coordinatively unsaturated sites "CUS") discussed by many authors over the years. Such vacancies are thought to be created upon reaction of hydrogen with a surface sulfide group, leading to the formation of H_2S and an "empty" coordination site. Since the original site was thermodynamically favorable for the presence of the metal bonded sulfide group, there would be a natural tendency to revert to that stable situation. This provides the driving force for the extrusion of sulfur from organosulfur compounds, thus conforming the basic "sulfur breathing" process represented by Eq. 1.2 that is ultimately responsible for HDS catalysis. The HDS activity would be thus related to the concentration of vacancies on the surface, and a wealth of experimental evidence seems to support this idea. Well founded theoretical models have also been advanced to explain the periodic trends in HDS activity for the different transition metals in terms of M-S bond energies (see section 1.2.2, p. 7). The results of these calculations predict with great accuracy the trends found experimentally.

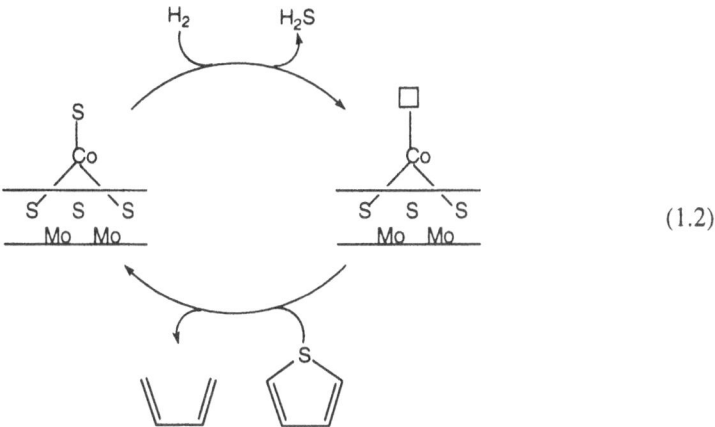

(1.2)

Also, despite Whitehurst's arguments for catalysts composed mostly of extremely small crystallites, several types of vacancies may be envisaged for many catalytic formulations, depending on the degree of sulfur removal from the original surface, and on the crystal plane being considered. This is related again to the idea of different sites with varying degrees of coordinative unsaturation being active for different reactions, since diverse modes of bonding from the various substrates then become possible. Some nice correlations with these observations can be found in organometallic HDS models as will be discussed in the following Chapters.

In the Co-Mo-S model incorporating a pentacoordinated Co, the apical sulfur atom on the square base pyramidal structure around the Co -which is the one extending away from the surface- is known (by EXAFS) to be loosely bound and could be easily removed to create the necessary vacancies for the catalysis to occur. Mechanistic models based on the vacancy approach indicate that in promoted catalysts a strong Co-Mo attractive interaction would result in a weaker metal-sulfur bond than in MoS_2 or Co_9S_8, and the application of such considerations produce remarkably good fits of experimental data (see ref. 15, p. 228). In the new version of the site deduced from STM measurements, tetrahedral Co would offer an intrinsically natural vacancy, resulting in a higher activity. However, this also leads to the question of what would be the driving force for sulfur extrusion leading to a presumably less stable pentacoordinated Co atom.

In spite of its popularity and definite usefulness, the concept of surface vacancies seems to be evolving into a more sophisticated -and probably more realistic- panorama. EXAFS measurements have indicated that the cobalt atoms in Co-Mo-S sites are capable of *variable* coordination numbers and oxidation states; this means that such cobalt atoms could have enough bonds to surrounding sulfurs that may be easily broken and reconstructed through low energy pathways.   Thus, the "vacancy" does not necessarily have to physically exist for an important length of time, but it would be created -through electron and sulfur mobility- only when needed, *i.e.* if and when an incoming sulfur-containing molecule is ready to take its place. Recent support for this extremely interesting concept has been provided by organometallic studies in solution using model Co-Mo-S containing metal clusters (see Section 4.2.2.3). Within this framework, there is no need to invoke physical "vacancies", but rather a resting state of the catalyst possessing what Curtis very adequately termed "latent vacancies" [36]. This refers to such *dynamic coordinative unsaturation*, a situation of lower energy than that corresponding to an "empty site" and intuitively more in line with nature's dislike for voids.

     Recent DFT calculations by Byskov *et al.* have also shown that if vacancies are created in MoS_2 surfaces, and relaxation is allowed to take place -including several layers of the solid in the calculations- surface reconstruction takes place. What is actually observed is the migration of sulfide ions from the lower layers to the surface so as to "fill", at least partially, the empty sites [51, 52]. Interestingly, similar considerations have begun to appear in many homogeneous catalytic reactions, where "empty sites" previously invoked in complexes are now more often thought of as being occupied by weakly bound solvent molecules or very labile ligands [53]. Startsev has also presented several theoretical and experimental arguments against the direct involvement of sulfur anion vacancies in the catalysis, pointing instead to "concerted mechanisms" that also seem to be in line with the concept of latent vacancies [14, 18]. Such a refinement in the idea of vacancies does not cause any inconsistencies with previous knowledge or theories. Instead it means a conceptual evolution requiring an intellectual change in the way of conceiving the acting catalytic surface or complex, in order to match it to the idea of both a more dynamic and at the same time a more stable system.

1.2.5. THE ACTIVATION OF HYDROGEN

Hydrogen is one of the key components in HDS as well as of all the other reactions implicated in hydrotreating, and therefore the way in which hydrogen reacts with the catalysts and the nature of the sites at which such activation takes place are of prime concern. Two main types of hydrogen activation processes have been considered [14, 15, 18]; one involves a simple homolytic splitting on surface $(S-S)^{2-}$ units to produce two - SH groups (Eq. 1.3), which some authors think is the major pathway for hydrogen activation:

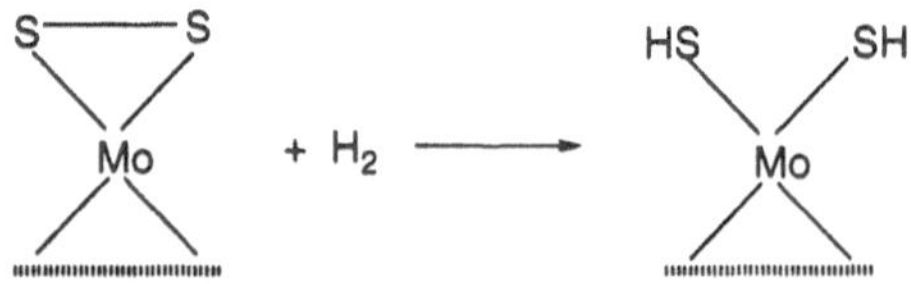

$$(1.3)$$

     A more elaborate and perhaps more appealing proposal advanced by several authors [54] involves the heterolytic activation of H_2 on *e.g.* a Mo=S or Mo-S-Mo bond to form Mo-H plus Mo-SH species. The metal hydride is sometimes though to take part in HDS catalysis, but also its oxidation by molybdenum to form a second -SH group can be envisaged as a facile process for a highly mobile H atom in such an sulfur rich surface (Eq. 1.4):

$$
\begin{array}{ccc}
\underset{\displaystyle \overset{S}{\diagdown}\ \overset{S}{\diagup}}{\underset{\displaystyle Mo}{}} + H_2 & \longrightarrow & \underset{\displaystyle Mo}{\overset{S\ \overset{H}{|}\ SH}{\diagdown|\diagup}} & \longrightarrow & \underset{\displaystyle Mo}{\overset{HS\ \ SH}{\diagdown\diagup}}
\end{array} \qquad (1.4)
$$

Although it has been difficult to obtain direct experimental proof of these reaction pathways, some particularly illustrative data have become available. For instance, the detection in one case by solid state NMR measurements of two types of surface protons on hydrogenated RuS_2 was interpreted in terms of the existence of both Ru-H and -SH groups; this view correlated well with hydrogenation activities [55, 56]. A theoretical study of the interaction of hydrogen with MoS_2 has also shown that heterolytic splitting of dihydrogen is a strong possibility [57]. For promoted catalysts some authors believe that even though the activation of the organosulfur compound takes place on the promoter Co (or Ni) atom, hydrogen may actually be dissociated on Mo-S edge sites and the resulting Mo-SH groups would be the principal hydrogen source in HDS. Others prefer to think that both H_2 and the sulfur-containing molecule are activated at the promoter atom, and/or that both metal-hydrides and M-SH groups serve as hydrogen sources for HDS [58]. Although this question of hydrogen activation on sulfide surfaces is the subject of continuing debate, it appears that the heterolytic mechanism is gaining more adepts since it correlates well with other experimental and theoretical data, and with the observed inhibition of hydrogenation and desulfurization reactions by H_2 and H_2S. Finally, oxidative addition of hydrogen on the Co (or Ni) sites of promoted catalysts has been claimed to be responsible for the high oxidation states - Co(III) or Ni(IV)- which some authors have detected and associated with catalytic activity.

The issue of hydrogen activation in connection with HDS has been very little addressed in organometallic modeling studies, perhaps because hydrogen activation in homogeneous catalysis is considered a well understood process. The main pathways frequently encountered in homogeneous hydrogenation systems [59, 60] involve either a concerted oxidative addition of H_2 to the metal center of the complexes to form metal dihydrides, as in the classical case of Wilkinson's catalyst, $RhCl(PPh_3)_3$ shown in Eq. 1.5, or a heterolytic activation leading to a metal monohydride and usually requiring an external base to remove the proton, as depicted in Eq. 1.6 for the reaction of $RuCl_2(PPh_3)_3$ with H_2 in the presence of triethylamine; the latter mechanism is analogous to the activation proposed on M=S or M-S-M units in which the sulfur would act as the base for proton abstraction.

$$
RhCl(PPh_3)_3 + H_2 \quad \longrightarrow \quad \overset{H}{\underset{H}{}\!\diagup}RhCl(PPh_3)_3 \quad \longrightarrow \quad Rh(H)_2Cl(PPh_3)_3 \qquad (1.5)
$$

$$
RuCl_2(PPh_3)_3 + H_2 \quad \xrightarrow{\ Et_3N\ } \quad RuHCl(PPh_3)_3 + Et_3N.HCl \qquad (1.6)
$$

Another well-documented way of activating hydrogen on metal complexes is the homolytic cleavage of dihydrogen on two adjacent metal centers which yields also monohydrides, as exemplified for $Co_2(CO)_8$ in Eq. 1.7.

$$Co_2(CO)_8 + H_2 \longrightarrow 2\, HCo(CO)_4 \qquad\qquad (1.7)$$

One further point probably related with the absence of coordination analogues of the hydrogen activation mechanisms invoked in HDS is the scarcity of adequate model complexes containing sulfide ligands [61] which has precluded further developments in this area; however, a few interesting examples of reactions of hydrogen with metal complexes with simple sulfur ligands have become available, and they represent very good models for the types of activation thought to occur on metal sulfide surfaces, as discussed in Chapter 5.

1.2.6. CHEMISORPTION OF ORGANOSULFUR COMPOUNDS

The fundamental mechanism by which thiophenes and other organosulfur compounds are activated by HDS catalysts is through the formation of a chemical bond to a metal center on the catalyst surface. This process, generally referred to as "chemisorption" in surface chemistry language, finds its parallel in organometallic chemistry in the idea of "coordination" of an incoming *ligand* (the molecular nomenclature for "adsorbate"). Whether one speaks of a solid surface or of a complex in solution, in order for a chemical bond to be formed between two entities, the appropriate orbitals must be available. Since it is generally known -or assumed- that the adsorbate (or ligand) is an electron donor (basic) species, the metal site must have the necessary "electronic room" (*i.e.* empty orbitals of adequate energy and symmetry) to accommodate the incoming electrons. This is what is known as *coordinative unsaturation* in metal complexes, whereas in the language of surface chemists this is referred to most usually as vacancies, although the terminology "coordinative unsaturated sites" (CUS) has gained considerable acceptance in recent times in the heterogeneous literature. The active sites in HDS catalysts have often been associated with sulfur anion vacancies - although as discussed in Section 1.2.4.2, their physical existence for any significant length of time is uncertain - and their concentration seems to be a function of catalyst formulation and reaction conditions. In any case, it is reasonable to assume that during HDS the adsorption of thiophene (as well as of other unsaturated molecules) takes place on such coordinatively unsaturated sites or latent vacancies. In promoted catalysts there seems to be a general agreement in that the organosulfur compounds are adsorbed and thereby activated *on the promoter atoms* (Co or Ni) rather than on Mo or W sites.

Another important question concerning reactivity is the degree of coordinative unsaturation that can be attained in a single site, since that in itself would determine the possible modes of bonding available for the various competing substrates. This bonding capabilities are in turn related to the issue of whether a unique type of site or a variety of them exist -differing in *e.g.* the number of CUS- or whether one single kind of site may be responsible for several different reactions; furthermore, selective poisoning effects which have often been observed in HDS of mixtures of substrates can also be related to this feature of the active sites [15, 16]. Fig. 1.6 shows the most commonly invoked modes of chemisorption of thiophenes on HDS sites: Experimental and theoretical data - obtained predominantly on clean surfaces- indicate that thiophene binds in different ways depending on the degree of coordinative unsaturation of the active sites available. Binding through the sulfur atom only -one point adsorption- has been claimed in many instances [*e.g.* ref. 62], and this would be the case particularly for singly coordinatively unsaturated sites (this is called $\eta^1 S$ bonding in the notation of organometallic chemistry, see Chapter 2). If doubly CUS's exist, coordination can occur through the sulfur and one double bond ($\eta^3 S, C=C$, "multi point adsorption" [63]); and if the site is triply unsaturated, then the whole π-ring of thiophene can interact in a "flat" manner (η^5) [64].

For benzothiophenes and dibenzothiophenes, besides the latter adsorption modes, flat binding through an arene ring (η^6) to a 3-CUS site is possible. It has been claimed that $\eta^1 S$ ("one point") adsorption leads to direct sulfur extrusion [62], while π-bonding ("multipoint") should be related to prehydrogenation of the thiophene ring [63] (See Section 1.2.7.2 below) but this point is far from definitely resolved. Steric effects are also important in binding to the surface, and this has been taken as an explanation, for instance, for the much lower activities observed in HDS of 2-methyl- or 2,5-dimethyl- substituted thiophenes in comparison with the unsubstituted molecule [65].

Also, some authors think that the great difficulty in desulfurizing 4,6-dimethyldibenzothiophene is in good part due to the fact that the methyl groups shield the sulfur from interacting with the surface active site. In this case, the preferred reaction pathway goes through hydrogenation of a benzene ring [16], most likely via a flat, π-bonded intermediate; upon saturation, the overall aromaticity of the molecule is reduced, and the ring twists and exposes the sulfur atom which can then be extruded by the catalyst (see Section 1.2.7.2). Considering the inherent difficulties in trying to experimentally determine the intimate details of chemical binding between a small molecule and a surface, even with the most sophisticated analytical techniques, this is an area in which organometallic chemistry has been particularly useful in identifying the various modes of bonding of thiophenes, as they occur in well defined and often structurally characterized metal complexes.

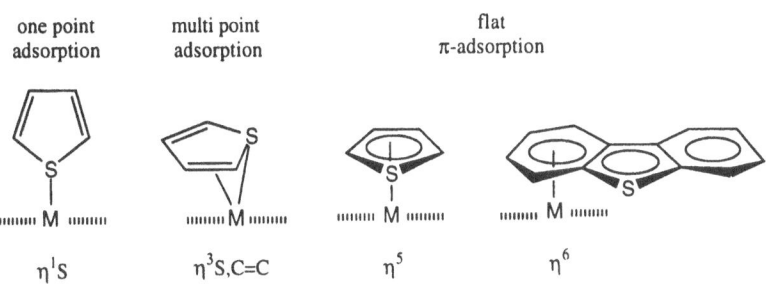

Fig. 1.6. Possible bonding modes for thiophenes on a surface metal site.

Additionally, extensive theoretical calculations on such complexes have provided a very deep understanding of the bonding between thiophenes and metal centers and this knowledge has been very helpful in distinguishing the most reasonable proposals to be envisaged in surface chemistry [21-34]. This subject is discussed in detail in Chapter 2.

1.2.7. HYDROGENATION AND HYDROGENOLYSIS OF ADSORBATES

Once an organosulfur compound and hydrogen have been chemisorbed on a catalytic surface they must interact with each other in order for desulfurization to be completed; therefore, all the major pathways that need to be considered involve hydrogenation of C=C bonds and/or hydrogenolysis of C-S bonds. The intimate details of how such elementary steps occur, and the way in which they are interlinked to form a catalytic cycle are not easy to establish on real operating systems, because of the intrinsic complexity of the catalyst and of the intervening reaction schemes; nevertheless, a number of very reasonable mechanistic proposals have been advanced by several research groups on the basis of kinetic studies, detection of intermediates, deuteration experiments, and other appropriate techniques, and the literature on this point is ample.

1.2.7.1. HDS of thiophenes.

As a first example of interest for our purposes, Fig. 1.7 presents a summary of the most commonly invoked reaction pathways for HDS of thiophene, which is the simplest of all model compounds of this family. Although some of the results have been obtained on unpromoted MoS_2 while others are from studies using Co-Mo catalysts or other alternative formulations, and the various experiments were carried out under a diversity of reaction conditions, rather than attempting a comprehensive coverage of the details of each paper it has been intended here to generalize the main conclusions and common features extracted from a perusal of the original literature. To avoid further complications, a simple graphic representation of a Co-Mo-S site is presented which is not meant to imply any particular structure, although it assumes thiophene bonding to the promoter Co atom.

Once thiophene has been adsorbed on the active site, several alternative reaction pathways may be envisaged (see Fig. 1.7): (i) The stepwise hydrogenolysis of both C-S bonds to yield 1,3-butadiene plus adsorbed sulfur. Subsequent hydrogenation of the primary products produces butenes/butane plus H_2S, thus regenerating the CUS. The presence of butadiene and the absence of tetrahydrothiophene as products of the mild HDS of thiophene support this pathway [66]. Startsev recently proposed the related idea of a concerted mechanism in which the intermediate butadiene is retained in the active site during the reaction so that the primary products are the butenes [18, 67]. (ii) Direct H_2S elimination ("dehydrodesulfurization") of thiophene to yield 1,3-butadiyne plus adsorbed H_2S; hydrogenation of the diacetylene and desorption of H_2S would lead to the same HDS products [68]. This possibility cannot be applied to alkyl substituted thiophenes which could not possibly generate the corresponding dialkynes, so its relevance is rather limited [69].

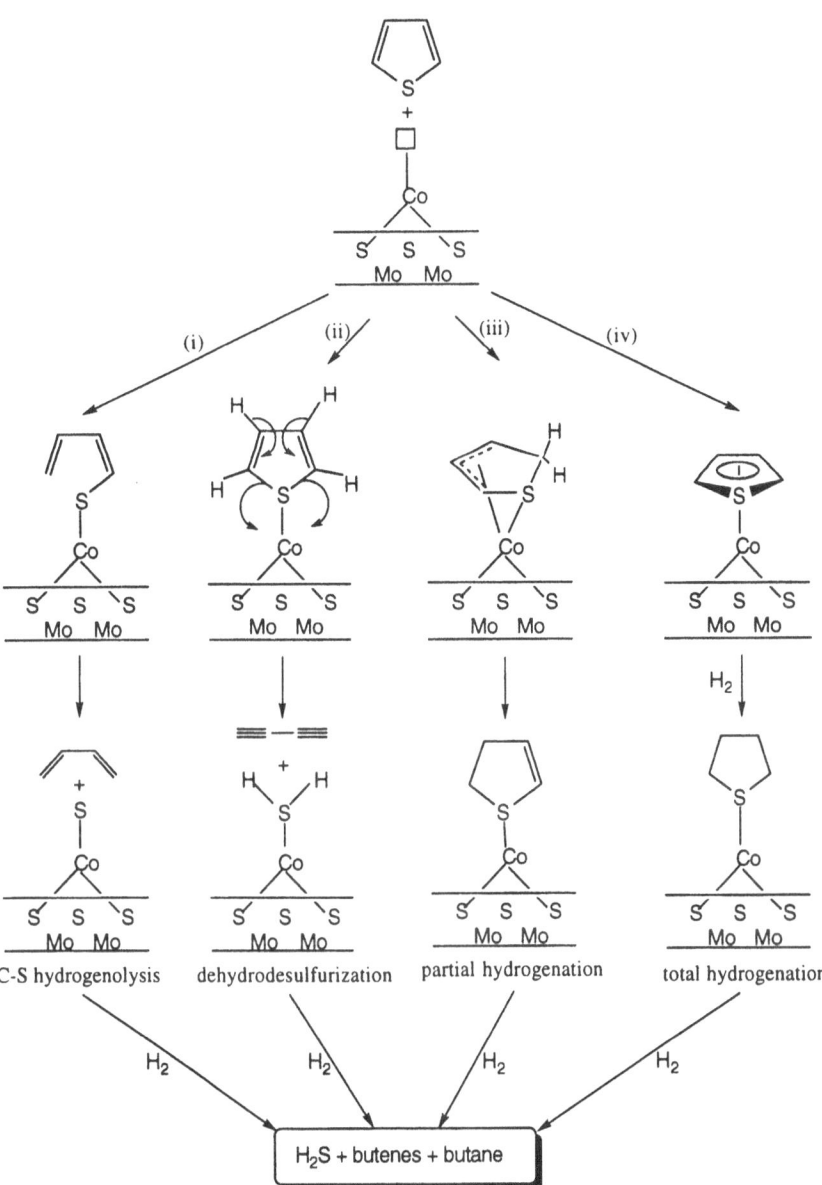

Fig. 1.7. Main reaction pathways in HDS of thiophene.

(iii) Partial hydrogenation of thiophene leading to the transient formation of 2,3-dihydrothiophene, followed by hydrogenolysis of the C-S bonds, which has been proposed on the basis of theoretical [70] as well as experimental [71, 72] studies, and (iv) Complete saturation of thiophene to yield tetrahydrothiophene, followed by hydrogenolysis of the thioether into butane plus H_2S, which is thought to occur at elevated pressures where tetrahydrothiophene has been observed as a major intermediate [73].

Thiophene itself and H_2S are known to inhibit the HDS reaction, and it is often thought that hydrogenation and hydrogenolysis occur at different active sites. It is also interesting to note that ring alkylation significantly affects the reactivity of thiophenes, but not in a straightforward manner; for instance it has been found that the reactivity of methylated thiophenes varies in the order:

thiophene > 2-methylthiophene > 2,5-dimethylthiophene

However, substitution by a third methyl group enhances the HDS rate showing that both steric and electronic effects are operative, and they may be acting upon different steps or upon more than one step, *e.g.* adsorption of the thiophene and/or the hydrogenation or hydrogenolysis step [19].

1.2.7.2 HDS of benzothiophenes and dibenzothiophenes and the deep desulfurization issue.

Besides thiophene, both benzothiophene and dibenzothiophene have been extensively used as model compounds by a number of groups over the years in numerous reactor and mechanistic studies, and a great deal of information has been derived in this manner, resulting in several proposals of the most important reaction pathways operating in sulfur removal from these polycyclic molecules. However, the new standards tending toward "zero sulfur" fuels have stimulated the detailed analysis of the sulfur containing species found in diesel fuels and middle distillates of crude oil which show that the types of compound remaining after conventional HDS has been performed are distributed non-uniformly, depending on the origin of the crude. Thiophenes are essentially absent from these cuts, and the polyaromatic sulfur compounds (PASC) present are mainly the benzothiophenes and dibenzothiophenes alkylated at different positions; in the case of benzothiophenes the alkyl substituents contain 1-7 carbon atoms while the dibenzothiophenes are substituted with C1-C5 alkyl chains. This clearly shows that these two families of compounds are extremely refractory; when the sulfur level is lowered from 0.20 to 0.05% the chemistry of gas oils is essentially the chemistry of the least reactive PASC which are the alkylated dibenzothiophenes, and particularly 4-methyldibenzothiophene and 4,6-dimethyldibenzothiophene [16, 19, 20]. Consequently, the most important challenge in HDS catalysis today is to find efficacious ways to remove or degrade such compounds from crudes or distillates in what is known as "deep desulfurization". It is therefore of prime importance to gain a detailed knowledge of the ways in which sulfur is extruded from such inert substrates, as a necessary first step toward the design of novel catalysts or catalytic processes.

The possible reaction pathways for HDS of benzothiophene are shown in Fig. 1.8, where some analogies to those described for thiophene become evident. The main routes may involve either the selective prehydrogenation of the thiophenic ring to yield the corresponding 2,3-dihydrobenzothiophenes, followed by sulfur removal through hydrogenolysis of the cyclic thioether; or alternatively, direct sulfur extrusion by the stepwise hydrogenolysis of the two C-S bonds of the heterocyclic ring of benzothiophene, followed by hydrogenation of the olefinic chain of the resulting styrene or even by complete hydrogenation also of the benzene ring. It is important to note that the C(2)=C(3) bond of benzothiophene has a more pronounced olefinic character than those of thiophene itself; this obviously makes hydrogenation to 2,3-dihydrobenzothiophene a much more facile process than hydrogenation of thiophene to either 2,3-dihydrothiophene or tetrahydrothiophene, which requires overcoming a higher energetic barrier due to the greater aromaticity of this molecule.

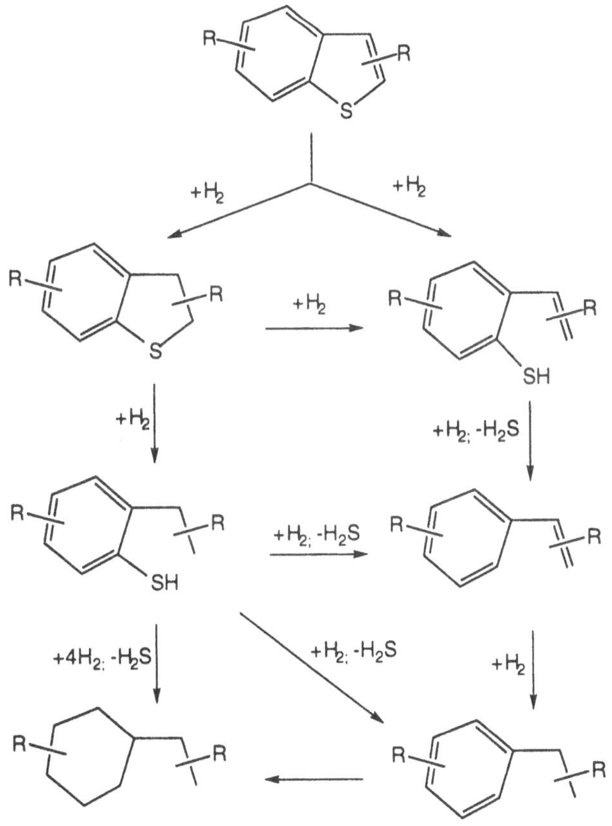

Fig. 1.8. Main reaction pathways in HDS of benzothiophene.

In general, hydrogenative routes are known to become more competitive for the larger S-containing ring systems such as benzonaphthothiophene. Nevertheless, it has been observed that methyl substitution of benzothiophene reduces the rate of partial hydrogenation in the order:

$$BT > 2\text{-MeBT} > 3\text{-MeBT} > 2,3\text{-Me}_2BT$$

This reactivity trend is attributed mainly to electronic rather than steric effects; as a consequence, the hydrogenolysis route becomes more favorable as the hydrogenation is inhibited. This is important in that, as explained above, C1-C7 alkyl substituted benzothiophenes are known to be among the main sulfur components in heavy fractions and distillates, and therefore direct sulfur extrusion rather than prehydrogenation may be the most important mode of reaction for the hydrodesulfurization of this class of compounds. Also interestingly, H_2S has been found to inhibit the hydrogenolysis but not the hydrogenation of benzothiophene [20], indicating either different active sites, or different requirements in terms of coordinative unsaturation, for the two reactions. In the latter case it is possible to envisage a unique active site for both reactions with at least a double coordinative unsaturation; sulfur extrusion may require both unsaturations in the chemisorption step, for instance the sulfur atom and the C=C bond, whereas C=C bond hydrogenation would need only one unsaturation for the olefinic bond to adsorb. Blockage of a single unsaturation by a H_2S molecule would thus inhibit hydrogenolysis but not hydrogenation.

Dibenzothiophenes can also be desulfurized via two different pathways: either prehydrogenation of one (or both) of the arene rings followed by hydrogenolysis of the heterocyclic moiety; or, alternatively, direct sulfur extrusion without any need for previous hydrogenation. This is shown in a simplified manner in the reaction Scheme depicted in Fig. 1.9. Perhaps the major difficulty in trying to interpret the various sets of data available in the literature for these substrates stems in the fact that the catalysts and the reaction conditions employed, as well as the kinetic models invoked, are quite different from one publication to another; furthermore, the experimental conditions reported are often very different from the operating conditions for real HDS catalysis. Detailed discussions on these studies are available in several recent reviews [16, 19, 20] and in the interesting series of papers by Vrinat and his coworkers devoted to the kinetics of the hydrodesulfurization of a variety of substituted dibenzothiophenes [74-77].

Both routes are known to operate in parallel during the hydrodesulfurization of unsubstituted dibenzothiophene and the preponderance of any one of them depends strongly on the catalyst formulation and on the reaction conditions. Alkyl substitution, particularly in the vicinity of the sulfur atom, as in 4-methyldibenzothiophene and 4,6-dimethyldibenzothiophene, results in a marked decrease in HDS rates, and in general, this is associated primarily with a decrease in the rate of the direct sulfur extrusion routes, which is known to be disfavored whenever there is steric congestion around the sulfur atom.

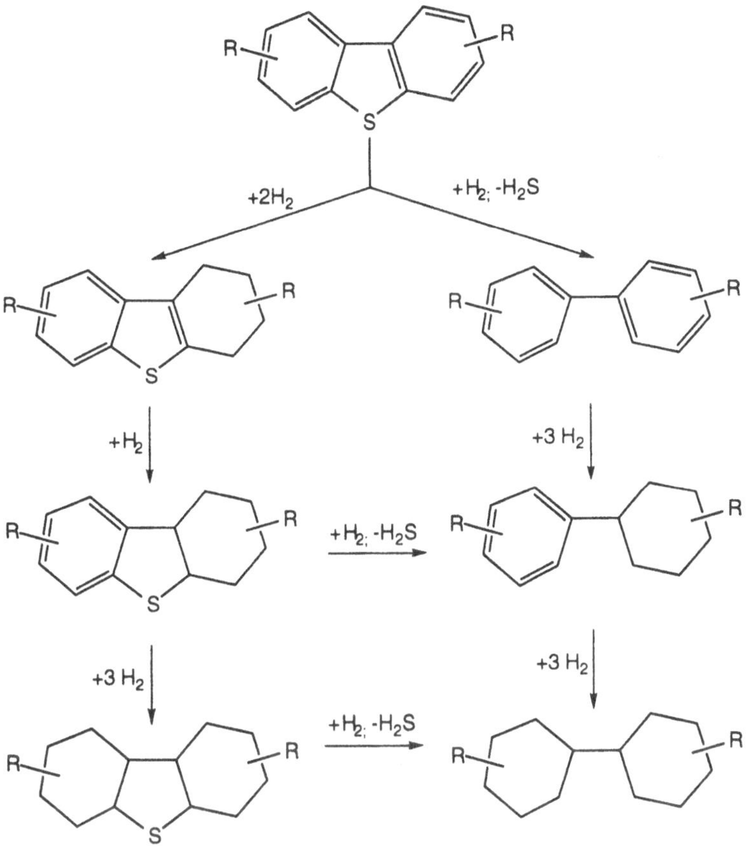

Fig. 1.9. Main reaction pathways in HDS of dibenzothiophenes.

This has been explained in two different ways: the first hypothesis, advanced by Houalla *et al.* [78] and by Mochida and coworkers [79], assumes that the rate of sulfur extrusion is limited by the adsorption step, which is thought to occur in a "sulfur only" $\eta^1 S$ manner; thus, the effect of the methyl substituent is essentially a steric hindrance for the sulfur atom to approach the surface active site; hydrogenation of one of the benzene rings allows a twisting of the molecule –as the sp^2 carbon atom becomes sp^3– which exposes the sulfur sufficiently for an effective interaction with the active site. Alternatively, Vrinat and his group have convincingly argued on the basis of extensive kinetic studies on the HDS of a wide range of individual dibenzothiophenes substituted with different alkyl groups in varying numbers and positions, as well as mixtures of them, that the adsorption of dibenzothiophenes occurs mainly via the π-electron cloud of one of the aromatic rings (η^6 bonding) and this type of bonding is not largely affected by the steric effects of the alkyl substituents in the 4- and/or 6- positions. In this model,

steric hindrance is thought to affect the sulfur elimination step instead [74-77].
Furthermore, Vrinat's mechanism, which is schematically shown in Fig. 1.10, assumes
that after adsorption of the dibenzothiophene has taken place, the molecule is first
partially hydrogenated to the corresponding *dihydro*dibenzothiophene, in accord with a
previous proposal by Singhal *et al.* [80]. This unstable partially reduced intermediate can
be subsequently transformed via two distinct parallel pathways on similar catalytically
active sites by hydrogenation into the tetrahydro- or hexahydrodibenzothiophenes, or
alternatively, by desulfurization via an elimination mechanism assisted by basic sites,
resulting in the corresponding biphenyl derivatives.

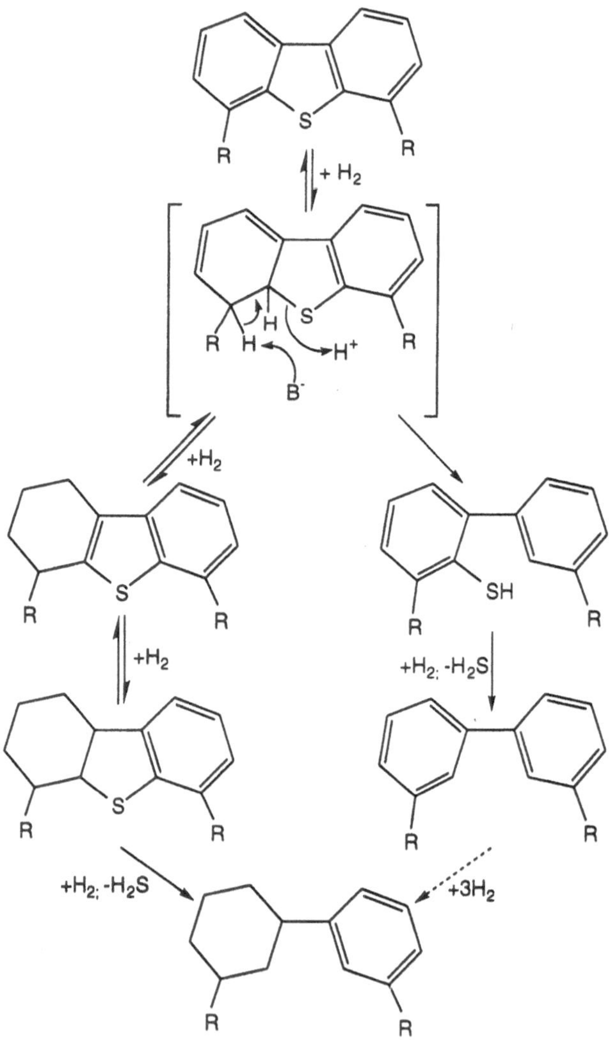

Fig. 1.10. Vrinat's mechanism for HDS of dibenzothiophenes (Adapted from ref. 75).

Despite the different mechanistic interpretations that have been advanced, it is clear from the experimental facts that the presence of alkyl substituents on the arene rings has a major inhibiting effect on the rate of the direct desulfurization route, and therefore the molecules of interest in the context of deep desulfurization are desulfurized predominantly via a hydrogenative pathway. This is very important in relation to the design of new effective HDS catalysts, which must display a high activity for the hydrogenation of arene rings.

In conclusion, an extensive literature is available on the reaction networks that are thought to operate in HDS of various types of thiophenic molecules; besides the great advances that have been made in direct studies on molybdenum sulfides and related catalysts, this is another area in which organometallic chemistry has made an impressive contribution to HDS catalysis, as a number of reaction pathways and mechanisms for the hydrogenation and hydrogenolysis of thiophenes on metal complexes in solution has been well established with the aid of a variety of physical techniques.

1.3. The Hydrodenitrogenation Reaction

Hydrodenitrogenation (HDN) is the name given to the removal of nitrogen from organonitrogen compounds by reaction with hydrogen over a catalyst to yield ammonia plus a hydrocarbon (Eq. 1.8). This is probably the most important reaction within the context of hydrotreating after hydrodesulfurization, since nitrogen is generally the second most abundant polluting heteroatom in crude oils [5, 13, 15, 20, 81-86].

$$[R\text{-}N] + H_2 \longrightarrow [R\text{-}H] + NH_3 \qquad (1.8)$$

The importance of this reaction arises from similar considerations to the ones invoked for the need for HDS, namely: (1) Environmental regulations and concerns, since combustion of nitrogen-containing fuels leads to the emission of highly polluting nitrogen oxides (NO_x), responsible for acid rain. (2) Possible poisoning effects due to the basic organonitrogen compounds that adsorb strongly on acid catalysts employed in subsequent refining steps, such as the ones used in fluid catalytic cracking (FCC) or hydrocracking (HCK). (3) Improving product quality since N-compounds are thought to be responsible for many of the problems related to coloring and gum formation during storage.

The nitrogen content of crudes is variable, but in general considerably lower than the sulfur content. Typically values of around 0.1 wt. % are found, although levels as high as 1 wt. % are not uncommon. The acceptable amounts according to present and future regulations are of the order of less than 0.001 wt. % in fuels and of less than 0.5 wt. % for FCC feeds. This type of nitrogen is present predominantly in the heavier fractions and in cracked fractions; the more easily degraded classes are the non aromatic N-compounds, mainly amines and nitriles, which are efficiently removed with standard catalysts and processes. On the other hand, the most highly refractory molecules that are relevant for HDN-modeling purposes are represented by the families depicted in Fig. 1.11, deriving from the basic pyridine (Py) and the less-basic pyrrole (Pyr) structures; these include the bicyclic quinolines (Q) and indoles (In), with one or several short-chain alkylsubstituents- as well as higher polycyclic homologues.

Due to the higher C=N bond strength (147 kcal mol^{-1} $vs.$ 114-128 kcal mol^{-1} for C=S bonds) [20, 87] and the smaller atomic radius of N (0.75 Å $vs.$ 1.09 Å for S), the removal of nitrogen from heteroaromatic molecules is indeed considerably more difficult to achieve than HDS of thiophenes. Consequently, typical operating conditions for industrial HDN (80-120 bar and 340-380 °C) are more forcing than those required for HDS. Furthermore, since hydrogenation steps of the heterocycle, the arene, or both rings are an important part of the overall reaction scheme, hydrogen consumption during HDN is high and therefore strongly hydrogenating catalysts perform the best. The most commonly used formulation is NiMo/Al$_2$O$_3$; ruthenium has been shown to be especially effective as a promoter in HDN catalysts. Nevertheless, hydrotreating processes are generally optimized for HDS rather than for HDN, since sulfur removal is the main problem that needs to be solved. The active hydrodenitrogenation sites are usually thought to be the same as -or very similar to- the HDS sites, $i.e.$ the promoter atom in the Co-Mo-S phase or in the analogous Ni-Mo-S phase (see Section 1.2.4.1).

1.3.1. CHEMISORPTION OF ORGANONITROGEN COMPOUNDS.

In general N-containing compounds, particularly the more basic ones are capable of strongly adsorbing on catalyst surfaces and they may actually act as inhibitors for other reactions implicated in hydrotreating [35]. Thus the information of the interactions between N-heterocycles and the catalyst surface sites is important not only in the context of HDN but also of that related to poisoning of HDS and hydrogenation activity of sulfide catalysts. The present knowledge on these interactions is much more limited than the one gained in the field of metal-thiophene binding, and a number of -often contradictory and scattered- proposals have been put forward in the HDN literature. No general trends can be safely invoked thus far to relate the different bonding situations of organonitrogen compounds in metal centers with particular reactivity patterns in the context of HDN catalysis.

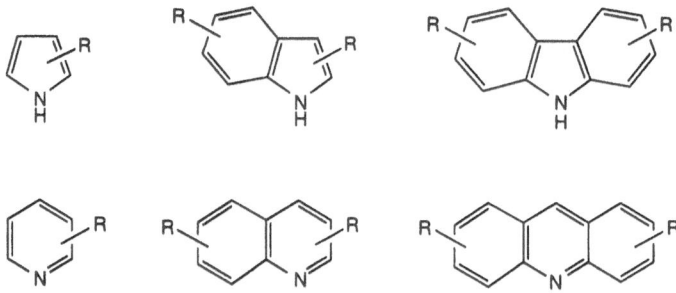

Fig. 1.11. Main classes of refractory organonitrogen compounds in crudes and distillates
(R is a short chain alkyl group).

Basic nitrogen heterocycles such as quinoline and related molecules interact with the metal sites on the catalyst surface in two principal modes: *end-on* through the nitrogen atom only (η^1N-coordination) or *flat* ($\pi-$ or η^6-coordination) through either the heterocycle or the arene ring (see Fig. 1.12). It is also generally agreed that in the absence of α-substituents that might produce serious steric congestion, quinoline uses preferentially the N-atom for interaction with the catalyst. In contrast, for non-basic rings such as pyrrole and its higher homologues, coordination through the N-atom only is, in principle, possible but it is disfavored over π-bonding. The delocalized type bonding is more frequently observed in the η^2–mode through the 2,3-C=C bond (or the tautomeric η^2(C=N)), (see Fig. 1.12); in the flat η^5 manner through the heterocycle, or η^6 through the entire carbocyclic ring.

No simple correlations between the bonding modes and reactivity patterns during HDN reaction networks are clear at present, but nevertheless, some interesting general ideas have been advanced in this respect. Hydrogenation of the heterocyclic rings may take place through initial η^1N-bonding, although some authors invoke some π-bonding, or even both types of coordination in equilibrium. Arene hydrogenation requires some kind of π-bonding, either η^4 or η^6 (see Chapter 3), while C-N bond rupture most probably requires a strong interaction between the N-atom and the active site, but this alone may not be sufficient to produce the bond scission; this point is far from being resolved.

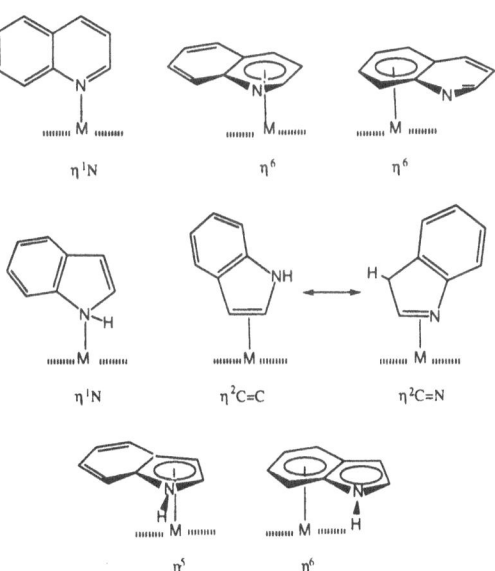

Fig. 1.12. Main adsorption modes of organonitrogen compounds.

1.3.2. MAIN REACTION PATHWAYS AND MECHANISMS IN HDN.

The mechanisms of heterogeneous HDN have been studied in some detail, albeit in much less depth than those of HDS, and the reader is referred to other reviews and to the original literature cited therein for further details [5, 13, 15, 20, 81-86]. The predominant reaction schemes are in essence different combinations of three major types of reactions that take place once the organonitrogen compound has been adsorbed:

-Hydrogenation of N-heterocycles
-Hydrogenation of arene rings
-Hydrogenolysis and cleavage of C-N bonds

This is illustrated in Fig. 1.13 for quinoline, one of the most frequently used model substrates for HDN mechanistic studies. Several points common to various proposals are worth mentioning: It is generally accepted that prehydrogenation of at least the N-heterocycle is mandatory for C-N bond breaking, in contrast to the case of HDS, where prehydrogenation of the S-ring is a possible pathway but not a prerequisite for C-S bond scission. It is clear that under operating conditions, hydrogenation reactions are reversible whereas C-N bond rupture is irreversible. Frequently the overall HDN rates are dictated by a hydrogenation rate rather than by the C-N bond cleavage rate and therefore HDN kinetics must always be thoroughly analyzed in order to avoid misleading conclusions. High pressures of hydrogen are needed in order to increase the equilibrium concentrations of saturated amines.

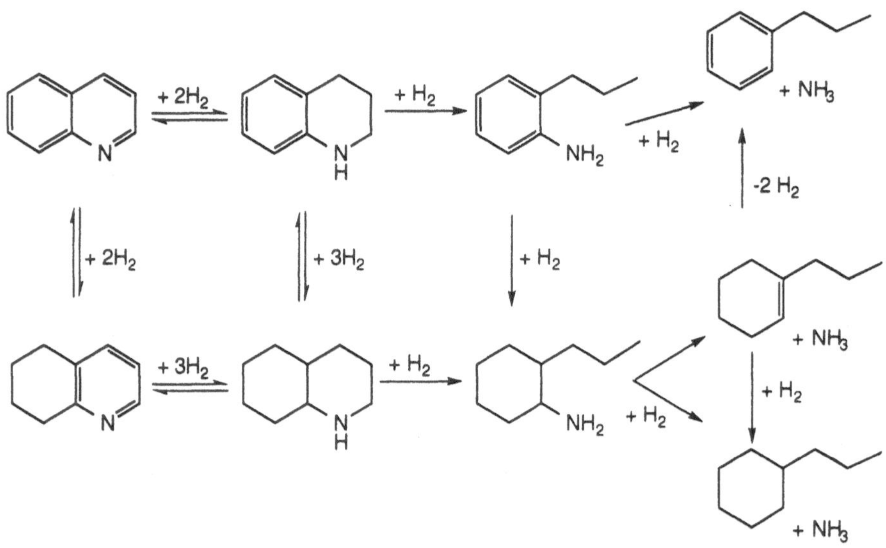

Fig. 1.13. Main reaction pathways in HDN of quinoline.

The preferred reaction pathway in Fig. 1.13 would be in principle the one involving pre-saturation of only the heterocyclic ring, prior to denitrogenation to yield *n*-propylbenzene, since this would consume less hydrogen and retain some aromaticity, thereby resulting in a lower decrease of the octane rating. However, many studies indicate that the first C-N bond that is broken in *e.g.* 1,2,3,4-tetrahydroquinoline (THQ) is the C(sp³)-N bond, leading to the corresponding *ortho*-propylaniline intermediate; in order to break the remaining C-N bond, the slow pre-saturation of the arene ring is the predominant pathway and thus the main product is *n*-propylcyclohexane. Nevertheless, denitrogenation without prehydrogenation has also been reported, while some authors consider that both rings are necessarily hydrogenated before any C-N bond is broken.

H₂S is known to inhibit hydrogenation reactions while at the same time promoting C-N bond breaking. Depending on the substrate and the reaction conditions, the net result may be a mild overall promotion or inhibition, or a compensation of both effects. The product distribution, however, is markedly –and reversibly- affected by the amount of H₂S present in the mixture, high concentrations leading mainly to denitrogenated aromatics instead of fully saturated hydrocarbons [15, 81, 82]. Two mechanisms have been proposed for the C-N bond breaking process based on acid-base interactions, as summarized in Fig. 1.14 [81, 83-86]. The first pathway corresponds to the classical Hoffmann degradation (Fig. 1.14 (a)) in which the N-atom is protonated and subsequently a base abstracts a proton from the β-carbon, causing elimination of NH₃ and the consequent generation of an olefin -which would be quickly hydrogenated under HDN conditions. Alternatively, the base can attack the α carbon of the quaternized amine in what is termed the nucleophilic substitution mechanism (Fig. 1.14 (b)) leading to elimination of NH₃ and formation of an intermediate containing the nucleophile. Applying these general mechanisms to the case of quinoline provides the reaction networks depicted in Fig. 1.15.

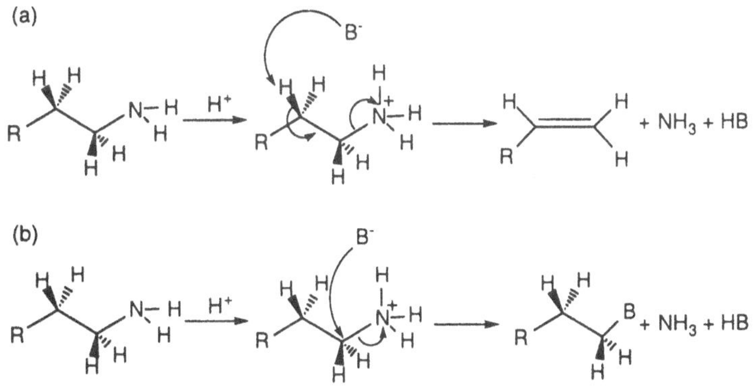

Fig. 1.14. Mechanisms of C-N bond breaking:
(a) Hoffmann elimination and (b) Nucleophilic substitution

In the Hoffmann-type elimination, a surface –SH group is the base required to remove the proton from the β-carbon to yield the 2-n-propenylaniline intermediate, which is subsequently hydrogenated to 2-n-propylaniline. Alternatively, Nelson and Levy [88] invoke quaternization of the nitrogen by a surface proton, followed by nucleophilic displacement of the –NH$_2$ moiety by an SH$^-$ group in the catalyst generating a transient thiol which is easily desulfurized under the reaction conditions, to yield the same product, 2-n-propylaniline. A study of the HDN of 1,2,3,4-tetrahydroisoquinoline revealed the exclusive formation of the aromatic denitrogenated product, which can only be generated by the latter route, since a Hoffmann elimination would lead to a fully saturated product [81, 88]. The promoting effect of H$_2$S on denitrogenation would be explained by an increase in the Brönstead acidity of the catalyst required to protonate the nitrogen base; surface –SH$^-$ groups would be the acting conjugate base in the reaction.

In an interesting variation, Laine has proposed a "modified Hoffmann mechanism" [83, 89, 90] in which the nitrogen is not protonated but instead it interacts strongly with a cationic metal site on the surface. As shown in Fig. 1.16, the metal first activates a C-H bond at the α position of e.g. piperidine leading to two possible hydride intermediates, viz. a metallaaziridinium species (pathway a) or a metal-iminnium complex (pathway b). Subsequent migration of the hydride to the α carbon in (a) or the N atom in (b) produces a metal alkyl or a metal alkylidene intermediate, respectively, both of which are further hydrogenated and ultimately decomposed through C-N bond scission.

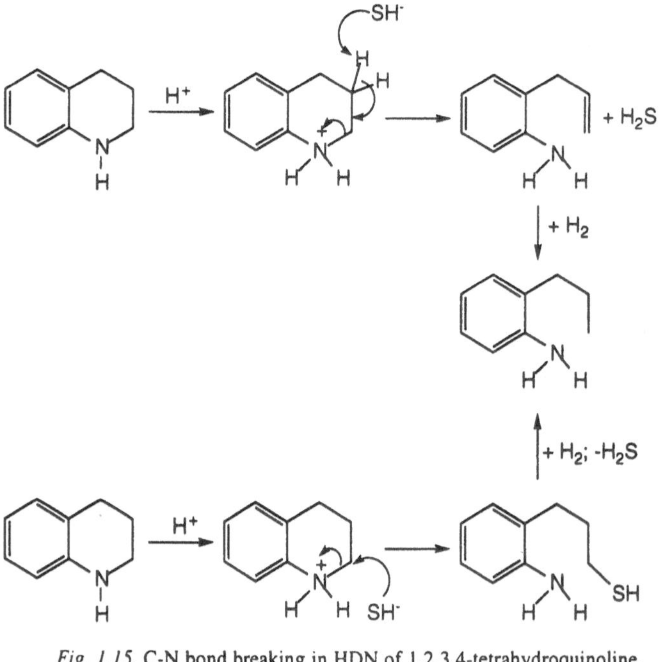

Fig. 1.15. C-N bond breaking in HDN of 1,2,3,4-tetrahydroquinoline. Hoffmann elimination (top) and nucleophilic substitution (bottom).

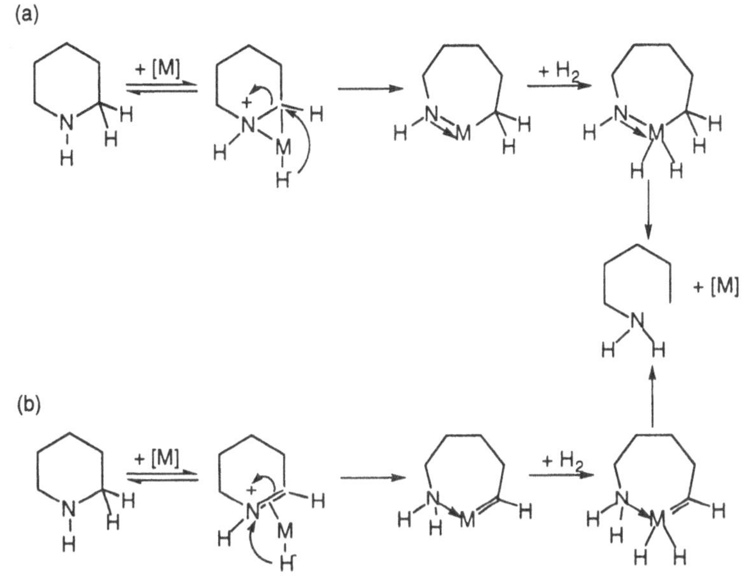

Fig. 1.16. Laine's mechanism for C-N bond breaking in HDN of piperidine.

The influence of H_2S in the overall HDN rate is explained in this model by a nucleophilic attack on a metal-*n*-pentylamine adduct causing the displacement of ammonia and the consequent formation of a thiolate intermediate which is readily desulfurized under the reaction conditions.

Concerning the organometallic chemistry related to HDN, the achievements reported to date are considerably less impressive than those related to HDS, and the relevant literature is rather scarce and scattered. Bianchini and coworkers have very recently published the only review so far available on this topic, including a short description of the main aspects of heterogeneous HDN [35]. The mechanisms of the hydrogenation of some N-heterocycles, notably quinoline to 1,2,3,4-tetrahydroquinoline in solution have been studied in some detail and are now well understood; this is discussed in Chapter 3. Although a large number of complexes containing aliphatic and aromatic amines have been known for a long time, very little is still known about the way in which C-N bonds can be broken, although some very interesting model systems have begun to appear. They will be discussed in Chapter 6.

In conclusion, this first Chapter attempts to provide the non specialist with and a general basis for understanding the importance and the major advances and drawbacks of heterogeneous HDS and HDN reactions, as well as the principal challenges that need to be addressed in the future. It is also meant to provide the context in which the organometallic chemistry related to these processes will be discussed in the rest of the book. Chapter 2 will be devoted to the description of metal complexes of the thiophenes, their syntheses, structures, bonding characteristics and reactivity patterns.

Chapter 3 deals with the metal complex catalyzed homogeneous hydrogenation of aromatic hydrocarbons, including nitrogen- and sulfur-heteroaromatic compounds, while Chapter 4 describes the major accomplishments reported in the literature on C-S bond rupture through stoichiometric and catalytic hydrogenolysis and related reactions of thiophenes and related substrates by transition metal complexes and clusters in homogeneous solutions. Despite the fact that in these cases the catalysts and the reaction conditions are generally very different from those employed in real HDS systems, the mechanisms deduced by use of this approach are sound and can be understood in great depth. Thus they are particularly useful in aiding this discussion and in allowing a distinction between the most reasonable proposals in heterogeneous catalysis and the less reasonable ones, which certainly represents a valid alternative method complementing the results obtained from analytical surface chemistry and heterogeneous catalysis and kinetic studies. In Chapter 5, the activation of hydrogen by metal complexes in relation with HDS will be discussed, including the possible roles of sulfur moieties in the H-H bond breaking process. Finally, the question of organometallic modeling of HDN will be addressed in Chapter 6, including binding modes of aromatic and aliphatic amines to metal centers in complexes, and some C-N bond breaking reactions of such amines and related molecules coordinated to metal containing molecular fragments, in the hope of stimulating further work in this hitherto somewhat neglected area of research. The book ends with a short epilogue in which an attempt is made to provide an overall connection between the wealth of information acquired from organometallic and coordination chemistry and the knowledge gained from surface chemistry and heterogeneous catalysis. This results in a rather provocative proposal of a possible generic structure of the active sites and a mechanistic picture for HDS and HDN on supported Co-Mo catalysts.

References

1.   P. C. H. Mitchell: *The Chemistry of Some Hydrodesulphurisation Catalysts Containing Molybdenum*, Climax Molybdenum Co. Ltd., London (1967).
2.   S. C. Schuman, H. Shalit: *Catal. Rev.* **4**, 245 (1970).
3.   O. Weisser, O. Landa: *Sulfide Catalysts. Their Properties and Applications*, Pergamon, Oxford (1973).
4.   F. E. Massoth: *Adv. Catal.* **27**, 265 (1978).
5.   B. C. Gates, J. R. Katzer, G. C. A. Schuit: *Chemistry of Catalytic Processes*, McGraw-Hill, New York (1979).
6.   P. Grange: *Catal. Rev.-Sci. Eng.* **21**, 135 (1980).
7.   J. G. Speight: *The Desulfurization of Heavy Oils and Residua*, Dekker, New York (1981).
8.   P. C. H. Mitchell, in C. Kemball, D. A. Dowden (eds.): *Catalysis, Specialist Periodical Reports*, Royal Society of Chemistry, London, Vol. 4, p. 175 (1981).
9.   H. Topsøe, B. S. Clausen: *Catal. Rev.-Sci. Eng.* **26**, 395 (1984).
10. M. Zdrazil: *Catalysis Today* **3**, 269 (1988).
11. R. Prins, V. H. J. de Beer, G. A. Somorjai: *Catal. Rev.-Sci. Eng.* **31**, 1 (1989).
12. B. Delmon: *Stud. Surf. Sci. Catal.* **53**, 1 (1990).
13. M. J. Girgis, B. C. Gates: *Ind. Eng. Chem. Res.* **30**, 2021 (1991).
14. A. N. Startsev: *Catal. Rev.-Sci. Eng.* **37**, 353 (1995).
15. H. Topsøe, B. S. Clausen, F. E. Masoth, in J. R. Anderson, M. Boudart, eds.: *Catalysis Science and Technology*, Vol. 11, Springer-Verlag, New York (1996).
16. D. D. Whitehurst, T. Isoda, I. Mochida: *Adv. Catal.* **42**, 345 (1998).
17. T. Weber, R. Prins, and R. A. van Santen, *Transition Metal Sulphides. Chemistry and Catalysis*, NATO ASI Series, Kluwer, Dordrecht, 1998.
18. A. N.Startsev, A. V. Kalinkin, I. I. Zakharov, D. G. Aksenov and V. N. Parmon: *J. Mol. Cat. A: Chem.* **151**, 171, (2000).

19. R. Shafi and G. J. Hutchings: *Catal. Today* **59**, 423 (2000).
20. M. V. Landau: *Catal. Today* **36**, 393 (1997).
21. R. J. Angelici: *Acc. Chem. Res.* **21**, 387 (1988).
22. R. J. Angelici: *Coord. Chem. Rev.* **105**, 61 (1990).
23. T. B. Rauchfuss: *Prog. Inorg. Chem.* **39**, 259 (1991).
24. B. C. Wiegand, C. M. Friend: *Chem. Rev.* **92**, 1 (1992).
25. R. A. Sánchez-Delgado: *J. Mol. Catal.* **86**, 287 (1994).
26. R. J. Angelici in R. B. King, ed.: *Encyclopedia of Inorganic Chemistry*, Wiley, New York, Vol. 3. P.1433 (1994).
27. R. J. Angelici: *Bull. Soc. Chim. Belg.* **104**, 265 (1995).
28. C. Bianchini, A. Meli in B. Cornils, W. A. Herrmann, eds.: *Applied Homogeneous Catalysis with Organometallic Compounds*, VCH, Weinheim, Vol. 2, p. 969 (1996).
29. C. Bianchini, A. Meli: *J. Chem. Soc. Dalton Trans.* 801 (1996).
30. R. J. Angelici: *Polyhedron* **16**, 3073 (1997).
31. C. Bianchini, A. Meli: *Acc. Chem. Res.* **31**, 109 (1998).
32. C. Bianchini, A. Meli, in B. Cornils, W. A. Herrmann, eds.: *Aqueous-Phase Organometallic Catalysis* VCH, Weinheim (1999).
33. R. A. Sánchez-Delgado, in I. T. Horváth, ed.: *Encyclopedia of Catalysis* John Wiley & Sons., New York (in press).
34. R. J. Angelici: *Organometallics* **20**, 1259 (2001).
35. C. Bianchini, A. Meli and F. Vizza: *Eur. J. Inorg. Chem.* **2001**, 46 (2001).
36. M. D. Curtis and S. H. Drucker: *J. Am. Chem. Soc.* **119**, 1027 (1997).
37. T. A. Pecoraro and R. R. Chianelli: *J. Catal.* **67**, 430 (1981).
38. J. P. R. Vissers, C. K. Groot, E. M. van Oers, V. H. J. de Beer and R. Prins: *Bull. Soc. Chim. Belg.* **93**, 813 (1984).
39. M. J. Ledoux, O. Michaud, G. Agostini and P. Panissod: *J. Catal.* **102**, 275 (1986).
40. P. Raybaud, G. Kresse, J. Hafner, H. Toulhoat: *J. Phys. Condens. Matter* **9**, 11085 (1997).
41. H. Toulhoat, P. Raybaud, S. Kasztelan, G. Kresse and J. Hafner: *Prep. Symp.Div. Petrol. Chem. Am. Chem Soc.* **42**, 114 (1997).
42. Y. Aray, J. Rodriguez, D. Vega and E. N. Rodriguez-Arias: *Angew. Chem. Int. Ed. Engl.*, in press (2000).
43. Y. Aray and J. Rodriguez, *CHEMPHYSCHEM* (in press).
44. S. Hatanaka, M. Yamada and O. Sadakane: *Ind. Eng. Chem. Res.* **36**, 1519 (1997).
45. S. Hatanaka, M. Yamada and O. Sadakane: *Ind. Eng. Chem. Res.* **36**, 5110 (1997).
46. S. Helveg, J. V. Lauritsen, E. Lægsgaard, I. Stensgaard, J. K. Nørskov, B. S. Clausen, H. Topsøe and F. Besenbacher: *Phys. Rev. Letters* **84**, 951 (2000).
47. J. V. Lauritsen, S. Helveg, E. Lægsgaard, I. Stensgaard, B. S. Clausen, H. Topsøe and F. Besenbacher: *J. Catal.* **197**, 1 (2001).
48. L. S. Byskov, J. K. Nørskov, B. S. Clausen and H. Topsøe: *J. Catal.* **187**, 109 (1999).
49. P. Raybaud, J. Hafner, G. Kresse and H. Toulhoat: *Surf. Sci.* **407**, 237 (1998).
50. P. Raybaud, J. Hafner, G. Kresse, S. Kasztelan and H. Toulhoat: *J. Catal.* **190**, 129 (2000).
51. L. S. Byskov, B. Hammer, J. K. Nørskov, B. S. Clausen and H. Topsøe, *Catal. Letters* **47**, 177 (1997).
52. L. S. Byskov, J. K. Nørskov, B. S. Clausen and H. Topsøe, in T. Weber, R. Prins, and R. A. van Santen, *Transition Metal Sulphides. Chemistry and Catalysis*, NATO ASI Series, Kluwer, Dordrecht, 1998, Vol. 60.
53. S. H. Strauss: *Chemtracts-Inorg.* **6**, 1 (1994).
54. S. Kasztelan and D. Guillaume: *Ind. Eng. Chem. Res.* **33**, 203 (1994), and references therein.
55. M. Lacroix, S. Yuan, M. Breysse, C. Dorémieux-Morin and J. Fraissard: *J. Catal.* **138**, 409 (1992).
56. M. Lacroix, C. Mirodatos, M. Breysse, T. Decamp and S. Yuan,, in: L. Guczi (Ed.): *Proc. 10th Int. Congr. Catal.*, Elsevier, Amsterdam, p. 34 (1992).
57. A. B. Anderson, Z. Y. Al-Saigh and K. W. Hall: *J. Phys. Chem.* **92**, 803 (1988).
58. I. I. Zakharov, A. N. Startsev, G. M. Zhidomirov and V. N. Parmon: *J. Mol. Cat. A: Chem.* **137**, 101, (1999).
59. B. R. James: *Homogeneous Hydrogenation*, John Wiley & Sons, New York (1982).
60. J. P. Collman, L. S. Hegedus, J. R. Norton and R. G. Finke: *Principles and Applications of Organo-Transition Metal Chemistry*, University Science Books, Ca. (1987).
61. G. R. Parkin: *Prog. Inorg. Chem.* **47**, 1 (1998).
62. J. M. J. G. Lipsch and G. C. A. Schuit: *J. Catal.* **15**, 179 (1969).
63. H. Kwart, G. C. A. Schuit and B. C. Gates: *J. Catal.* **61**, 128 (1980).
64. S. Harris and R. R. Chianelli: *J. Catal.* **86**, 400 (1984).
65. W. X. S. O'Brien, J. W. Chen, R. V. Nayak, G. S. Carr: *Ind. Eng. Chem. Process Res. Dev.* **25**, 221 (1986).
66. K. F. McCarthy and G. L. Schrader: *J. Catal.* **103**, 261 (1987).
67. N. Startsev, V. A. Burmistrov, Yu. I. Yermakov: *Appl. Catal.* **45**, 191 (1988).
68. R. J. Mikovsky, A. J. Sivestri, H. Heinemann: *J. Catal.* **34**, 324 (1974).
69. M. Zdrazil: *Appl. Catal.* **4**, 107 (1982).
70. F. Ruette and E. Ludeña: *J. Catal.* **67**, 266 (1981).
71. N. N. Sauer, E. J. Markel, G. L. Schrader and R. J. Angelici: *J. Catal.* **117**, 295 (1989).

72. Delmon and J.-L. Dallons: *Bull. Soc. Chim. Belg.* **97**, 473 (1988).
73. H. Schulz and D.-V. Do: *Bull. Soc. Chim. Belg.* **93**, 645 (1984).
74. V. Lamure-Meille, E. Schulz, M. Lemaire and M. Vrinat: *Appl. Catal. A* **131**, 143 (1995).
75. V. Meille, E. Schulz, M. Lemaire and M. Vrinat: *J. Catal.* **170**, 9 (1997).
76. V. Meille, E. Schulz, M. Lemaire and M. Vrinat: *Appl. Catal. A* **87**, 179 (1999).
77. M. Macaud, A. Milenkovic, E. Schultz, M. Lemaire and M. Vrinat: *J. Catal.* **193**, 255 (2000).
78. M. Houalla, D. Broderick, A. V. Sapre, N. K. Nag, V. J. H. de Beer, B. C. Gates and H. Kwart: *J. Catal.* **61**, 523 (1980).
79. X. Ma, K. Sakanishi, T. Isoda and I. Mochida: *Pap. Am. Chem. Prepr. Div. Petrol. Chem.* **39**, 622 (1994).
80. G. H. Singhal, R. L. Espino, J. E. Sobel and G. A. Huff: *J. Catal.* **67**, 457 (1981).
81. G. Perot: *Catal. Today* **10**, 447 (1991).
82. S. Kasztelan, T. Des Courières and M. Breysse: *Catal. Today* **10**, 433 (1991).
83. R. M. Laine: *Catal. Rev.-Sci. Eng.* **25**, 459 (1983).
84. T. C. Ho: *Catal. Rev.-Sci. Eng.* **30**, 117 (1988).
85. J. R. Katzer and R. Sivasubramanian: *Catal. Rev.-Sci. Eng.* **20**, 155 (1979).
86. R. Prins, M. Jian and M. Flechsenhar: *Polyhedron* **16**, 3235 (1997).
87. T. L. Cottrell: *The Strengths of Chemical Bonds*, 2nd. Ed., Butterworth, London (1958).
88. N. Nelson and R. B. Levy: *J. Catal.* **58**, 485 (1979).
89. Eisenstadt, C.M. Giandomenico, M. F. Frederick and R. M. Laine: *Organometallics* **4**, 2033 (1985)
90. R. M. Laine: *J. Mol. Catal.* **21**, 119 (1983).

CHAPTER 2
COORDINATION AND ACTIVATION OF THIOPHENES
IN METAL COMPLEXES

2.1. Introduction

Understanding HDS reactions requires a good comprehension of the various steps involved in the overall process; taking into account the generally accepted mechanisms, one very important goal is to try to define the principal ways in which thiophenic molecules are activated through their bonding interactions with metal centers on catalytically active surfaces, a phenomenon commonly referred to as *chemisorption* in the language of surface chemistry. Thiophenes are usually envisaged as aromatic compounds regarding their structures and reactivities, although precise structural microwave data for thiophene itself indicate that electron delocalization is considerably lower than that for *e.g.* benzene [1, 2]. In terms of bonding capacity, the likely sites for coordination to metal centers are clearly the areas of highest electron density, *viz.* the C=C bonds and the lone pairs at the sulfur atom. As mentioned in Section 1.2.6, a number of proposals have been put forward over the years for the modes in which thiophenes can interact with surfaces, the most important ones being the so called "one point adsorption", in which the strong interaction is thought to occur between the sulfur atom and vacancy on the metal ion in the surface, and the "multi-point adsorption" involving also one or both of the C=C bonds in a delocalized π-bonding situation (See Fig. 1.6) [1].

While it is very difficult to experimentally obtain detailed information on the bonding of thiophenes to metal centers in extended solids, the use of routine techniques such as NMR spectroscopy and single crystal X-ray diffraction has allowed the unambiguous identification of a variety of coordination modes of thiophenes in metal complexes [2-7]. The most important bonding situations authenticated to date are summarized in Fig. 2.1, together with the common nomenclature for organometallic compounds based on "hapticity", a notation that continues to gain adepts in the surface chemistry literature. Complexes are known in which thiophenes coordinate to metal centers through the sulfur only ($\eta^1 S$), through a C=C bond ($\eta^2(C=C)$), through four carbons (η^4) or through the entire thiophene ring (η^5). Benzothiophenes are known to bind to metals through the sulfur atom ($\eta^1 S$), through the 2,3-C=C bond ($\eta^2(C=C)$), through the sulfur atom and the 2,3-C=C bond ($\eta^3(S,C=C)$), through four carbon atoms of the benzene ring (η^4) and through the six carbon atoms of the benzene ring (η^6). Dibenzothiophenes in turn coordinate through the sulfur atom ($\eta^1 S$), through four carbon atoms of the benzene ring (η^4) and through the six carbon atoms of the benzene ring (η^6).

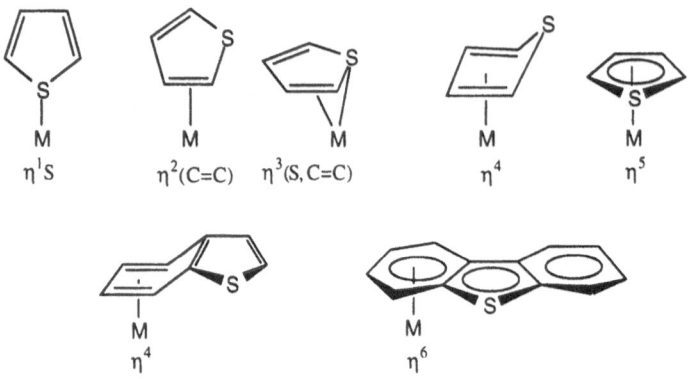

Fig. 2.1. Bonding modes of thiophenes known in metal complexes

On comparing the structures represented in Fig. 2.1 with those contained in Fig. 1.6 (Chapter 1), the analogies between the known bonding modes in metal complexes and the proposed modes of chemisorption on catalytic solids become evident; furthermore, other possibilities not previously envisaged on solids emerge on extrapolating the knowledge of molecular chemistry to surface situations. Most of the organometallic compounds known have been well characterized by NMR spectroscopy and chemical methods, and a good number of X-ray structures of complexes containing thiophenic ligands are now available, providing a sound basis for the understanding of chemisorption of such entities. Besides the experimental data at hand, extensive theoretical calculations have been performed on metal thiophene complexes including Fenske-Hall [8-10], CNDO [11, 12], and more recently *ab initio* [13-15] methods, which have contributed in an important manner to a better understanding of the electronic structures of such compounds.

Perhaps more importantly for our purposes, some general patterns have begun to emerge concerning the reactivity associated with thiophenes in each particular bonding situation. This allows some interesting parallels to be drawn with the ways in which thiophenes are known or thought to be degraded after being chemisorbed on the active sites of heterogeneous HDS catalysts. The discussion in this Chapter has been organized in terms of the different coordination modes, describing in each Section the most relevant aspects of the syntheses, the structures, the bonding and the reactions of metal complexes of the thiophenes; also, an attempt has been made, whenever it seemed possible, to correlate this molecular information with what is known or believed to take place in heterogeneous HDS catalysis. Some excellent earlier reviews on some of these aspects are available, notably those by Angelici [2, 4-6] and Rauchfuss [3]. Although some overlap with those papers –hopefully not an excessive one- was unavoidable, an update of this chemistry covering the literature up to the middle of the year 2001, organized in the manner described seemed appropriate here.

2.2. η^1S-Bonded metal thiophene complexes

In the η^1S binding mode thiophenes may be formally envisaged as neutral 2-electron ("L-type") donor ligands occupying a single coordination site on the metal center. Some of the earliest proposals for a heterogeneous HDS mechanism involved the adsorption of thiophene through the sulfur atom only, and thus the successful synthesis and full characterization of η^1S complexes, as well as the theoretical studies on their bonding characteristics have provided a sound basis for the understanding of the so called "one point adsorption" on solid catalysts [16-20]. Table 2.1 provides a fairly comprehensive list of known η^1S complexes of the thiophenes [21-34]; NMR data have been most useful in their characterization and a number of them have also been studied crystallographically. An interesting feature is the fact that in all cases where X-ray structures are available, the sulfur atoms may be described as pyramidal, corresponding to an approximate sp^3 hybridization although the ring itself is planar and perturbed only by a slight lengthening of the C-S bonds. This indicates that this type of coordination might not result in a very large activation of the heterocycle.

TABLE 2.1. Known η^1S complexes of the thiophenes.

Complex	Characterization	Refs.
Cr(CO)$_5$(Th)	NMR	21
(Th = 2,5-Me$_2$T, BT, 3-MeBT)		
Cr(CO)$_5$(DBT)	NMR + X-rays	21
Mo(CO)$_3$[2,5-(Ph$_2$PCH$_2$CH$_2$)$_2$T]	NMR + X-rays	32
Mo(CO)$_5$(DBT)	NMR	21
[Me$_2$Si(C$_5$Me$_4$)Mo(DBT)]	NMR + X-rays	22
W(CO)$_5$(Th)	NMR	21
(Th = 2,5-Me$_2$T, BT)		
W(CO)$_5$(DBT)	NMR + X-rays	21
CpMn(CO)$_2$(Th)	NMR	21
(Th = 2,5-Me$_2$T, BT)		
CpMn(CO)$_2$(DBT)	NMR + X-rays	21
Cp*Re(CO)$_2$(T)	NMR + X-rays	23
[CpRe(NO)(PPh$_3$)(Th)]BF$_4$	NMR	
(Th = T, 2-MeT, 2,5-Me$_2$T, BT, 2-MeBT, 3-MeBT)		24, 25
[CpFe(CO)$_2$(T)]BF$_4$	NMR	26, 27
[CpFe(CO)$_2$(2,5Me$_2$T)]PF$_6$	NMR	28
[CpFe(CO)$_2$(BT)]BF$_4$	NMR	27
[CpFe(CO)$_2$(DBT)]BF$_4$	NMR + X-rays	27
[C$_5$H$_4$CH$_2$-2-C$_4$H$_3$S)Ru(PPh$_3$)$_2$]BF$_4$	NMR + X-rays	29
RuCl$_2$(Ph$_2$P-DBT)$_2$	NMR + X-rays	30
RuCl$_2$(Ph$_2$P-DBT)$_2$(CO)	NMR + X-rays	30
[CpRu(CO)(PPh$_3$)(2-MeT)]BF$_4$	NMR + X-rays	31
[CpRu(CO)(PPh$_3$)(Th)]BF$_4$	NMR	
(Th = T, 2,5-Me$_2$T, 3-MeT, BT, Me$_4$T, DBT)		31
{Co(CO)$_2$[2,5(Ph$_2$PCH$_2$CH$_2$)$_2$T]}BPh$_4$	NMR	32
{Rh(CO)[2,5-(Ph$_2$PCH$_2$CH$_2$)$_2$T]}BPh$_4$	NMR + X-rays	32
[Ir(H)$_2$(PPh$_3$)$_2$(Th)]PF$_6$	NMR + X-rays	33
(Th = T, THT, BT, DHBT, DBT)		
Cp*IrCl$_2$(DBT)	NMR + X-rays	34

Cp = C$_5$H$_5$; Cp* = C$_5$Me$_5$;T = thiophene; BT = benzothiophene; DBT = dibenzothiophene

The values of the angle θ between the ring plane and the vector defined by the M-S bond (Fig. 2.2) clearly demonstrate that thiophene is invariably tilted away from a perpendicular arrangement. Values ranging from 16° in [Me$_2$Si(C$_5$Me$_4$)Mo(DBT)] [22] to 61° in [CpRu(CO)(PPh$_3$)(2-MeT)]BF$_4$ [31] have been observed and the differences have been ascribed predominantly to steric effects. This is in agreement with elegant spectroscopic studies on the adsorption of thiophene on clean surfaces which have also concluded that thiophene was tilted away from perpendicularity at an angle of 30-40° [35-38], and this is now accepted as a general phenomenon both in metal complexes and on surfaces. This is an important advance in connection with the early view of the "one point adsorption", in which the thiophene was assumed to be located perpendicular to the catalyst surface, as was then known to be the case for adsorbed pyridine.

The bonding in η^{1}S complexes of thiophenes has been studied in detail by several theoretical methods, mainly by the groups of Harris [8, 10], Sánchez-Delgado [11-13] and Sargent [14]. In this binding mode the thiophenes may act as both σ and π donors and also as π acceptors; calculations show that the interaction is predominantly a ligand to metal donation of electron density from orbitals concentrated primarily on the sulfur lone pairs, in agreement with a small effect of coordination on the C-S bonds or on the ligand as a whole. However, the π acceptor ability of thiophene becomes important under certain conditions, since increasing the electron density on the metal weakens the C-S bond through an enhanced back donation into the corresponding antibonding π* orbital; this provides an adequate pathway for the activation of the coordinated thiophene. Indeed, it will be shown in Chapter 4 that electron rich metal fragments are particularly prone to promote ring-opening reactions of thiophenes through C-S bond activation. Tilting of the ring away from a perpendicular binding avoids a strong repulsive interaction between filled ligand π-orbitals and an occupied d$_\pi$ metal orbital [8, 10, 11, 13]. As an example, Fig. 2.3 shows the energy level diagram calculated by Harris for Cp*(CO)$_2$Re(η^{1}S-T) [10].

Examples of S-bonded thiophene complexes are known for a wide variety of metals, *viz.* Mn, Re, Cr, Mo, W, Fe, Ru, Co, Rh and Ir, as shown in Table 2.1; the procedures used to synthesize such compounds are usually straightforward, involving addition of the thiophene to an unsaturated metal precursor or displacement of a labile ligand. This type of binding has generally turned out to be weak, resulting in rather unstable compounds; the stability increases along the trend   thiophenes < benzothiophenes < dibenzothiophenes.

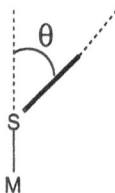

Fig. 2.2. Thiophene tilt angle (θ) in η^{1}S complexes.

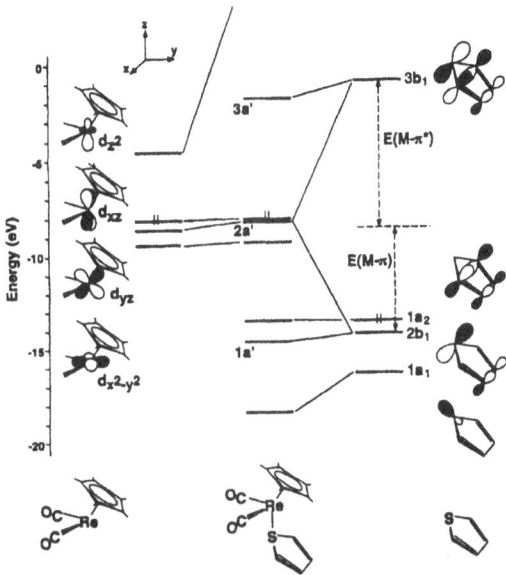

Fig. 2.3. Energy level diagram for Cp*(CO)₂Re(η¹S-T) (Reproduced with permission from Ref. 10).

Some early examples of S-bonded Ru-thiophene complexes provided by Rauchfuss were stabilized by attaching the thiophenic moiety to a strongly coordinating group such as a phosphine (**a**) [30] or a cyclopentadienyl(**b**) [29] but even in these cases dissociation of the thiophene was apparent from the NMR spectra.

(a)   (b)

Other attempts have been made to promote η¹S coordination by incorporating the thiophenic moiety in a polydentate ligand structure. By use of the ligand 2,5-bis(2-diphenylphosphinoethyl)-thiophene (P₂S), capable of chelating a metal by use of the thiophene sulfur plus the two phosphorus donor atoms, Mathieu and coworkers have prepared the complexes $Mo(CO)_3(P_2S)$ (**c**), $[Co(CO)_2(P_2S)]BPh_4$ (**d**) and $[Rh(CO)(P_2S)]BPh_4$ (**e**), the only isolable complexes of these three metals containing an S-bonded thiophene [32]. Thiophene rings have also been attached to 2,2-bipyridine, and the resulting N₂S ligand can bind to *e.g.* Ru(II) in hexacoordinated $[Ru(N_2S)_2Cl]^+$ in which one chelating ligand is coordinated through the three donor atoms and the other through the two nitrogens only [39].

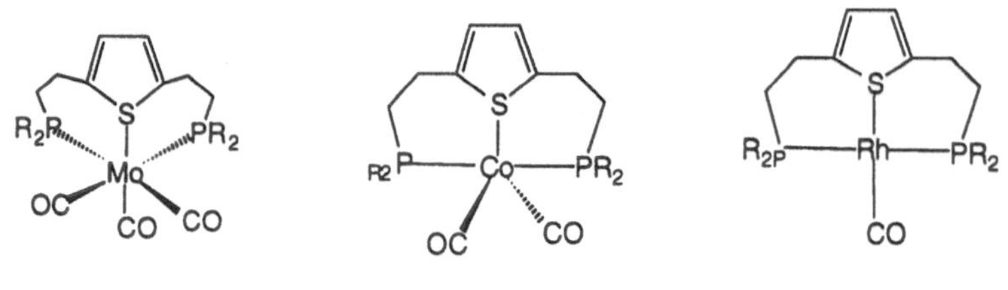

(c)   (d)   (e)

Replacement of a pyrrole ring by a thiophene in a porphyrin structure produces thioporphyrin ligands which bind to metals through the sulfur plus the remaining three nitrogen donors [40, 41]. Thus, S-binding is normally very weak, and as a consequence, in complexes of simple thiophenes even poorly coordinating solvents like dichloromethane or nitromethane have been reported to displace the S-heterocycle [6, 7]. This has limited the study of the reactivity of η^1-coordinated thiophenes, since ligand dissociation and/or exchange processes rather than reactions of the thiophene molecule itself dominate the behavior of such compounds in solution. Some exceptions to this behavior have been observed: for instance, the complex $Cp^*(CO)_2Re(\eta^1S\text{-}T)$ reacts with $Fe_2(CO)_9$ to yield the new thiophene-bridged heterobimetallic derivative $Cp^*(CO)_2Re(\mu\text{-}T)Fe(CO)_3$ ($Cp^* = C_5Me_5$) in which the thiophene ligand remains η^1S-bonded to Re and also binds in a η^4 fashion to the $Fe(CO)_3$ fragment [42] (Eq. 2.1). Since free thiophene fails to react with $Fe_2(CO)_9$, it is evident that S-bonding to rhenium makes it more susceptible to reactions with electrophiles.

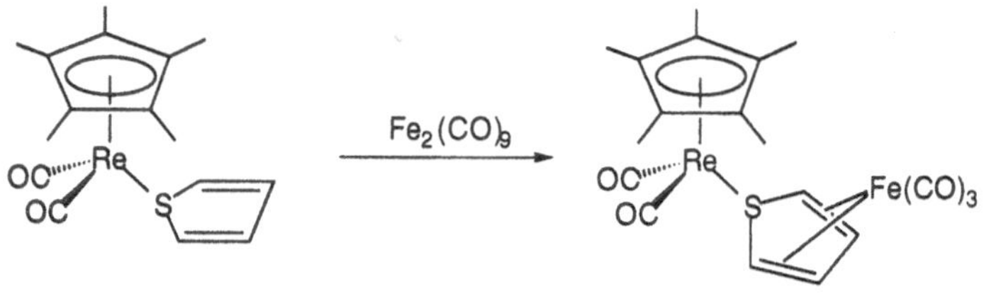

(2.1)

Also interestingly, bases such as OH⁻ or Et_3N induce the activation of the C-H bond α to S in $[Cp^*(NO)(PPh_3)Re(\eta^1S\text{-}T)]^+$ (Eq. 2.2); this C-H activation can be related to the mechanism of the α-H/D exchange known to occur in thiophene over solid HDS catalysts [43].

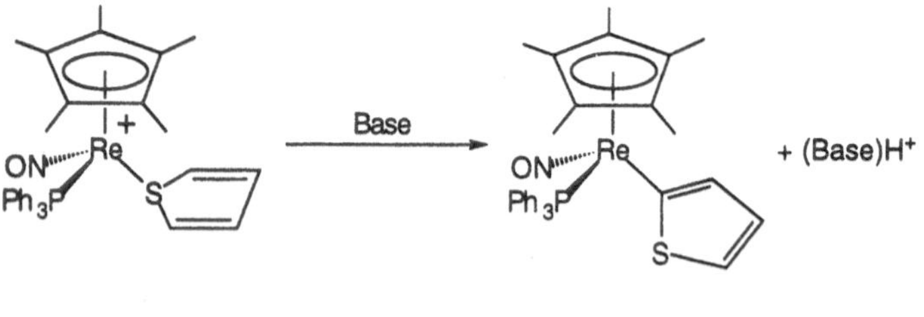

(2.2)

NMR measurements of the equilibrium constants for a series of ruthenium complexes $[Cp(CO)(L)Ru(\eta^1 S\text{-Th})]^+$ (Cp = cyclopentadienyl; L = CO, PPh_3; Th = T, 2-MeT, 3-MeT, $2,5\text{-Me}_2 T$, $Me_4 T$, BT, DBT)) (Eq. 2.3) have been performed by Angelici [44], showing that increased methylation on the thiophene results in stronger binding to the metal center and that steric effects are important for the stability of the complexes. $\eta^1 S$-BT and $\eta^1 S$-DBT complexes are also more stable than their T analogues. Tetrahydrothiophene (THT) was found to bind 7.1×10^6 times more strongly than thiophene.

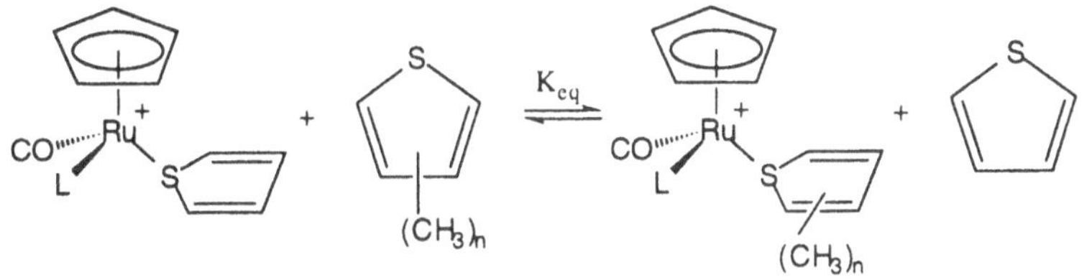

$$(2.3)$$

relative K values:
L = CO: T (1.0) < 2-MeT (3.30) < 3Me-T (4.76) < $2,5\text{-Me}_2 T$ (20.7) < BT (47.6) < $Me_4 T$ (887)
L = PPh_3: T (1.0) < $2,5\text{-Me}_2 T$ (2.76) < 2-MeT (4.11) < 3-MeT (6.30) < $Me_4 T$ (57.4) < BT (29.96) < DBT (74.1)

This is interesting in that it is in parallel with trends previously reported for the adsorption and desulfurization of thiophenes on $Co\text{-}Mo/Al_2 O_3$ catalysts [44]. Also important in connection with HDS mechanisms, Jones has convincingly argued that the S-bonded complex $Cp^*(\eta^1 S\text{-T})Rh(PMe_3)$ is the key –unobserved- precursor toward insertion of the metal into the C-S bond to form a thiametallacycle (Eq. 2.4, see also Chapter 4) [45]; a theoretical analysis subsequently carried out by Sargent and Titus is in agreement with this idea, and clearly shows that this is indeed a very reasonable reaction pathway for the activation of the thiophene toward ring opening on electron rich d^8 metal fragments [14].

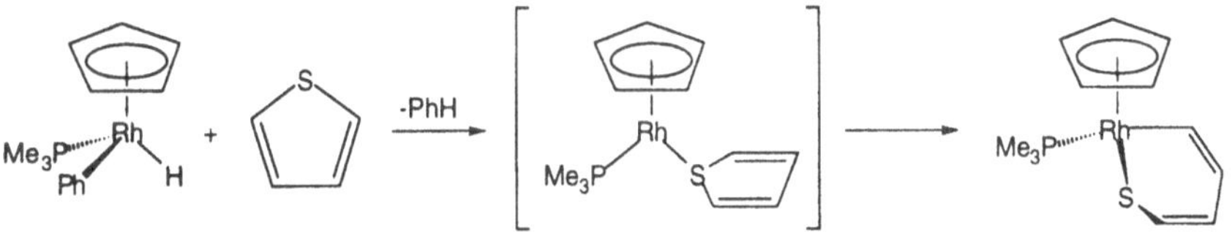

$$(2.4)$$

An interesting series of stable d^6 Ir(III) compounds $[Ir(H)_2(PPh_3)_2(Th)_2]PF_6$ (Th = T, BT, DBT, THT, DHBT) containing two thiophenes and two hydrides on the same metal atom was obtained by the mild hydrogenation of the diene in the complex $[Ir(COD)(PPh_3)_2]PF_6$ (COD = 1,5-cyclooctadiene) in the presence of the appropriate organosulfur ligand [33], as exemplified for thiophene in Eq. 2.5. The existence of these complexes demonstrates that a single highly unsaturated and electrophilic metal center is capable of simultaneously activating a hydrogen molecule plus two thiophenes; this is an intriguing possibility, even if it may arguably be an unlikely event in a surface containing several neighboring active centers. Whitehurst's proposal of the nature and reactivity of the active Co-Mo-S sites actually considers this possibility of thiophene and H_2 activation on the same metal center [46] (see Chapter 1, Fig. 1.4).

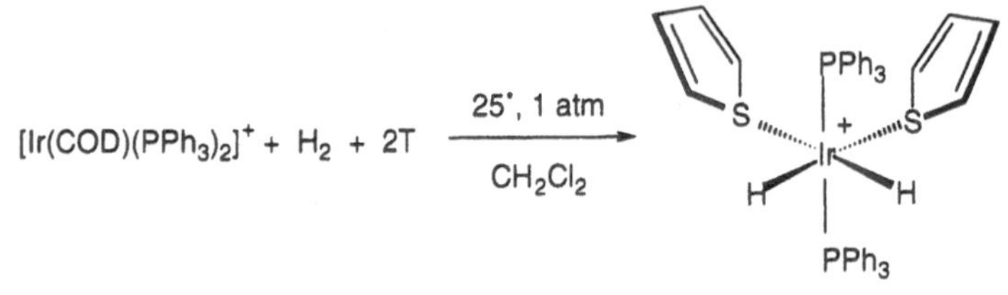

$$[Ir(COD)(PPh_3)_2]^+ + H_2 + 2T \xrightarrow[CH_2Cl_2]{25°, \text{ 1 atm}} \qquad (2.5)$$

Furthermore, the unusually high stability of these Ir(III) hydrido-thiophene complexes allowed some interesting studies to be carried out, involving the transformation of the coordinated thiophene through intramolecular migration of a coordinated hydride to the C=C bond (Eq. 2.6). This led to a particularly rare example of the mild homogeneous hydrogenation of thiophene to tetrahydrothiophene described in detail in Chapter 3.

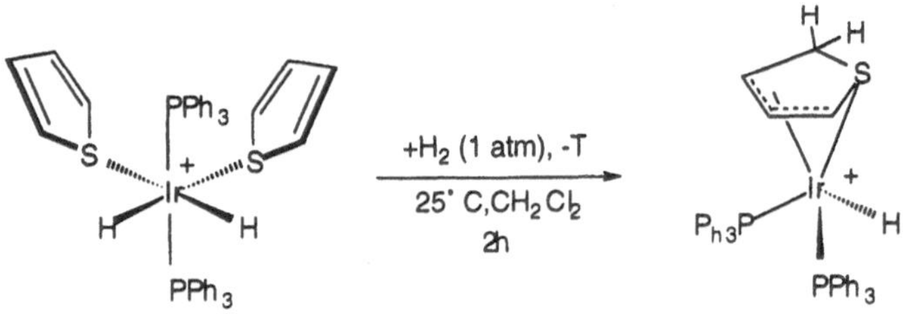

$$\xrightarrow[\substack{25° C, CH_2Cl_2 \\ 2h}]{+H_2 \text{ (1 atm), -T}} \qquad (2.6)$$

Until recently, one notable deficiency in the organometallic modeling of HDS-related structures had been the lack of well defined thiophene complexes of the metals commonly used in practical heterogeneous catalysts. A theoretical study carried out by the group of Sánchez-Delgado predicted as early as 1989 that the complex $Mo(CO)_5(\eta^1S\text{-}T)$ -as well as the related π-bonded $Mo(CO)_3(\eta^5\text{-}T)$ species- should be stable [11], but it was only ten years later that Angelici and his coworkers reported the spectroscopic (NMR) characterization of $Mo(CO)_5(\eta^1S\text{-DBT})$ as well as some evidence for the formation of its analogues containing $2,5\text{-Me}_2T$ and BT. These elusive compounds were obtained through UV photolysis of the metal hexacarbonyl in the presence of an excess of the appropriate thiophene (see e.g. Eq. 2.7). They were found to be indeed extremely labile, rapidly decomposing in solution through loss of the thiophene ligand; this instability made it impossible to isolate them in a pure form, thus precluding further reactivity studies. A number of chromium and tungsten analogues $M(CO)_5(\eta^1S\text{-Th})$ (M = Cr, Mo; Th = $2,5\text{-Me}_2T$, BT and DBT) were synthesized in an similar manner and they turned out to be considerably more stable, which allowed their spectroscopic characterization as well as the determination of the X-ray structures of both DBT derivatives [21].

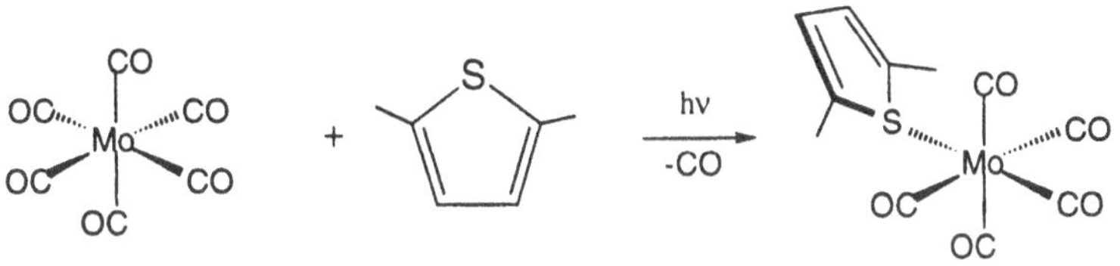

$$(2.7)$$

Almost simultaneously to Angelici's report, the first structural characterization of a molybdenum complex containing η^1S-bonded thiophenic ligand was achieved in Parkin's group, that of $ansa$-Cp$_2$''Mo(η^1S-DBT) [Cp$_2$'' = Me$_2$Si(C$_5$Me$_4$)$_2$]. This interesting compound was prepared by photolysis of the dihydride Cp$_2$''Mo(H)$_2$ in the presence of DBT (Eq. 2.8) [22]; its X-ray structure displayed the usual features of η^1S-bonded derivatives with the peculiarity of showing the smallest tilt angle θ thus far observed (16°). This was ascribed to steric interactions between the coordinated DBT and the methyl substituents on the Cp ligands. The reaction can be reversed by treatment of the DBT adduct with H$_2$ at 80 °C and atmospheric pressure:

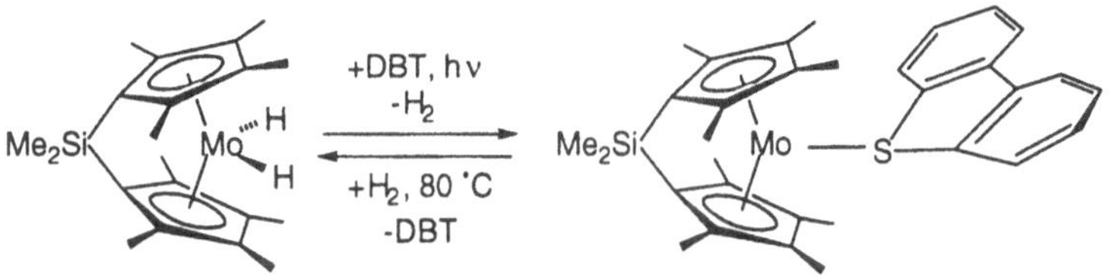

$$(2.8)$$

Interestingly this S-bonded DBT complex did not react further so as to cleave the C-S bond, whereas T and BT reacted photochemically with the dihydride under similar conditions by a ring opening C-S scission, without any η^1S intermediates being observed (See Chapter 4) [22]. No examples of stable well defined η^1S-bonded thiophene complexes of Co or Ni, the other important metals in industrial HDS catalysts have been reported to date.

2.3. η^2(C=C)-Bonded metal thiophene complexes

Examples of stable η^2(C=C) complexes which incorporate thiophenes as olefin-like 2-electron donor (L-type) ligands are very scarce, having been characterized only in a few cases. Taube reported the magnesium reduction of [Os(NH$_3$)$_5$(OTf)]$^{2+}$ in the presence of thiophene to yield {Os(NH$_3$)$_5$[η^2(C=C)-T]}$^{2+}$, which was characterized by NMR spectroscopy [47]. Several analogues with a variety of thiophenes, $viz.$ {Os(NH$_3$)$_5$[η^2(C=C)-Th]}$^{2+}$ (Th = 2-MeT, 3-MeT, 2,5-Me$_2$T, 2-MeOT, and BT), were isolated in a pure form by Spera and Harman from similar reactions and characterized by NMR spectroscopy [48]. These complexes readily add electrophiles to the sulfur atom and the resulting adducts react further with a variety of nucleophiles (H⁻, CN⁻, OAc⁻, py,

PrNH$_2$, N$_3^-$, PPh$_3$, PhO$^-$, PhS$^-$) to yield the corresponding ring-opened η^2-4-(alkylthio)-1,3-butadiene complexes (Eq. 2.9) [49, 50]:

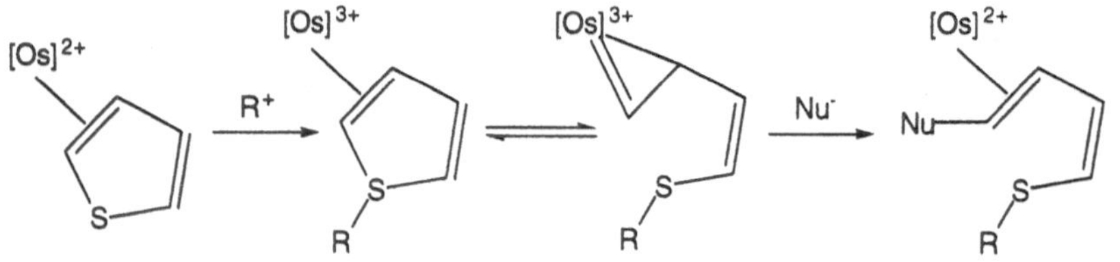

$$(2.9)$$

In contrast, protonation of {Os(NH$_3$)$_5$[η^2(C=C)-Th]}$^{2+}$ with triflic acid proceeds via addition to the exo side of the ring to give a η^2-2H-thiophenium product {Os(NH$_3$)$_5$[η^2(C=C)-Th.H]}$^{2+}$ (Eq. 2.10) [48]. The analogous {Os(NH$_3$)$_5$[η^2(C=C)-BT]}$^{2+}$ complex is protonated at the S- atom but the product was not stable enough to be isolated. Interestingly, the latter complex could be desulfurized with Raney Ni to yield the corresponding styrene derivative {Os(NH$_3$)$_5$[η^2(C=C)-styrene]}$^{2+}$ [51].

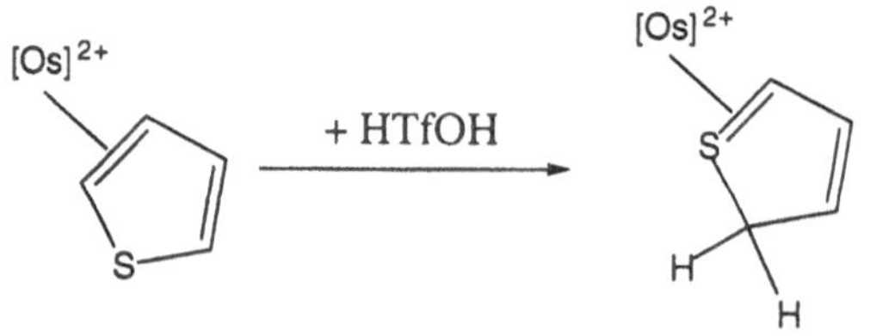

$$(2.10)$$

Angelici and coworkers detected by NMR spectroscopy the complexes Cp'Re(CO)$_2$[η^2(C=C)-BT] (Cp' = Cp, Cp*) which coexist with their η^1S isomers in solution in slowly interconverting mixtures ($t_{1/2}$ = 1.7-13 min at 22° C) (Eq. 2.11); the η^2(C=C) form is favored by the presence of the more electron-donating Cp* ligand. Introducing steric congestion and a stronger donor ability of the sulfur atom, as in 2-Me- and 3-MeBT leads to the formation of the S-bonded isomers only [52, 53]. The existence of a more stable analogous selenophene complex Cp'Re(CO)$_2$[η^2(C=C)-Sel] was also demonstrated by this group [54].

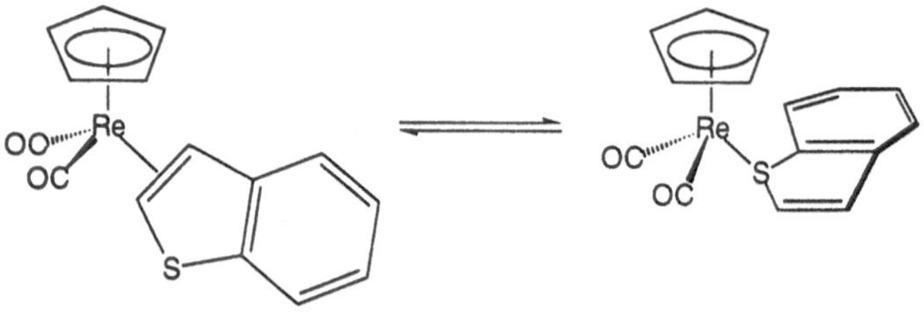

$$(2.11)$$

Furthermore, if the BT ligand is previously η^6-coordinated to the electron-withdrawing Cr(CO)$_3$ group, reaction with Cp(CO)$_2$Re(THF) leads exclusively to the

$\eta^2(C=C)$ isomer of the Re complex, as a result of the greater π-acceptor ability of the olefin in the Cr complex (see Eq. 2.12):

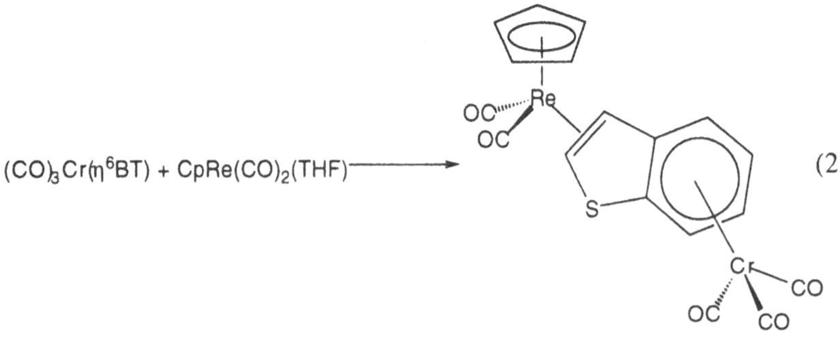

$$(CO)_3Cr(\eta^6BT) + CpRe(CO)_2(THF) \longrightarrow \qquad (2.12)$$

On the other hand, when the η^1-η^2 mixtures of $Cp'Re(CO)_2[\eta^2(C=C)\text{-}BT]$ are allowed to react with electrophiles, only the S-atom coordinates to the incoming group, e.g. Me_3O^+ produces $Cp'Re(CO)_2[\eta^2(C=C)\text{-}BT.Me]$ while $W(CO)_5$ yields the S-adduct exclusively (see Eq. 2.13) [53]:

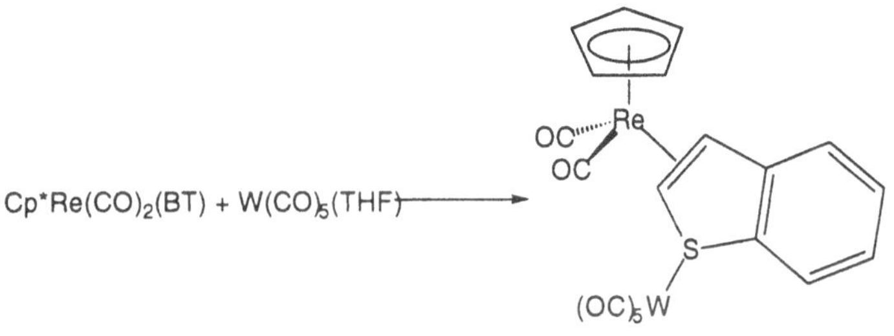

$$Cp^*Re(CO)_2(BT) + W(CO)_5(THF) \longrightarrow \qquad (2.13)$$

Similarly, reaction of $Cp'Re(CO)_2[BT]$ with $Cp'Re(CO)_2[THF]$ gives a product containing the BT ligand bound η^1S to one $Cp'Re(CO)_2$ group and $\eta^2(C=C)$ to the other. If the reaction is carried out between $Cp^*Re(CO)_2[BT]$ and $CpRe(CO)_2[THF]$, the only product isolated contains the BT ligand bound η^1S to the $CpRe(CO)_2$ group and $\eta^2(C=C)$ to $Cp^*Re(CO)_2$, consistent with the more electron rich metal fragment binding to the π-accepting olefin while the less electron rich one accepts electron density from the sulfur donor. In this way, bonding to the two Re centers is mutually favored, since π-back bonding from $Cp^*Re(CO)_2$ provides extra electron density to the sulfur, which strengthens its bonding to $CpRe(CO)_2$.

Concerning theoretical studies of the $\eta^2(C=C)$-coordination of thiophenes to metals, semi-empirical calculations were carried out by Sánchez-Delgado et al. for the model complex $\{(PH_3)_2(H)_2Rh[\eta^2(C=C)\text{-}BT]\}^+$ -the PPh_3 analogue of which is an excellent catalyst for the hydrogenation of BT. These calculations indicate that the bonding between the $\eta^2(C=C)$-BT and Rh(III) closely resembles the coordination of simple olefins, and consequently BT is activated mainly through a weakening of the C=C bond attached to the metal; this renders it susceptible to the transfer of the coordinated hydrides, eventually leading to the saturation of the C=C bond [12]. Clearly, this olefin-like bonding mode is favored for electrophilic metal centers capable of forming strong bonds to the C=C moiety; also η^2(C=C)-binding is favored for thiophene ligands not substituted at the α and β carbon atoms which would cause important steric impediments. Finally, the energy barrier for the η^1-S $\leftrightarrow$ η^2(C=C) interconversion is low, and therefore, the equilibrium described by Angelici [52, 53] is now accepted as a standard necessary transformation in all the mechanistic schemes proposed for the metal complex-catalyzed homogeneous hydrogenation of benzothiophene to 2,3-dihydro-benzothiophene (see Chapter 3, Section 3.3.1).

Another interesting point was the suggestion made by Jones that the olefin-like complex Cp*[η^2(C=C)-T]Rh(PMe$_3$) is a key intermediate in the C-H activation of thiophene (at the α-carbon atom) whereas its η^1S isomer would be responsible for C-S activation [45]; nevertheless, recent DFT calculations by Sargent and Titus indicate that in the η^2(C=C)-T complex, back bonding from Rh to T is actually more pronounced than ligand to metal electron donation, and more interestingly, they demonstrate that the C-H activation process is not restricted to the reaction pathway involving the π-olefin bonding mode. Reasonable routes leading to both C-S and C-H activated products can be envisaged from the η^1S-bonded isomer [14]. What does seem clear is that C-S bond scission is unlikely to occur directly from η^2(C=C) bonded thiophenes.

2.4. η^3(S,C=C)-Bonded metal thiophene complexes

Bianchini and Sánchez-Delgado have provided the only example thus far available of a well-characterized (^{1}H, ^{13}C and ^{31}P NMR) metal complex containing benzothiophene in the unique 4-electron combination of η^1S + η^2(C=C) L$_2$-type bonding at a single metal center [55]. The stable complex {(triphos)Ir[η^3(S,C=C)-BT]}$^+$ (triphos = MeC(CH$_2$PPh$_2$)$_3$) was prepared through displacement of the η^4-coordinated benzene in [(triphos)Ir(C$_6$H$_6$)]$^+$ by benzothiophene at room temperature (Eq. 2.14). This is a very interesting bonding situation in that it requires two CUS's and also it coincides with the structure suggested in some early "multi point adsorption" proposals in heterogeneous catalysis [56].

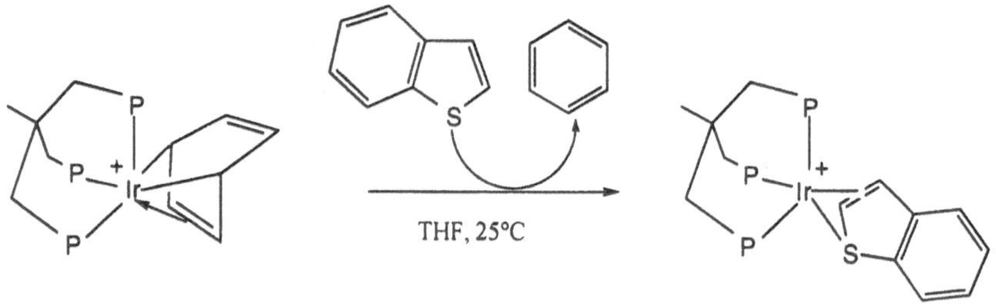

$$\text{THF, 25°C} \tag{2.14}$$

Such multi point adsorption has been previously related to a desulfurization route proceeding through prehydrogenation of the ring rather than through direct sulfur extrusion [56] but in fact, on gentle warming (refluxing THF) of the $\eta^3(S,C=C)$ Ir(I) complex, further reaction takes place readily through insertion of the metal atom into the C-S bond to form the corresponding thiametallacycle, as represented in Eq. 2.15 (see also Chapter 4, section 4.2.1.2):

$$(2.15)$$

This demonstrates that this type of S + C=C coordination -occupying 2 CUS- can be an important way of activating thiophenes toward ring opening in those cases where the reactive molecular fragment or surface site contains no more than 14 valence electrons, a likely possibility in working HDS catalysts. Recent DFT calculations by Rincón *et al.* indicate that this $\eta^3(S,C=C)$ bonding is not only more stable than the $\eta^1 S$ or η^4 coordination modes of thiophene in $[(T)Ir(PH_3)_3]^+$ model complexes, but it is also more adequate for promoting C-S bond scission for d^8 ML_3 14e fragments [57] (see also Section 4.2.1.2).

2.5. η^4-Bonded metal thiophene complexes

The diene-type η^4-bonding of Th ligands to transition metals (Th = thiophene and its methylated derivatives), in which the thiophenes act as 4-electron donors (L_2-type ligands occupying 2 CUS), is comparatively infrequent. A relatively small number of complexes of ruthenium and iridium containing η^4-bonded thiophenes, together with cyclopentadienyl or arene co-ligands, have been synthesized, mainly by the groups of Angelici and Rauchfuss [58-64]. A good proportion of these compounds has been fully characterized by single crystal X-ray diffraction techniques. Examples of η^4-bonded complexes of benzothiophenes and dibenzothiophenes are limited to two examples, *viz.* Cp*Ir(η^4-BT) and Cp*Ir(η^4-DBT); these two compounds proved to be extremely air-sensitive and this precluded any characterization other than by *in situ* NMR spectroscopy. For the same reason, their reactivity has not been further studied (see Table 2.2).

One of the most successful preparative routes for η^4-bonded metal complexes of thiophenes has been the chemical or electrochemical reduction of the corresponding 18e η^5 Ru, Rh, and Ir precursors [58-64]; upon addition of two extra electrons to complexes like Cp*Ir(η^5-Th), the thiophenic ligand necessarily transforms from a 6e-donor into a 4e-donor situation in order to avoid oversaturation at the metal center, as exemplified for Cp*Ir(2,5-Me$_2$T) in Eq. 2.16.

TABLE 2.2. Known η^4-complexes of the thiophenes.

Complex	Characterization	Refs.
Cp*Ir(Th) (Th = T, 2-MeT, 3-MeT, BT, DBT)	NMR	58
Cp*Ir(2,5-Me$_2$T)	NMR + X-rays	59
Cp*Ir(2,5-Me$_2$T.A) (A = BH$_3$, CH$_3^+$, CS$_2$, Ru(η^6-C$_6$H$_6$)Cl$_2$, Fe(CO)$_4$, Co$_4$(CO)$_{11}$, Ru$_3$(CO)$_{11}$, Re$_2$(CO)$_9$)	NMR +/or X-rays	63
Cp*Rh(Me$_4$T)	NMR + X-rays	60
Cp*Rh[Me$_4$T.Fe(CO)$_4$]	NMR + X-rays	60
Cp*Rh(Me$_4$T.O)	NMR + X-rays	64
(η^6-C$_6$Me$_6$)Ru(Th) (Th = T, 2-MeT, 2,5-Me$_2$T, Me$_4$T)	NMR	61
(η^6-C$_6$Me$_6$)Ru[η^4-T.Mo(CO)$_5$]	NMR + X-rays	61, 62
(η^5-Me$_4$T)Ru(η^4-Me$_4$T)	NMR	61, 62
(η^5-Me$_4$T)Ru(η^4-Me$_4$T.Fe(CO)$_4$)	NMR + X-rays	61, 62

$$(2.16)$$

Similarly, reduction of Cp*Rh(η^5-Me$_4$T) with Cp$_2$Co leads to the stable Cp*Rh(η^4-Me$_4$T), while one ring of the Ru bis(tetramethylthiophene) derivative [Ru(η^5-Me$_4$T)$_2$]$^{2+}$ is reduced in an analogous manner to yield a rather unstable mixed η^5-η^4 species [(η^5-Me$_4$T)Rh(η^4-Me$_4$T)]$^{2+}$, which may be stabilized by coordination of the sulfur atom to Fe(CO)$_4$ through reaction with Fe(CO)$_5$. Other examples noted in Table 2.2 have been produced by this type of reductive procedure.

The single crystal X-ray structures available for η^4-Th metal complexes invariably show that the thiophene ring is highly distorted with the four thiophene carbon atoms coordinated to the metal center, and the sulfur atom notably bending out of the plane, as exemplified for Cp*Ir(η^4-2,5-Me$_2$T) in Fig. 2.4. Another interesting structural feature in this type of compound is that the C-S distances in η^4-Th derivatives are in general significantly longer (e.g. average 1.77 Å in Cp*Ir(η^4-2,5-Me$_2$T) [59]) than those measured in free thiophenes (1.714(1) Å [2]), clearly indicating a significant perturbation of the thiophenic ligand. It seems clear that the η^4-Th type of bonding in metal complexes is imposed essentially by the electron count at the metal center, even if there are no particularly important spatial restrictions to accommodate the ligand in a flat η^5-bonding fashion. In fact, most η^4-complexes can be easily oxidized by e.g. Cp$_2$Fe$^+$ to yield the corresponding η^5-derivative [51] in the reverse reaction shown in Eq. 2.16.

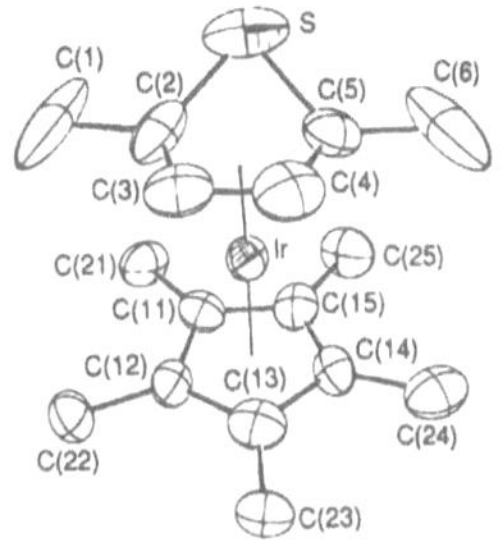

Fig. 2.4. Structure of Cp*Ir($\eta^4$2,5-Me$_2$T) (Reproduced with permission from Ref. 59).

Two-electron redox processes are facile in metal sulfide surfaces, and thus M-S bond breaking and re-making in such extended arrays may provide an energetically favorable pathway to modify the electron count at a particular metal center. Although η^4-bonded thiophenes have not been invoked in heterogeneous HDS mechanisms, η^5 $\leftrightarrow \eta^4$ transformations such as the ones defined in organometallic complexes could be envisaged to take place at the active sites of HDS catalysts, as possible routes for thiophene activation.

Concerning reactivity patterns, η^4-coordination renders the organosulfur molecules highly reactive; in particular, the uncoordinated sulfur atom becomes strongly nucleophilic and, besides the ease of oxidation already mentioned , this feature dominates the chemistry of the complexes. Consequently, a common reaction of η^4-thiophene metal complexes is the formation of adducts with a variety of Lewis acids which bind to the sulfur atom (see Table 2.2). More interesting in relation to HDS is the fact that in some instances η^4-coordination proceeds other reactions leading to C-S bond breaking, in accord with the observed lengthening of the C-S bonds. Some particularly interesting examples related to HDS include Rauchfuss' report on the protonation of (η^6-C$_6$Me$_6$)Ru(η^4-T) to give a transient cationic thioallylic intermediate which evolves by C-S bond scission into a butadienethiolate ligand [65] (Eq. 2.17):

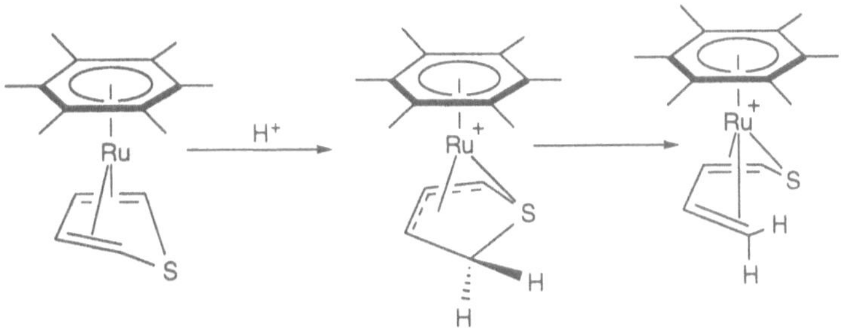

(2.17)

Angelici observed a remarkable ring opening of thiophenes in $Cp*Ir(\eta^4\text{-Th})$ catalyzed by basic alumina or by triethylamine, or promoted by UV light [58] (Eq. 2.18), although the mechanisms for such transformations are not clear. It was also shown that reaction of $Cp*Ir(\eta^4\text{-Th})$ with H_2 promotes ring opening, even though hydride transfer to the thiophene was not observed (Eq. 2.19) [66].

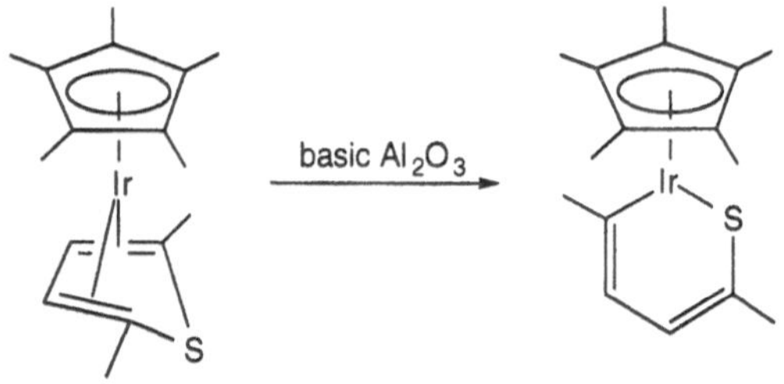

(2.18)

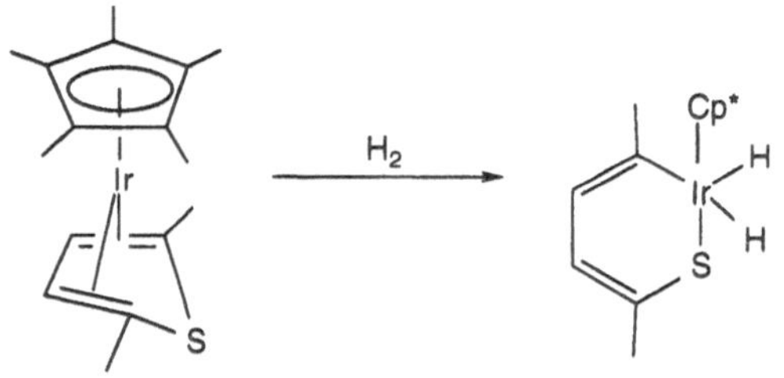

(2.19)

The detailed bonding in these molecules, as well as the possible reasons for the enhanced nucleophilicity of the S atom, and the tendency of η^4 complexes to isomerize to the ring-opened form have been discussed at length by Harris [8, 10]. In many ways this coordination mode resembles the bonding of simple dienes to metals but additionally it incorporates a substantial antibonding metal-sulfur interaction; thus, breaking one C-S bond may be viewed as a way of relieving the repulsive metal-sulfur interaction.

2.6. η^5-Bonded metal thiophene complexes

Metal complexes containing η^5-bonded thiophenes, where the ring formally donates 6 electrons and occupies three coordination sites (L_3-type ligand) are the most numerous and stable of the transition metal-thiophene derivatives, examples being available for Cr, Mn, Re, Fe, Ru, Rh, and Ir. Curiously, the synthesis of the first π-thiophene metal complex, viz. $Cr(CO)_3(\eta^5\text{-T})$ reported by Fischer as early as 1958 [67] represents still today the only example available for a Group 6 metal π-bonded to thiophene; its X-ray structure was solved by Dahl in 1965 albeit with a strong rotational disorder for the thiophene ligand [68]. In 1989 Ruette *et al.* advanced theoretical arguments [11] based on CNDO calculations that predicted the stability of $Mo(CO)_3(\eta^5\text{-}T)$, but attempts to synthesize π-bonded thiophene complexes of the HDS-active metals Mo and W have so far met with no success; no examples of stable η^5-thiophene derivatives of the promoter metals Co or Ni have been reported to date either. A list of known complexes with this type of thiophene bonding is presented in Table 2.3 [67-86]; more comprehensive discussions on the synthetic and structural aspects of this chemistry can be found in earlier reviews [2-7].

The principal structure and bonding features of these compounds, extensively discussed by Harris [8], resemble those of the well known cyclopentadienyl (Cp) analogues which is not surprising since the frontier orbitals in both T and Cp ligands are quite similar. The main differences between them are that T is a poorer electron donor but a better acceptor than Cp, due to the relative orbital energies. Also the presence of the larger sulfur atom in thiophene causes a tilting of the ring relative to a perfectly horizontal disposition (and in some cases a ring slip); consequently, thiophenes are in general less strongly bound to metals than Cp ligands.

TABLE 2.3. Known η^5 complexes of the thiophenes.

Complex	Characterization	Refs.
$Cr(CO)_3(T)$	NMR + X-rays	67, 68
$Cr(CO)_3(Th)$	NMR	3, 69, 70
(Th = T, 2-MeT, 3-MeT, 2,5-Me$_2$T, Me$_4$T, ...)		
$[Mn(CO)_3(Th)](OTf)$	NMR	71-73
(Th = T, 2-MeT, 2,5-Me$_2$T, Me$_4$T)		
Cp'Fe(Th)	NMR	74-76
(Cp' = Cp, Th = T, 2-MeT, 3-MeT, 2,5-Me$_2$T, Me$_4$T; Cp' = Et-Cp, Th = 2,5-Me$_2$T, Me$_4$T;)		
$[Fe(Me_4T)_2](PF_6)_2$	NMR	77
$[Cp'Ru(Th)]X$	NMR	78-81
(Cp' = Cp, X = BF$_4$, Th = T, 2-MeT, 3-MeT, 2,5-Me$_2$T, 2,3,5- Me$_3$T, Me$_4$T; Cp' = Cp*, X = PF$_6$, Th = T, 3-MeT, 2,5-Me$_2$T)		
$[Ru(Th)_2](X)_2$	NMR	82
((Th)$_2$ = (Me$_4$T)$_2$, X = BF$_4$; Th$_2$ = T, 2-MeT, Me$_4$T, X = OTf)		
$[Ru(Th)_2](BF_4)_2$	NMR + X-rays	83
$[Ru(Me_4T)(p\text{-cymene})](BF_4)_2$	NMR	84
$[Ru(Me_4T)Cl_2]_2$	NMR + X-rays	82, 83
$Ru(Me_4T)Cl_2(PR_3)$	NMR	82
$Ru(Me_4T)Cl_2(H_2NTol)$	NMR	82
$[(Me_4T)Ru(Cl)]_3S(BF_4)$	NMR + X-rays	82
$[(Me_4T)Ru(L)_3](OTf)_2;$	NMR + X-rays	83
(L = H$_2$O, NH$_3$)		
$[Rh(T)(PPh_3)_2]PF_6$	NMR + X-rays	85
$[Rh(Th)(diene)]PF_6$	NMR	84
(Th = 2,5-Me$_2$T Me$_4$T, diene = COD, NBD)		
$[Cp*Rh(Th)][PF_6]_2$; Th = T, Me$_4$T	NMR	84, 86
$[Cp*Ir(Th)][X]_2$	NMR	82, 84, 86
(Th = T, 2-MeT, 2,5-Me$_2$T, Me$_4$T, X = PF$_6$, BF$_4$)		

Most of the η^5-bonded thiophene complexes known correspond to d^6 hexacoordinated metal centers, a particularly stable situation; as an example of the bonding characteristics of these complexes, the energy level diagram for $Cr(CO)_3(\eta^5$-T) is shown in Fig. 2.5 together with the one corresponding to $[Cr(CO)_3Cp]^-$ for comparison [8]. The $2b_1$, $1a_2$, and $1b_1$ filled orbitals of T act as electron donors into the vacant 2e and $2a_1$ orbitals in the metal fragment, whereas the empty $3b_1$ and $2a_2$ ligand orbitals are acceptors; in the Cp complex, the main interactions are those of the donor ligand orbitals e_1'' and a_2'' with the vacant 2e and $2a_1$ orbitals in the $Cr(CO)_3$ unit, and the vacant e_2'' with the filled 1e orbitals. As mentioned above, the presence of the large S atom results in the tilting of the ring with respect to the plane parallel to the three C atoms of the CO ligands; this distortion is accompanied by a small slip in the ring thus optimizing the bonding betwen the metal and the S and C atoms.

An interesting exception to the case of d^6 complexes is $[(\eta^5$-T)Rh(PPh$_3)_2]$PF$_6$; this rare example of a 5-coordinated Rh(I) d^8 η^5-T complex was prepared by Sánchez-Delgado et al. through the mild hydrogenation of the coordinated cyclooctadiene in the precursor $[(COD)Rh(PPh_3)_2]$PF$_6$, in the presence of excess thiophene (see Eq. 2.20) [85].

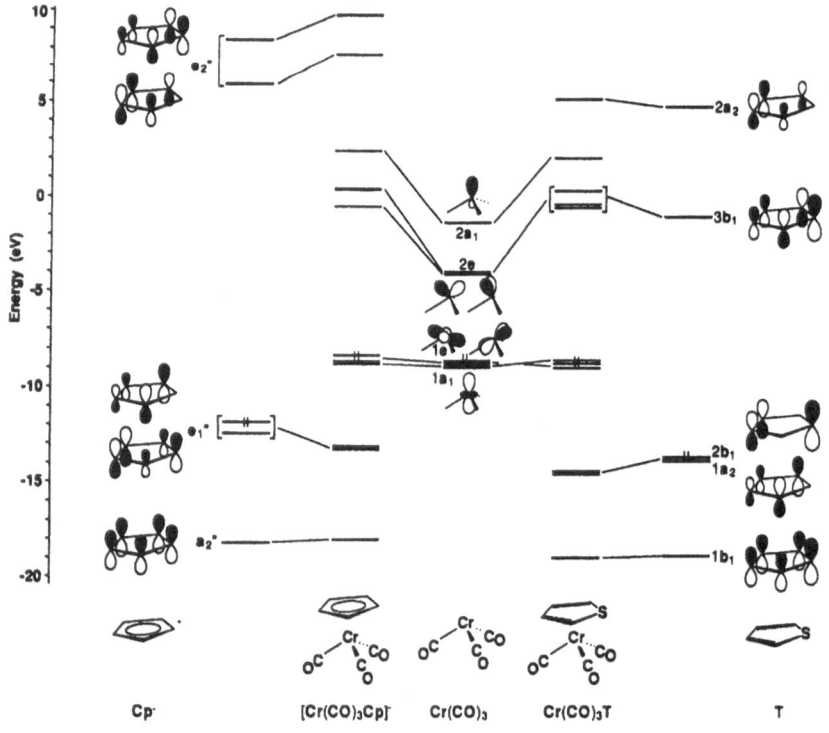

Fig. 2.5. Energy level diagrams for $[Cr(CO)_3(Cp)]^-$ and $[Cr(CO)_3(\eta^5$-T)] (Reproduced with permission from Ref. 8).

$$[Rh(COD)(PPh_3)_2]^+ + H_2 + T \xrightarrow[CH_2Cl_2]{25°, 1\ atm} \quad + C_8H_{16} \quad (2.20)$$

Once the diene hydrogenation product (cyclooctane) is released, the remaining highly unsaturated fragment "$[Rh(PPh_3)_2]PF_6$" (12e, 3 CUS) readily picks up thiophene, which is the only molecule with efficient ligation capability in the reaction medium, as a 6e-donor. This rhodium compound provided the first non-disordered X-ray structure of a π-bonded metal thiophene complex [85]. The thiophene ring in this case is not only tilted as usual, but also slightly bent (see Fig. 2.6) in what may be viewed as a unique case of an intermediate structure between a "normal η^5" and an η^4 mode, in which the sulfur tends to be located away from the metal, but nevertheless remains coordinated to Rh in order to attain an 18e configuration. The explanation for these interesting features are found in the energy level diagram for the PH_3 analogue of this complex, calculated by Harris [8] which is shown in Fig. 2.7.

The main bonding component in this case involves electron donation from the filled $1b_1$, $1a_2$ and $2b_1$ thiophene orbitals to the $3a_1$, $1b_2$ and $2b_1$ orbitals in the $[Rh(PH_3)_2]^+$ fragment. Additionally there is a substantial interaction between a filled thiophene $2b_1$ orbital and a filled metal $1b_1$ orbital, which is strongly antibonding. Mixing of the bonding Rh-T b_1 interaction partially relieves the repulsive 4-electron interaction, which is further reduced by the observed slight bending of the sulfur atom out of the ring plane. Although the net result of this situation is a destabilized HOMO with antibonding character between the metal and the thiophene sulfur atom, such a destabilization is considerably smaller than it would be if the ring maintained its planarity.

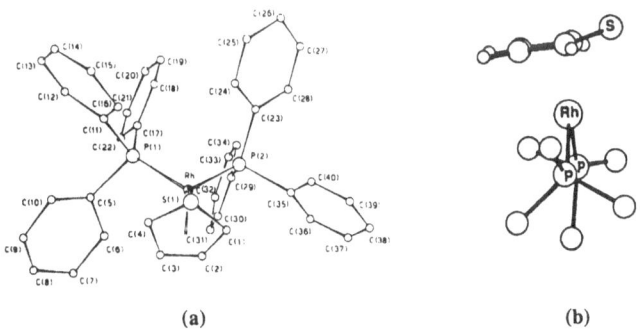

(a)　　　　　　　　　　　　(b)

Fig. 2.6. Structure of $[(\eta^5\text{-}T)Rh(PPh_3)_2]^+$
(Reprinted with permission from from (a) Ref. 85 and (b) Ref. 8).

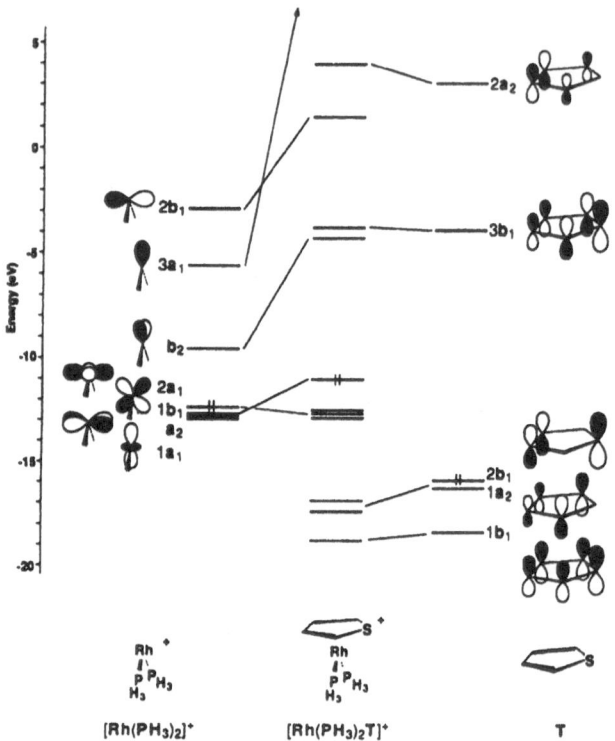

Fig. 2.7. Energy level diagram for $[(\eta^5\text{-}T)Rh(PH_3)_2]^+$ (Reproduced with permission from Ref. 8).

Methyl substituted thiophenes are known to be better σ-donors than the unsubstituted ligand and therefore they can be expected to bind more strongly to transition metal centers; Angelici and coworkers have nicely illustrated this trend through an extensive NMR study of a series of ligand exchange reactions of CpRu(η^5-Th) complexes at 50° C in acetone-d_6, as represented by Eq. 2.21. It was found that the corresponding equilibrium constant increases by a factor of about 6 for each methyl group added on the thiophene [80], *viz.* $K_{eq} = 1$ (T), 6 (2-MeT), 7 (3-MeT), 35 (2,5-Me$_2$T), 50 (2,3-Me$_2$T), 200 (2,3,4-Me$_3$T), 300 (2,3,5-Me$_3$T), 1300 (Me$_4$T).

$$\text{(2.21)}$$

This trend is interesting in that it is consistent with the one observed for the relation between adsorption constants and HDS activities of thiophenes on Co-Mo/Al$_2$O$_3$ catalysts; nevertheless, a similar trend was also observed in the case of the η^1-S bonded complexes [Cp(CO)$_2$Ru(η^1S-Th)]$^+$ also studied by Angelici [44] (see Section 2.2) and therefore no clear conclusions can be safely drawn from these organometallic model compounds as to a preferred mode of surface adsorption (η^1S or η^5) being related to HDS activity.

It was also shown by Angelici that benzene, benzothiophene and dibenzothiophene are capable of readily displacing thiophene under the conditions of Eq. 2.21. In the case of benzothiophenes and dibenzothiophenes η^5-bonding through the heterocycle to triply unsaturated metal centers is possible, but it seems to be largely disfavored in comparison with the alternative very stable η^6-coordination through a benzene ring which also occupies triply unsaturated sites. The existence of a η^5-bonded benzothiophene complex [(η^5-BT)Rh(PPh$_3$)$_2$]$^+$ was suggested by Sánchez-Delgado *et al.* on the basis of some *in situ* NMR data, and supported by semi-empirical MINDO calculations [87], but neither this compex nor any other η^5-BT or η^5-DBT derivative has been well characterized to date. The preference for η^6-bonding in BT and DBT complexes is further evidenced by calorimetric studies which show that in Cr complexes Cr(CO)$_3$(Th), displacement of T is 2.6 kcal/mol more favorable than the displacement of 2,5-Me$_2$T, and 4.2 kcal/mol more favorable than the displacement of benzene [88]. This was also found to be consistent with kinetic studies of the displacement of π-ligands by phosphines from Cr(CO)$_3$(π-L) which indicate that the rate of displacement increases in the same order mentioned above: benzene << 2,5-Me$_2$T < T [89].

Interestingly, these observations can be related with the known inhibition effect of benzene and other aromatic hydrocarbons on HDS of BT over solid catalysts [90]; this has been ascribed to a much stronger adsorption of benzene on metal sulfide surfaces as compared to common sulfur heterocycles.

Concerning reactivity features possibly related to HDS, the π-bonded thiophene ring, particularly in cationic complexes, is activated in ways which can be expected from the lowering of the electron density resulting by donation from the ligand to the metal center. For instance, H-D exchange is rapidly observed at the H(2,5) positions of the bound thiophene in CpRu(η^5-Th) complexes (Th = T or its methylated derivatives) when they are dissolved in CD$_3$OD in the presence of OD$^-$, while H(3,4) exchange takes place much more slowly; this is in agreement with an increased acidity of thiophenes as a result of complexation to the metal, and it is an interesting parallel to the trends observed for thiophene H-D exchange when D$_2$ is passed over conventional HDS catalysts [79].

Another appealing feature in relation to HDS mechanisms is the activation of π-bonded thiophenes in cationic metal complexes toward nucleophilic attack, particularly by hydrides, which ultimately leads to hydrogenation of the C=C bonds (Eq. 2.22) [72] or to ring opening through C-S bond breaking (Eq. 2.23) [91] of the thiophene. These interesting reactions are discussed in more detail in Chapters 3 and 4. On the other hand, neutral η^5-thiophene complexes do not suffer attack by nucleophiles but can be deprotonated with strong bases to yield thienyl products.

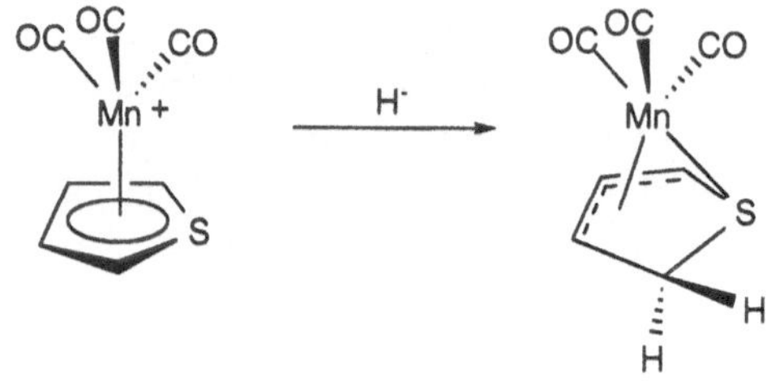

$$(2.22)$$

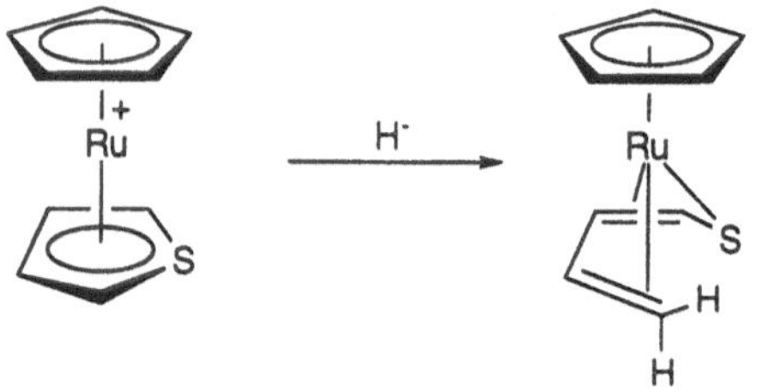

$$(2.23)$$

In contrast to Eq. 2.22, carbanions add to $[Mn(CO)_3(\eta^5T)]^+$ exclusively at the sulfur atom, even if the 2- and 5-positions are substituted with methyl groups, demonstrating that the sulfur in this complex is indeed highly electrophilic (Eq. 2.24). This suggests the interesting idea that the sulfur atom might be the initial site of attack for most nucleophiles, which would subsequently migrate to an adjacent carbon atom. In the particular case of the carbanions, the C-S bond formed would be kinetically inert while a –S-H bond would be weaker (and/or more labile), allowing the migration to occur.

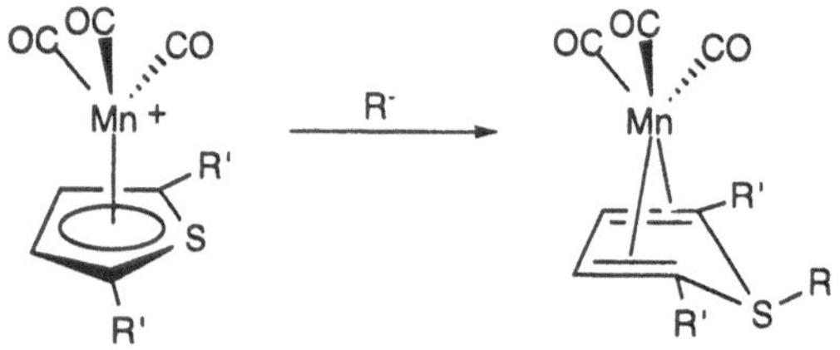

$$(2.24)$$

This extensive organometallic chemistry involving η^5-coordination of thiophene to metal centers followed by nucleophilic attack to the ring has been used by Angelici and coworkers in conjunction with some detailed heterogeneous reactor studies [92-94] to propose the mechanism for HDS of thiophene depicted in Fig. 2.8 [95]. In this scheme, a heterolytic activation of hydrogen is invoked, as well as π-bonding of the thiophene to an unsaturated metal site; this is followed by a nucleophilic attack of a metal hydride to the carbon atom adjacent to sulfur and subsequent proton addition to C(3) producing a 2,3-dihydrothiophene intermediate, which rearranges to the 2,5-dihydrothiophene isomer before being desulfurized to yield free 1,3-butadiene plus a surface metal sulfide.

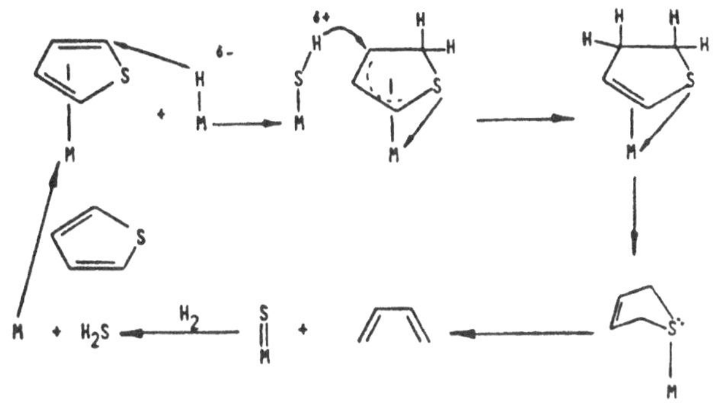

Fig. 2.8. Angelici's mechanistic proposal for HDS of thiophene (reproduced with permission from ref. 95).

Nevertheless, Harris has argued very convincingly [8] that η^5-coordination does not produce any significant alteration in the charge distribution on the thiophene ring, or any important weakening of the C-S bonds. Since nucleophilic attack is observed only in positively charged complexes it seems that it is the overall charge on the metal derivative rather than significant changes in the electronic structure of the bound thiophenes, which leads to the activation toward nucleophiles.

Furthermore, it is becoming clear from the wealth of information now available on organometallic HDS models that most of the reactions of metal-thiophene complexes with hydrogen leading to ring saturation or C-S bond breaking, and particularly the homogeneous *catalytic* hydrogenolysis and desulfurization reactions (see Chapter 4), are associated with other types of thiophene bonding requiring only 1 or 2 CUS, specifically $\eta^1 S$, $\eta^2(C=C)$, and -less frequently- $\eta^3(S,C=C)$ [7-15].

Thus it appears from the knowledge gained from the chemistry of metal complexes of π-bonded thiophenes that η^5 adsorption on highly (triply) unsaturated metal sites, which has been frequently invoked in the heterogeneous literature, may in itself be a peripheral situation, rather than a crucial phenomenon directly related to the actual HDS reactions taking place, since a number of other pathways involving the more abundant singly or doubly CUS' are available under desulfurization conditions.

2.7. η^6-Bonded metal benzothiophene and dibenzothiophene complexes

As mentioned in the preceding Section, η^6 coordination of arenes to 12e metal centers is a particularly stable situation in general, and this type of bonding is well documented, as is the consequent activation of such fragments for further reactions. This information can be found in specialized treatises [96, 97], as well as in major textbooks in organometallic chemistry [98, 99], so there is no need for any further discussion on those points here. A number of transition metal complexes containing BT and DBT as ligands π-coordinated through the benzene ring are known, indicating a strong preference of these two ligands for this type of 6-electron (L_3 type) arene bonding to triply CUS. Examples are known for Cr, Mn, Fe, Ru, Co, Rh, and Ir, as shown in Table 2.4.

TABLE 2.4. Known η^6 complexes of the benzo- and dibenzo-thiophenes.

Complex	Characterization	Refs.
Cr(CO)$_3$(Th) (Th = BT, DBT, BNT)	NMR	100, 101
[Mn(CO)$_3$(Th)]BF$_4$ (Th = 7-MeBT, 7-EtBT)	NMR	101-105
[CpFe(Th)]PF$_6$ (Th = BT, DBT)	NMR	101, 106
trans-[(CpFe)$_2$(DBT)$_2$](PF$_6$)$_2$	NMR	106
[CpRu(Th)]PF6 (Th = BT, DBT)	NMR	101, 107
trans-[(CpRu)$_2$(DBT)$_2$](PF$_6$)$_2$	NMR, X-rays	108
[(C$_6$Me$_6$)Ru(BT)$_2$](BF$_4$)$_2$	NMR	101, 105, 109
Co$_4$(CO)$_9$(DBT)	NMR	110
Cp*M(Th) (M = Rh, Ir) (Th = BT, 2-MeBT, 3-MeBT, 2,3-Me$_2$BT)	NMR	86

Such structures do not seem to be of great general relevance as HDS models, since they produce activation predominantly of the coordinated *benzene* ring towards nucleophilic attack [107, 108]. For instance, [Mn(CO)$_3$(η^6-BT)]BF$_4$ reacts with nucleophiles (H⁻, Me⁻, Et⁻, Ph⁻, CH$_2$CN⁻) by addition to either the 4- or 7-position of the carbocyclic ring, without any indication of attack to the heterocycle [111].

However, some other aspects of the chemistry of η^6 bonded metal BT and DBT complexes could be of importance in connection with HDS. It is now known, for instance, that 4-methyl- and 4,6-dimethyldibenzothiophene are desulfurized mainly through prior hydrogenation of one or both of the arene rings [46, 112-115], and therefore the study of model η^6-DBT complexes could be of use in developing efficient homogeneous, heterogeneous or hybrid catalysts for reducing the arene moieties in this type of molecule either as a pretreatment for conventional HDS, or as an added functionality in novel multimetallic catalyst formulations; this possibility is further discussed in Chapter 3.

Also of interest are Sweigart's recent reports [101-103, 105, 109] showing that some η^6-BT or DBT complexes are actually activated toward C-S bond scission by a second metal center, as exemplified in Eq. 2.25.

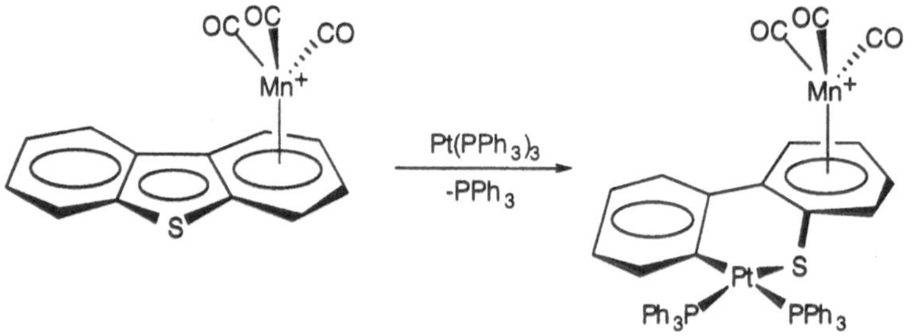

$$(2.25)$$

These reactions, described in some detail in Chapter 4, illustrate an interesting concept which could be of use in trying to develop bifunctional catalysts specially adapted to the difficult case of desulfurization of methylated dibenzothiophenes and related molecules.

2.8. Conclusions and further comments

Despite the fact that only 15 years ago metal thiophene complexes were considered a rarity, this field is today as a well developed area of coordination and organometallic chemistry. Synthetic methods have become available for a large variety of metals, thiophenes, and stoichiometries, and the general structural trends have been well established, mainly through extensive use of NMR spectroscopy and numerous X-ray structure determinations. The various types of bonding between thiophenic ligands and metal centers have been studied in detail by several theoretical methods of varying degrees of sophistication and a deep understanding of the interactions involved and the factors affecting them has been reached. The number of Mo and W thiophene complexes is still small but it will probably grow in the near future; also, examples of well defined stable Co and Ni derivatives containing coordinated thiophenes are still needed and will hopefully become available in the coming few years. This would add to this already extensive area of organometallic chemistry in a particularly useful way, since modeling HDS-related species and reactions with complexes of the metals actually used in practical catalysts would bring this field a step closer to real situations.

It is possible now to begin drawing clear parallels between the coordination modes of thiophenes authenticated in metal complexes and the ways in which such molecules are adsorbed on catalytic surfaces. Also, the very interesting idea of each type of binding being associated with a particular kind of reaction of the activated substrate has been advanced in this Chapter and will be further developed in the subsequent ones.

References

1.   H. Topsøe, B. S. Clausen, and F. E. Massoth, in J. R. Anderson and M. Boudart, eds.: *Catalysis Science and Technology*, Springer-Verlag, New York, Vol. 11 (1966).
2.   R. J Angelici: *Coord. Chem. Rev.* 105, 61 (1990).
3.   T. B. Rauchfuss: *Prog. Inorg. Chem.* **39**, 259 (1991).
4.   R. J Angelici: *Bull. Soc. Chim. Belg.* **104**, 265 (1995).
5.   R. J Angelici, in K. R. Bruce, ed.: *Encyclopedia of Inorganic Chemistry*, Vol. 3, John Wiley & Sons, New York, p. 1433 (1994)
6.   R. J Angelici, *Organometallics* **20**, 1259 (2001), and references contained therein.
7.   R. A. Sánchez-Delgado, in I. T. Horvath, ed.: *Encyclopedia of Catalysis*, John Wiley & Sons, New York, in press.
8.   S. Harris: *Organometallics*, **13**, 2628 (1994).
9.   M. Palmer, K. Carter and S. Harris: *Organometallics*, **16**, 2448 (1997).
10.   S. Harris: *Polyhedron*, **16**, 3219 (1997).
11.   F. Ruette, N. Valencia and R. A. Sánchez-Delgado: *J. Am. Chem. Soc.* **111**, 40 (1989).
12.   R. A. Sánchez-Delgado, V. Herrera, L. Rincón, A. Andriollo, G. Martín: *Organometallics* **13**, 553 (1994).
13.   L. Rincón, J. Terra, D. Guenzburger and R. A. Sánchez-Delgado: *Organometallics* **14**, 1292 (1995).
14.   A. L. Sargent and E. Titus: *Organometallics* **17**, 65 (1998).

15. M. S. Palmer, S. Rowe and S. Harris: *Organometallics* **17**, 3798 (1998).
16. J. M. J. G. Lipsch and G. C. A. Schuit: *J. Catal.* **15**, 179 (1969).
17. F. Ruette and E. V. Ludeña: : *J. Catal.* **67**, 266 (1981).
18. R. P. Diez, A. H. Jubert: : *J. Mol. Catal.* **83**, 219 (1993).
19. G. V. Smith, C. C. Hinckley and F. Behbahany: *J. Catal.* **30**, 218 (1973).
20. S. Harris and R. R. Chianelli: *J. Catal.* **86**, 400 (1984).
21. M. A. Reynolds, I. A. Guzei, B. C. Logsdon, L. M. Thomas, R. A. Jacobson and R. J. Angelici: *Organometallics* **18**, 4075 (1999).
22. D. G. Churchill, B. M. Bridgewater and G. Parkin: *J. Am. Chem. Soc.* **122**, 178 (2000).
23. M. -G. Choi and R. J. Angelici: *Organometallics* **10**, 2436 (1991).
24. C. J. White and R. J. Angelici: *Organometallics* **13**, 5132 (1994).
25. M. J. Robertson, C. J. White and R. J. Angelici: *J. Am. Chem. Soc.* **116**, 5190 (1994).
26. N. Kuhn and H. Schumann: *J. Organomet. Chem.* **276**, 55 (1984).
27. J. D. Goodrich, P. N. Nickias and J. P. Selegue: *Inorg. Chem.* **26**, 3424 (1987).
28. D. Catheline and D. Astruc: *J. Organomet. Chem.* **248**, C9 (1983).
29. M. Draganjac, T. B. Rauchfuss and C. J. Ruffin: *Organometallics* **4**, 1909 (1985).
30. S. M. Bucknor, M. Draganjac, T. B. Rauchfuss, C. J. Ruffin, W. C. Fultz and A. L. Rheingold: *J. Am. Chem. Soc.* **106**, 5379 (1984).
31. J. W. Benson and R. J. Angelici: *Organometallics* **11**, 922 (1992).
32. M. Alvarez, N. Lugan, B. Donnadieu and R. Mathieu: *Organometallics* **14**, 365 (1995).
33. R. A. Sánchez-Delgado, V. Herrera, C. Bianchini, D. Masi and C. Mealli: *Inorg. Chem.* **32**, 3766 (1993).
34. K. M. Rao, C. L. Day, R. A. Jacobson and R. J. Angelici: *Inorg. Chem.* **30**, 5046 (1991).
35. J. Stohr, J. Gland, E. B. Kollin, R. J. Koestner, A. L. Johnson, E. L. Muetterties and F. Sette: *Phys. Rev. Lett.* **53**, 2161 (1984).
36. J. F. Lang and R. I. Masel: *Surf. Sci.* **183**, 44 (1985).
37. B. A. Sexton: *Surf. Sci.* **163**, 99 (1985).
38. F. Zaera, E. B. Kollin and J. L. Gland: *Surf. Sci.* **184**, 75 (1987).
39. E. Constable, R. P. G. Hanney and D. A. Tocher: *J. Chem. Soc. Dalton Trans.* 2335 (1991).
40. L. Latos-Grazynski, J. Lisowski, P. Chmielewski, M. Grzeszczuk, M. Olmsted and A. Balch: *Inorg. Chem.* **33**, 192 (1994).
41. P. Chmielewski, L. Latos-Grazynski, E. Pacholska: *Inorg. Chem.* **33**, 1992 (1994).
42. M. -G. Choi and R. J. Angelici: *J. Am. Chem. Soc.* **111**, 8753 (1989).
43. C. J. White and R. J. Angelici: *Organometallics* **13**, 5132 (1994).
44. J. W. Benson and R. J. Angelici: *Organometallics* **12**, 680 (1993).
45. L. Dong, S. B. Duckett, K. F. Ohman and W. D. Jones: *J. Am. Chem. Soc.* **114**, 151 (1992).
46. D. D. Whitehurst, T. Isoda and I. Mochida: *Adv. Catal.* **42**, 345 (1998).
47. R. Cordonne, W. D. Harman and H. Taube: *J. Am. Chem. Soc.* **111**, 5969 (1989).
48. M. L. Spera and W. D. Harman: *Organometallics* **14**, 1559 (1995).
49. M. L. Spera, W. D. Harman: *J. Am. Chem. Soc.* **119**, 8843 (1997).
50. M. L. Spera and W. D. Harman: *Organometallics* **18**, 2988 (1999).
51. W. D. Harman: *Chem. Rev.* **97**, 1953 (1997).
52. M. -G. Choi, M. J. Robertson and R. J. Angelici: *J. Am. Chem. Soc.* **113**, 4005 (1991).
53. M. -G. Choi and R. J. Angelici: *Organometallics* **11**, 3328 (1992).
54. M. -G. Choi, M. J. Robertson and R. J. Angelici: *J. Am. Chem. Soc.* **113**, 5651 (1991).
55. C. Bianchini, A. Meli, M. Peruzzini, F. Vizza, S. Moneti, V. Herrera and R. A. Sánchez-Delgado: *J. Am. Chem. Soc.* **116**, 4370 (1994).
56. H. Kwart, G. C. A. Schuit and B. C. Gates: *J. Catal.* **61**, 128 (1980).
57. H. Diez y Riega, L. Rincón and R. Almeida: *J. Mol. Struct. (Theochem)* **493**, 259 (1999).
58. J. Chen, L. M. Daniels and R. J. Angelici: *Organometallics* **8**, 2277 (1989).
59. J. Chen, and R. J. Angelici: *J. Am. Chem. Soc.* **112**, 199 (1990).
60. S. Luo, A. E. Ogilvy, T. B. Rauchfuss A. L. Rheingold and S. R. Wilson: *Organometallics* **10**, 1002 (1991).
61. S. Luo, T. B. Rauchfuss and S. R. Wilson: *J. Am. Chem. Soc.* **114**, 8515 (1992).
62. S. Luo, T. B. Rauchfuss and S. R. Wilson: *Organometallics* **11**, 3497 (1992).
63. J. Chen and R. J. Angelici: *Coord. Chem. Rev.* **206-207**, 63 (2000).
64. A. E. Skaugset, T. B. Rauchfuss and C. L. Stern: *J. Am. Chem. Soc.* **112**, 2432 (1990).
65. S. Luo, T. B. Rauchfuss and Z. Gan: *J. Am. Chem. Soc.* **115**, 4943 (1993).
66. J. Chen, L. M. Daniels and R. J. Angelici: *Polyhedron* **9**, 1883 (1990).
67. E. O. Fischer and K. Öfele: *Chem. Ber.* **91**, 2395 (1958).
68. M. F. Bailey and L. F. Dahl: *Inorg. Chem.* **4**, 1306 (1965).
69. K. Öfele: *Chem. Ber.* **99**, 1732 (1966).
70. M. J. Sanger and R. J. Angelici: *Organometallics* **13**, 1821 (1994).
71. H. Singer: *J. Organometal. Chem.* **9**, 135 (1967).
72. D. A. Lesch, J. W. Richardson, Jr., R. A. Jacobson and R. J. Angelici: *J. Am. Chem. Soc.* **106**, 2901 (1984).
73. S. C. Huckett, N. N. Sauer and R. J. Angelici: *Organometallics* **6**, 591 (1987).

74. V. Guerchais and D. Astruc: *J. Organometal. Chem.* **316**, 335 (1986).
75. P. Bachmann and H. Singer: *Z. Naturforsch.*, **31b**, 525 (1976).
76. C. C. Lee, M. Iqbal, U. S. Gill and R. G. Sutherland: *J. Organometal. Chem.* **288**, 89 (1985).
77. D. M. Braitsch and R. Kumarappan: *J. Organometal. Chem.* **84**, C37 (1975).
78. G. H. Spies and R. J. Angelici: *J. Am. Chem. Soc.* **107**, 5569 (1985).
79. N. N. Sauer and R. J. Angelici: *Organometallics* **6**, 1146 (1987).
80. J. W. Hachgenei and R. J. Angelici: *Organometallics* **8**, 14 (1989).
81. B. Chaudret, F. Jalón, M. Pérez-Manrique, F. Lahoz, F. J. Plou and R. A. Sánchez-Delgado: *Nouv. J. Chim.* **14**, 331 (1990).
82. J. R. Lockemeyer, T. B. Rauchfuss, A. L. Rheingold and S. R. Wilson: *J. Am. Chem. Soc.* **111**, 8828 (1989).
83. E. A. Ganja, T. B. Rauchfuss and S. R. Wilson: *Organometallics* **10**, 270 (1991).
84. M. J. H. Russell, C. White, A. Yates and P. M. Maitlis: *J. Chem. Soc. Dalton Trans.*, 857 (1978).
85. R. A. Sánchez-Delgado, R.-L. Márquez-Silva, J. Puga, A. Tiripicchio and M. Tiripicchio-Camellini: *J. Organometal. Chem.* **316**, C35 (1986).
86. S. C. Hucket, L. L. Miller, R. A. Jacobson and R. J. Angelici: *Organometallics* **7**, 686 (1988).
87. R. A. Sánchez-Delgado, V. Herrera, L. Rincón, A. Andriollo and G. Martín: *Organometallics* **13**, 553 (1994).
88. S. Zhang, J. K. Shen, F. Basolo, T. D. Ju, R. F. Lang, G. Kiss and C. D. Hoff: *Organometallics* **13**, 3692 (1994).
89. J. A. S. Howell, P. C. Yates, N. F. Ashford, D. T. Nixon and R. Warren: *J. Chem. Soc. Dalton Trans.* 3959 (1996).
90. M. J. Girgis and B. C. Gates: *Ind. Eng. Chem. Res.* **30**, 2021 (1991) and references therein.
91. G. H. Spies and R. J. Angelici: *Organometallics* **6**, 1897 (1987).
92. E. J. Markel, G. L. Schrader, N. N. Sauer and R. J. Angelici: *J. Catal.* **116**, 11 (1989).
93. N. N. Sauer, E. J. Markel, G. L. Schrader and R. J. Angelici: *J. Catal.* **117**, 295 (1989).
94. J. W. Benson, G. L. Schrader and R. J. Angelici: *J. Mol. Catal. A: Chem.* **96**, 283 (1995).
95. R. J. Angelici: *Acc. Chem. Res.* **21**, 387 (1988).
96. E. W. Abel, F. G. A. Stone and G. Wilkinson, eds.: *Comprehensive Organometallic Chemistry*, Pergamon, London (1982).
97. E. W. Abel, F. G. A. Stone and G. Wilkinson, eds.: *Comprehensive Organometallic Chemistry II*, Pergamon, London (1995).
98. J. P. Collman, L. S. Hegedus, J. R. Norton and R. G. Finke: *Principles and Applications of Organo-Transition Metal Chemistry*, University Science Books, Ca. (1987).
99. R. H. Crabtree: *The Organometallic Chemistry of the Transition Metals*, 2nd Ed., John Wiley & Sons, N. Y. 1994).
100. E. O. Fischer, H. A. Goodwin, C. G. Kreiter, H. D. Simmons, Jr., K. Sonogashira and S. B. Wild: *J. Organomet. Chem.* **14**, 359 (1968).
101. C. A. Dullagham, X. Zhang, D. L. Greene, G. B. Carpenter, D. A. Sweigart, C. Camiletti, E. Ralaseelan: *Organometallics* **17**, 3316 (1998).
102. C. A. Dullagham, S. Sun, G. B. Carpenter, B. Weldon and D. A. Sweigart: *Angew. Chem. Int. Ed. Engl.* **35**, 5604 (1997).
103. C. A. Dullaghan, X. Zhang, D. Walther, G. B. Carpenter, D. A. Sweigart and Q. Meng: *Organometallics* **16**, 5604 (1997).
104. J. D. Jackson, S. J. Villa, D. S. Bacon, R. D. Pike and G. B. Carpenter: *Organometallics* **13**, 3972 (1994).
105. X. Zhang, C. A. Dullaghan, E. J. Watson, G. B. Carpenter and D. A. Sweigart: *Organometallics* **17**, 2067 (1998).
106. C. C. Lee, B. R. Steele and R. G. Sutherland: *J. Organometal. Chem.* **186**, 265 (1980).
107. S. C. Hucket and R. J. Angelici: *Organometallics* **7**, 1491 (1988).
108. C. -M. Wang and R. J. Angelici: *Organometallics* **9**, 1770 (1990).
109. S. Sun, C. A. Dullaghan and D. A. Sweigart: *J. Chem. Soc. Dalton Trans.* 4493 (1996).
110. J. Chen and R. J. Angelici: *Organometallics* **18**, 5721 (1999).
111. S. S. Lee, Y. K. Chung and S. W. Lee: *Inorg. Chim. Acta* **253**, 39 (1996).
112. V. Lamure-Meille, E. Schulz, M. Lemaire and M. Vrinat: *Appl. Catal. A* **131**, 143 (1995).
113. V. Meille, E. Schulz, M. Lemaire and M. Vrinat: *J. Catal.* **170**, 9 (1997).
114. V. Meille, E. Schulz, M. Lemaire and M. Vrinat: *Appl. Catal. A* **187**, 179 (1999).
115. M. Macaud, A. Milenkovic, E. Schultz, M. Lemaire and M. Vrinat: *J. Catal.* **193**, 255 (2000).

CHAPTER 3
HYDROGENATION REACTIONS

3.1   Introduction

Hydrogenation reactions play a major role in catalytic hydrotreating processes not only as a direct way of removing unwanted olefins and diolefins from the feeds, or of increasing the H/C ratio of products through the breakdown of the resins and asphaltenes, but also as a means of achieving the low levels of aromatic compounds allowed in gasoline, diesel and other refined products by partially reducing the polyaromatics present in those fractions [1]. Also, hydrogenation pathways have been frequently mentioned as crucial steps in HDS and HDN reaction networks [1-6]; as described in Chapter 1 it appears that in many hydrotreating reactions prehydrogenation of the heterocyclic moiety occurs prior to C-S or C-N bond breaking. In particular, a kinetic analysis by Whitehurst *et al.* for the HDS of 4,6-dimethyldibenzothiophene shows that the favored reaction pathways go through hydrogenative routes involving saturation of one or both of the *benzene* rings rather than through direct sulfur abstraction, while the opposite is true for unsubstituted dibenzothiophene [2]. Also, a detailed study of the HDS of a series of alkyl substituted dibenzothiophenes over Ni/Mo catalysts by Vrinat and coworkers [3-6] indicates that partial prehydrogenation of the dibenzothiophenes is a common initial step for the various possible HDS mechanisms of these highly refractory compounds. Thus, hydrogenation is of prime importance in connection with the deep desulfurization issue (see Section 1.2.7.2, p. 20), and with the generally accepted HDN mechanisms (see Section 1.3.2, p. 28). Hydrogenation of aromatics requires higher pressures of hydrogen than those needed for olefin hydrogenation on typical Co/Mo, Ni/Mo and related W-based HDS catalysts, mainly because of the high energetic barrier arising from resonance stabilization of the conjugated system that needs to be overcome. In polyaromatic molecules, the outer ring (less resonance stabilized) is always hydrogenated first, but as the hydrogenation proceeds the remaining rings become more benzene-like and are thus more difficult to reduce. Since very thorough discussions of the hydrogenation schemes accompanying HDS reactions over metal sulfide catalysts can be found in *e.g.* refs. 1-6, there is no need to duplicate that information here.

Instead we will concentrate on the field of *homogeneous* hydrogenation, which has developed explosively over the last four decades and it constitutes today one of the most frequently used and best understood applications of coordination and organometallic chemistry. Of particular interest for our purposes is the metal complex catalyzed hydrogenation of aromatic and heteroaromatic hydrocarbons, a subject that has attracted considerable attention over a number of years; a few catalytic systems are now fairly well mastered, although the number of efficient catalysts and the level of understanding of the reaction mechanisms is far lower than that which has been accumulated for other substrates like *e.g.* olefins. Thus this Chapter reviews some of the most relevant aspects of the homogeneous metal complex catalyzed hydrogenation of aromatic and hetero-aromatic hydrocarbons as well as their possible relations with the HDS problem. First the hydrogenation of simple mononuclear and polynuclear hydrocarbon molecules will be discussed, and subsequently the reduction of the corresponding sulfur- and nitrogen-containing heteroaromatic compounds will be presented, together with associated aspects of the pertinent coordination chemistry. The known mechanistic details will be placed in context, so that, whenever possible, parallels will be drawn with some of the major issues concerning hydrogenation reactions within the overall mechanistic networks of HDS catalysis.

3.2. Homogeneous Hydrogenation of Aromatic Hydrocarbons

3.2.1. THE CATALYSTS

Despite the impressive advances achieved in the general field of homogeneous hydrogenation over the last several decades, metal complex catalyzed hydrogenation of simple aromatic hydrocarbons remains rather undeveloped and unclear, in contrast with heterogeneous systems which have long been known to effect the hydrogenation of aromatic compounds with relative ease (*e.g.* Sabatier's early work on phenol and aniline hydrogenation [7] or on benzene reduction [8] by Ni metal). Such heterogeneous hydrogenation reactions on metallic catalysts are now commonly practiced in industry and in research laboratories, and the principal reaction pathways are fairly well understood [9,10]. On the other hand, comparatively few organometallic complexes are capable of effecting the hydrogenation of arenes to the corresponding partially or fully reduced products; a fairly comprehensive list of known homogenous catalysts for the hydrogenation of aromatic hydrocarbons is presented in Table 3.1 [11-57].

As can be seen from the data contained in Table 3.1, a wide variety of different metals have been found to be active for the homogeneous hydrogenation of aromatics. Most of the work has concentrated on cobalt, rhodium and ruthenium compounds but other late (Ni, Pt), early (Nb, Ta), and intermediate (Cr, Mo, W, Mn) metals have been employed. Commonly, the reaction conditions required to effect the reduction of arene rings with metal complexes are rather mild compared to those used in heterogeneous systems although in many cases they are considered quite drastic in comparison with the homogeneous hydrogenation of other substrates, *e.g.* olefins. Moreover these liquid phase catalysts are in general not efficient enough to be of practical industrial use, with perhaps the only notable exception being the IFP Process for benzene hydrogenation [24]. Even though new work on the homogeneous reduction of aromatics has continued to appear sporadically in the literature, no major breakthrough has been recently reported, and until today homogenous catalysts for this type of reaction generally suffer from either low activities and/or short lifetimes which makes them unattractive for practical applications. Although most arene reduction reactions have been carried out under conventional homogenous conditions (*i.e.* in solution under low H_2 pressures),

TABLE 3.1. Homogeneous catalysts for the hydrogenation of aromatic hydrocarbons

Catalyst	Substrates	T (°C)	P(H$_2$) (atm)	Ref.
Co$_2$(CO)$_8$	polyaromatics	135-185	230-270[a]	11-12
(η^3-C$_3$H$_5$)Co[P(OR)$_3$]	benzenes polyaromatics	25	1	13-19
M(acac)$_2$ (Co, Ni) + LiAlH$_4$	benzene	30	1	20
(η^6-CH$_3$C$_6$H$_5$)Ni(C$_6$F$_5$)$_2$	benzenes		35	21
Metal alkoxides, acac, or carboxylates + AlR$_3$	benzenes polyaromatics	150-210	70	22-25
Co(acac)$_2$+ AlR$_3$	benzenes	30	1	26
(η^3-cyclooctenyl)Co[(Cy$_2$P)$_2$(CH$_2$)$_3$]	benzene	25	1	27
Rh /phenylanthranylic acid	Benzene			28-30
[Cp*RhCl$_2$]$_2$	benzenes anthracene		50	31, 32
(μ-H)$_2$Rh[P(O^{i}Pr)$_3$]$_4$	benzenes	25	1	33
Rh(acac)[P(OPh)$_3$]$_2$	benzenes	80	10	34
HRh[P(NC$_4$H$_4$)$_3$]$_4$	benzenes	25	5	35
HRh(CO)[P(NC$_4$H$_4$)]$_3$				
[Rh(diphos)(MeOH)$_2$]$^+$	anthracenes	50-75	1	36
[RhCl(diene)]$_2$+ [NR$_4$]X	benzenes[b] naphthalene	25	1	37
Rh or Pt-isocyanide complexes on SiO$_2$-supported metals (Pd, Pt, Ru)	benzenes[c] naphthalene	40	1	38-39
(η^6-C$_6$Me$_6$)Ru(η^4-C$_6$Me$_6$)	benzenes	90	2-3	40-41
[(η^6-C$_6$Me$_6$)$_2$Ru$_2$(μ-H)$_2$(μ-Cl)]Cl$_2$	benzenes	50	50	42
Ru(H)$_3$(PPh$_3$)$_3$	anthracenes	50-100	5	43-45
Ru(H)$_2$(H$_2$)(PPh$_3$)$_3$				
(η^6-C$_6$Me$_6$)$_4$Ru$_4$H$_4$	benzenes[d]	90	60	46- 48
(η^6-C$_6$Me$_6$)$_2$Ru$_2$Cl$_4$				
Metal carbonyls (Fe, Co, Mn, Rh, Ru, W, Mo, Cr)	polyaromatics[a, e]	180	25	49, 50
Fe(CO)$_5$ + ammonium salt	anthracene[b]	150	35	51
Ta(OR)$_2$(H)$_3$(PMe$_2$Ph)$_2$	benzenes	80-100	3-100	52-54
M(OAr)$_3$Cl$_2$ + ^{n}BuLi (Nb, Ta)	benzenes polyaromatics			
Early metal complexes on oxides	benzenes[c] polyaromatics	100-120	70-90	55-56
PtCl$_2$(CH$_3$CN)$_2$ + H$_2$O.BF$_3$	benzene naphthalenes			57

[a](H$_2$/CO); [b]phase transfer catalysis; [c]supported metal complexes; [d]liquid biphasic catalysis; [e]CO/H$_2$O

other interesting systems have been employed, such as CO/H$_2$O (water gas) or CO/H$_2$ (syngas) as reducing mixtures [49, 50], phase transfer catalysis [37], and more recently, aqueous [46, 47] and non-aqueous ionic liquid [48] biphasic catalysis which offer more promise for practical uses. Some interesting examples of metal complexes grafted onto oxides [55, 56] or supported metals [38, 39] as arene hydrogenation catalysts have been provided.

With regard to possible relations between homogeneous hydrogenation systems and heterogeneous HDS catalysts, polynuclear aromatic molecules are easier to reduce by metal complexes than isolated benzene rings, since the partial saturation of fused-ring aromatic compounds does not suffer from as dramatic a loss of resonance stabilization;

this reflects essentially a thermodynamic limitation and thus this characteristic is shared with heterogeneous catalysts for which similar trends have been observed.

Since MoS_2 and WS_2 are known to be good hydrogenation catalysts it is important to note that molybdenum and tungsten complexes have not proved to be active in the *homogeneous* hydrogenation of aromatics except in the case of the low-valent carbonyls under water gas shift or syngas conditions [49, 50], systems that are rather far from those encountered in HDS reactions. This is an important drawback in considering the homogeneous systems as models of heterogeneous hydrogenation linked to HDS reaction networks catalyzed by unpromoted metal sulfides. On the other hand, cobalt compounds (and, to a lesser extent nickel complexes) are among the most active and best understood homogeneous catalysts for arene reduction. This is interesting in connection with the idea that the active sites for binding the reducible molecules in HDS catalysis are associated with the promoter (Co or Ni) atoms in the Co-Mo-S or Ni-Mo-S type structures. Hydrogen dissociation is sometimes though to take place predominantly at the MoS_2 edge sites [see *e.g.* ref. 1, p. 229], but also oxidative addition at the promoter metal sites is considered important (see Chapter 1, section 1.2.5). However, the conditions under which homogeneous reductions take place are so dramatically different from those employed in heterogeneous reactions that direct comparisons of activities for the various metals must be taken with caution. The value of coordination chemistry and homogeneous hydrogenation catalysis lies predominantly in the mechanistic considerations that will be described in the remaining parts of this Chapter, as well as in Chapter 5 where the activation of hydrogen on metal complexes with sulfur ligands is discussed in detail.

3.2.2. REACTION MECHANISMS

The mechanistic knowledge of the homogeneous hydrogenation of aromatics is far poorer than that accumulated over the years for other homogeneous hydrogenation reactions, olefins in particular. Some excellent monographs are available, for instance the ones by James [58, 59] and by Chaloner *et al.* [60] covering the general aspects of homogeneous hydrogenation and including early work on aromatics; this latter subject has been also reviewed by Fish [61]. Therefore, rather than attempting a comprehensive discussion on this topic, only the most recent work, together with some selected earlier cases that seem particularly pertinent will be presented here, in an attempt to relate the known details of homogeneous reactions to heterogeneous hydrogenation of aromatic hydrocarbons within the context of HDS catalysis.

Three major classes of reaction mechanisms can be identified: (1) Radical processes for cobalt and manganese carbonyl complexes that give the expected products with little stereoselectivity [11, 12, 49, 50]. Although some claims have appeared of the intervention of radicals in HDS-related hydrogenation, most of the evidence points to other types of surface mechanisms which can be better related to coordinative mechanisms; therefore no further mention will be made of radical reactions in this Chapter. (2) Reactions involving Ziegler-type catalysts (a transition metal complex mixed with an alkyl aluminum co-catalyst); these poorly defined systems have proved to be difficult to study in detail [20, 22-25], and they appear rather unrelated to HDS-active solids. (3) Coordinative catalysis with well defined metal complexes, which has allowed detailed investigations of some particularly efficient systems by use of the standard methods of organometallic chemistry, notably by NMR spectroscopy. For our purposes only the latter need to be considered. The main classes of catalysts known will be described in some detail in the following sections.

3.2.2.1. Allyl metal complexes.

The best understood catalysts for homogeneous arene hydrogenation are no doubt Muetterties' allyl cobalt complexes such as $(\eta^3\text{-}C_3H_5)Co[P(OMe)_3]_3$ and related compounds, which are able to reduce benzene to cyclohexane at exceedingly mild conditions, *viz.* 25°C and 1 atm of hydrogen pressure, albeit with dramatically short life-times (*ca.* 25 turnovers) [13-19]. Besides benzenes, the allylcobalt catalysts fully hydrogenate naphthalene, anthracene, and phenanthrene at pressures of 1-3 atm, although at markedly reduced rates; this is in contrast to other homogeneous and most heterogeneous catalysts which are known to be more efficient for the reduction of fused ring systems than for isolated benzene rings. Pyrene, carbon black and asphaltenes derived from coal or petroleum were not hydrogenated by this allylcobalt derivative. Alkyl substitution of the benzene ring has a pronounced inhibiting effect; thus, the hydrogenation rate falls in the order:

benzene > toluene > xylenes > mesitylene > 1,2,4,5-tetramethylbenzene > 1,2,3-trimethylbenzene > 1,2,3,4-tetramethylbenzene > hexamethylbenzene

Electronically, the expected order would be reversed; hence, the inhibition can be safely attributed to steric impediments. Terminal olefins are generally hydrogenated at faster rates than benzene and effectively block the reaction of arenes if the two substrates are hydrogenated simultaneously. Similar catalysts containing bulkier phosphite or phosphine ligands instead of trimethylphosphite generally led to an increase in catalytic activity in the following order: $P(OCH_3)_3 < P(OC_2H_5)_3 < P(CH_3)_3 < P(O\text{-}i\text{-}C_3H_7)_3$. However, this desirable change was invariably associated with a marked decrease in the catalyst life.

In their earlier publications, Muetterties and coworkers observed that the allyl group could be easily lost as propane from $(\eta^3\text{-}C_3H_5)Co[P(OMe)_3]_3$ by reaction with hydrogen, and they associated complete loss of the allyl group with cessation of catalytic hydrogenation. This led them to believe that the allyl ligand remained bound to cobalt throughout the catalytic cycle, and that a η^3/η^1 allyl isomerization was an important part of the mechanism, since it allowed the opening of a coordination site when needed, and stabilized the complex by π-coordination when no substrate was available [13-17]. However, further studies carried out on the analogous $(\eta^3\text{-}C_8H_{13})Co[P(OR)_3]_3$ by the same group [18, 19] showed that the η^3-cyclooctenyl ligand is very quickly lost as cyclooctane, yet the catalytic hydrogenation proceeded for an extended period. This suggests that the actual catalytically active species must be an allyl-free hydride complex, most likely $HCo[P(OR)_3]_2$; in the presence of free phosphite, this highly unsaturated complex will successively form $HCo[P(OR)_3]_n$ (n = 3,4), and further interaction with hydrogen also produces $H_3Co[P(OR)_3]_3$; these three complexes are inactive for arene hydrogenation. In the absence of donor molecules, $HCo[P(OR)_3]_2$ decomposes through a first step of dimerization, but in the presence of donor molecules (L) like olefins, dienes, and arenes, relatively stable $HCo[P(OR)_3]_2L_{1,2}$ adducts are formed, which enter a productive catalytic cycle. This set of reactions is summarized in Fig. 3.1.

Perhaps what can be viewed as the key feature of this catalytic system is the possibility, in the productive cycle, of coordination of the benzene ring to the cobalt atom through two of the double bonds to form $HCo[P(OR)_3]_2(\eta^4-C_6H_6)$. Hydrogen transfer to this diene-like η^4-coordinated arene then occurs intramolecularly, via allylic intermediates like $(\eta^3-C_6H_7)Co[P(OR)_3]_2$ and subsequent hydrogen addition produces $(\eta^3-C_6H_7)Co(H)_2[P(OR)_3]_2$, as depicted in Fig. 3.2 [13-19] to generate the corresponding 1,3-cyclohexadiene bound to Co. Repeated oxidative addition of hydrogen followed by hydride transfer transforms the carbocycle into cyclohexene and finally cyclohexane. If a terminal olefin like 1-hexene is present in the reaction medium, it will compete with cyclohexene in the last step of the cycle, so that the final product contains varying amounts of cyclohexene and cyclohexane; in the absence of added olefins, the only reaction product is the fully saturated hydrocarbon.

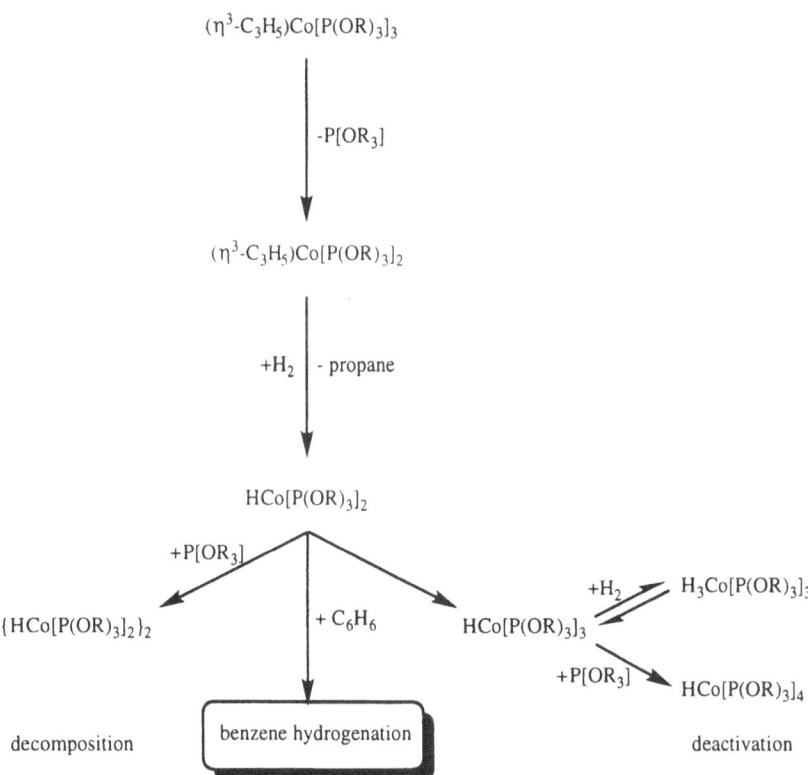

Fig. 3.1. Formation, deactivation and degradation of the active species for arene hydrogenation from $(\eta^5-C_3H_5)Co[P(OMe)_3]_3$.

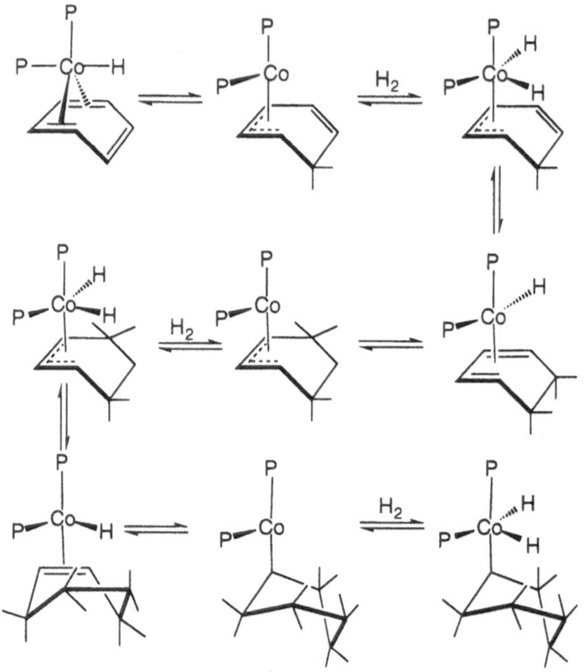

Fig. 3.2. The mechanism of benzene hydrogenation catalyzed by $(\eta^3$-$C_3H_5)Co[P(OMe)_3]_3$.

Some features of the structure and reactions of this catalyst are interesting in relation to the discussion on surface active sites for analogous reactions on HDS catalysts: (1) The starting allyl complex is a saturated (18 electron) Co(I) species which may be envisaged either as a *tetrahedral* array of the three phosphorus atoms and the centroid of the π-allyl ligand, or a *square-based pyramidal* structure with the allyl occupying two sites of the square base and the phosphorus atoms on the remaining three; this can be related with the geometries proposed for the active Co-Mo-S sites which are indeed square-based pyramidal or tetrahedral Co atoms (see Chapter 1, Section 1.2.4.1). (2) The active homogeneous species for arene hydrogenation, *viz.* $HCo[P(OR)_3]_2$ is readily formed by elimination of the allyl ligand through its hydrogenation to propane, in a way reminiscent of the manner in which surface sulfides are thought to be eliminated through hydrogenation as H_2S in order to generate an "anionic vacancy" in HDS catalysts (see Chapter 1, Section 1.2.4.2). (3) Also interestingly, the active homogeneous intermediate, $HCo[P(OR)_3]_2$, corresponds to Co(I) in a 14e configuration -with *two vacant coordination sites*; this species is thus capable of binding arenes -as pseudodienes- to yield coordinatively saturated intermediates. This availability of two CUS's –which has also been invoked in solid catalysts- seems to be the key feature of this catalyst for arene hydrogenation activity. On the other hand, $HCo[P(OR)_3]_3$ (16e), also formed under the reaction conditions by coordination of an extra phosphorus ligand, has only *one vacant site* and it is therefore very well adapted to *olefin* but not arene binding; the latter would necessarily be coordinated in a η^1-fashion, generally a less stable situation than simple alkene coordination. Consequent $HCo[P(OR)_3]_3$ is capable of preferentially hydrogenating olefins in mixtures containing also arenes, or, in other words, olefins selectively block the hydrogenation activity for arenes but continue themselves to be reduced by the catalyst; this is clearly in parallel with the poisoning of aromatic hydrogenation or HDS sites by olefins, sometimes observed in Co-Mo-S catalysts. Binding of a fourth phosphite is also possible, yielding $HCo[P(OR)_3]_4$, which is a coordinatively saturated 18-electron complex and therefore catalytically inactive for the hydrogenation of either arenes or alkenes.

In summary, this homogeneous cobalt chemistry is surprisingly in line with the currently accepted models of Co-Mo-S phases and some of the ways in which active sites are thought to be formed and to react. Hydrogenation of olefins and aromatics are sometimes claimed to take place on different types of sites, in fact related to different degrees of coordinative unsaturation [1, 2]. Such transformations, well identified in the homogeneous cobalt case, could be easily achieved in metal-sulfide surfaces by simple low energy processes resulting from changes in the number or the type of M-S and S-S bonds.

Allyl rhodium phosphite analogues of the cobalt catalysts have also been used as catalysts for arene hydrogenation; it has been shown that the rhodium complexes undergo rapid loss of propene in the presence of hydrogen which makes them too short-lived and thus not adequate for reduction of arenes. However, a product of hydrogenation of a related allyl complex containing triisopropylphosphite, identified as a triply hydride-bridged dimer $[L_2HRh(\mu-H)_3RhL_2]$ (where $L = P(O^iPr)_3$) is a good catalyst precursor for mild homogeneous arene hydrogenation (20 °C, 1 atm H_2); the rate of hydrogenation decreases with increasing methyl substitution, similarly to the cobalt catalyst. Also, olefins are reduced by this rhodium system at rates that are about 10^5 greater than those observed for arene hydrogenation. Although mechanistic details in this case have not been fully investigated, presumably the dimer is cleaved to yield monomeric species like $(H)_3RhL_2$ which could easily lose dihydrogen to generate $HRhL_2$, analogous to $HCo[P(OR)_3]_2$, and thus capable of operating by a similar mechanism to the one described for the cobalt catalysts [33].

3.2.2.2. Cyclopentadienyl and arene metal complexes.

Maitlis and coworkers found that the complex $[Cp^*RhCl_2]_2$ (Cp^* = pentamethyl-cyclopentadienyl) is active for arene hydrogenation to the corresponding cyclohexanes in the presence of a base, under moderate reaction conditions (50° C, 50 atm) in 2-propanol as the solvent. Ether, ester, keto, and amino substituents did not affect the activity but -OH or -COOH groups inhibited the reaction. Although the detailed mechanism of action of this catalyst was not established, the evidence points toward the heterolytic activation of hydrogen in the reaction mechanism –hence the need for a base in the reaction medium. Also, the observed high *cis* stereoselectivity in the reduction of disubstituted benzenes and the formation of 1,2,3,4-tetrahydroanthracene as the major product of anthracene hydrogenation were taken as suggestive that similarly to the Co-allyl catalysts, η^4-coordination of the substrate is important also in this case. The analogous iridium complex was also capable of hydrogenating arenes, but the activities were much lower, *ca.* 28% of that observed for rhodium [31, 32].

Another interesting example of an efficient catalyst for arene hydrogenation, also discovered by the Muetterties group, is bis(hexamethylbenzene)ruthenium(0), an 18 electron species which contains a conventional η^6-arene ligand and a nonplanar η^4-arene ligand. Unlike the allylcobalt- and allylrhodium-phosphite systems, this is a long-lived catalyst for arene hydrogenation at 80-90°C and 1-3 atm H_2 although its stereoselectivity is lower that that observed for the allylcobalt systems. Also in contrast with the cobalt and rhodium catalysts, considerable amounts of cyclohexenes are produced together with the corresponding cyclohexanes when this ruthenium catalyst is used [40, 41]. Detailed NMR studies indicate that the fluxional character observed for this molecule can be interpreted in terms of an exchange involving $Ru(\eta^4\text{-}C_6Me_6)_2$, an excited state containing two non-planar η^4-arene ligands; this intermediate is also the key species for the reaction with dihydrogen in the catalytic hydrogenation cycle to yield $H_2Ru(\eta^4\text{-}C_6Me_6)_2$. Since hexamethylbenzene itself is not hydrogenated by this complex, arene exchange to introduce the reducible substrate takes place subsequently to H_2 addition. Although this was not observed experimentally, it happens conceivably through an "unzipping" series of $\eta^6 \leftrightarrow \eta^4 \leftrightarrow \eta^2$ transformations as depicted in Fig. 3.3, accompanied at some stage by a "zipping up" of the incoming arene in a series of comparable processes.

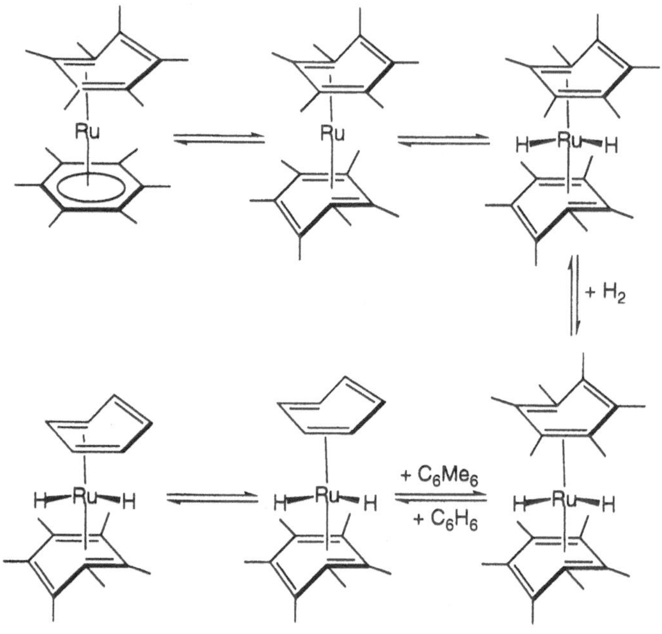

Fig. 3.3. Initial steps in the hydrogenation of benzene catalyzed by $(\eta^6\text{-}C_6Me_6)Ru(\eta^4\text{-}C_6Me_6)$. (Adapted from refs. 18, 40, 41).

By contrast, some ruthenium hydride clusters reported by Süss-Fink are thought to hydrogenate arenes through η^6-bonded intermediates, as represented in Fig. 3.4. The proposed mechanisms, however, have not yet been clearly established and more detailed studies under *catalytic* conditions [46-47] would be welcome. These ruthenium clusters have also been shown to be active for arene hydrogenation in a biphasic medium using ionic liquids as the solvent, where the catalysts is retained, allowing its easy recovery and recycling [48].

3.2.2.3. Other metal complexes.
A number of mononuclear and dinuclear Ru complexes with phosphine ligands such as $RuCl_2(PPh_3)_3$ [49, 50], $Ru(H)_2(H_2)(PPh_3)_3$ [43-45], and $(\eta^6\text{-}C_6Me_6)_2Ru_2(\mu\text{-}H)_2(\mu\text{-}Cl)]Cl_2$ [42], as well as some cationic rhodium diphosphine derivatives like $[Rh(MeOH)_2(diphos)]^+$ (diphos = 1,2-bis(diphenylphosphino)ethane) [36] are known to catalyze the hydrogenation of polynuclear aromatic hydrocarbons under mild reaction conditions. They are thought to be mechanistically related to Muetterties' system [18] in that they all presumably operate through intermediates containing η^4-coordinated arenes. Halpern's detailed kinetic study of anthracene hydrogenation by cationic rhodium diphosphine complexes [36] led to the mechanism summarized in Eqs. 3.1-3.3; it is also assumed that a key intermediate in the catalytic cycle is the η^4-bonded Rh(III) $[Rh(H)_2(diphos)(anthracene)]^+$. This mechanism is in accord with a second order rate law, $-d[arene]/dt = k[Rh(diphos)(arene)^+][H_2]$, with $k(59.7° \text{ C}) = (9.0 \pm 1.0) \times 10^{-2} \text{ M}^{-1}\text{s}^{-1}$, $\Delta H^{\ddagger} = 16.6 \pm 2$ kcal/mol, and $\Delta S^{\ddagger} = -14 \pm 5$ cal/mol K.

A very different set of interesting d^0 niobium and tantalum hydride catalysts containing bulky aryloxide ligands recently reported by Rothwell [52-54] appear very promising. For instance, naphthalene and anthracene are reduced at 80° C and 3-100 atm H_2 by $[Ta\{OC_6H_3(C_6H_{11})_2\text{-}2,6\}_2(H)_3(PMe_2Ph)]$ to produce mainly tetralin or 1,2,3,4-tetrahydroanthracene, respectively.

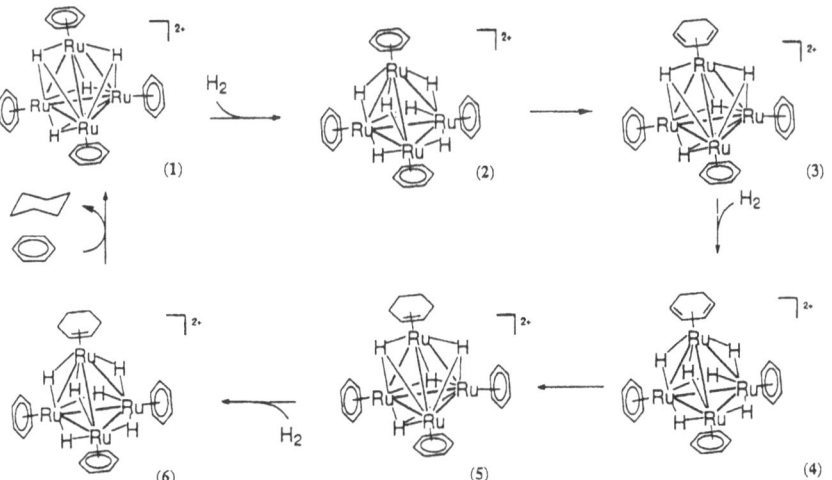

Fig. 3.4. Hydrogenation of benzene catalyzed by $[(\eta^6\text{-}C_6Me_6)_4Ru_4H_4]^{2+}$
(Reproduced with permission from ref. 46).

$$[Rh(diphos)(A)]^+ + H_2 \xrightarrow{\text{rds}} [Rh(diphos)(MeOH)_2]^+ + 1,2\text{-}AH_2 \qquad (3.1)$$

$$1,2\text{-}AH_2 + H_2 \xrightarrow{[Rh(diphos)(MeOH)_2]^+} 1,2,3,4\text{-}AH_4 \qquad (3.2)$$

$$[Rh(diphos)(MeOH)_2]^+ + A \underset{\text{fast}}{\overset{}{\rightleftharpoons}} [Rh(diphos)(A)]^+ \qquad (3.3)$$

(A = anthracene)

Since the synthetic methods to these catalysts are not straightforward, an alternative way of generating them is to use chloride precursors [Nb(OAr)$_3$Cl$_2$] activated *in situ* by 2-3 eq. of n-BuLi, followed by exposure to H$_2$. The hydrogenation mechanism involves metal dihydrido complexes containing η^6-bonded arenes that are reduced stepwise through cyclohexadienyl and cyclohexadiene intermediates. These systems are expected to be further developed and better understood in the near future, since they represent a very original alternative for arene hydrogenation. However, they seem to be quite far from any direct connection with the type of sites or reactions involved in hydrogenation of aromatics in the context of HDS.

3.2.2.4. Tethered complexes on supported metals.

The combination of homogeneous-heterogeneous catalysts developed by Angelici through the grafting of metal complexes onto conventional heterogeneous supported metal catalysts not only has the advantage of simple recovery and recycle, but also functions by a sinergistic action of the two components in the catalytic reactions (grafted complex and supported metal) [38, 39]. For example, the rhodium complexes [Rh(COD)(L)]BF$_4$ (where L is either a pyridylphosphine or a bipyridyl ligand containing alkoxysilane groups) were grafted onto silica-supported palladium metal to give the corresponding hybrid catalysts; these solids display activities for arene hydrogenation at 70° C and 4 atm H$_2$ that are higher than any other homogeneous or immobilized complex catalyst. Simple arenes such as toluene, as well as substituted ones like PhCO$_2$Me and PhOH are readily hydrogenated, but reduction of polynuclear aromatics was not reported. A possible explanation advanced by these authors for the high activities observed involves considering Pd-SiO$_2$ as the site for hydrogen dissociation, which is followed by spillover onto the silica, while the tethered complex would be the site for arene activation. This proposal is somewhat similar to some mechanisms that have been invoked in HDS and related reactions, through bimetallic interactions and spillover of hydrogen, but further work is needed in order to confirm this hypothesis.

No other detailed studies on the mechanisms of hydrogenation of aromatics have been reported to date and are clearly desirable. Although this area of research has remained relatively dormant, in comparison with other homogeneous hydrogenation reactions, the current urgent need for better processes aimed at reducing the aromatic contents in fuels will no doubt stimulate additional efforts to develop novel efficient catalysts as well as to better understand the underlying chemistry.

3.2.3. CONCLUSIONS AND PERSPECTIVES

Large scale applications such as fuel upgrading are generally not very adequate for homogeneous processes which require costly catalyst separation or regeneration and recycling, but the advent of new technologies, such as liquid biphasic catalysis using water soluble complexes or ionic liquids as solvents has already found a place in industrial chemistry [62] and offers exciting new possibilities for aromatic hydrogenation processes [46-48]. Also, anchoring organometallic complexes on solid supports or on supported metals [38, 39] or metal sulfides appears as a possible means of obtaining new effective catalysts.

Some examples of such systems for the hydrogenation of arenes are already available. A particularly appealing possibility for practical applications in the case of deep desulfurization would be the development of a two stage process incorporating a highly active novel coordination catalyst (liquid biphasic or supported metal complex) for the hydrogenation of one or both of the benzene rings in *e.g.* 4,6-dimethyldibenzothiophene, followed by a mild HDS of the saturated product which is known [1, 2] to be easier on conventional Co-Mo-S catalysts, as shown in Eq. 3.4.

Besides the possibility of practical uses, such as the one depicted in Eq. 3.4, or the direct hydrogenation of simple aromatics in fuels, a deeper understanding of homogeneous hydrogenation reactions of aromatic hydrocarbons is desirable as it will certainly shed additional light on the reaction networks which operate under HDS conditions. From the knowledge accumulated so far, some parallels can be drawn in relation to the nature of the active sites for hydrogenation reactions on Co-Mo-S and related catalysts. A crucial issue in hydrogenation of aromatics is the character of the metal-arene bonding which can occur through (A) the whole aromatic ring (η^6) occupying *three coordination sites*, (B) a diene portion of the ring (η^4) taking up *two coordination sites* or (C) one C=C bond (η^2) requiring only *a single empty coordination site*. Interestingly, η^6-coordination of arenes in complexes does not appear to be in general, favorable for hydrogenation, perhaps because the metal-π-benzene interaction represents too stable a situation; this type of binding could, however, be related to poisoning effects of aromatics, particularly of highly unsaturated (3 CUS) HDS active sites.

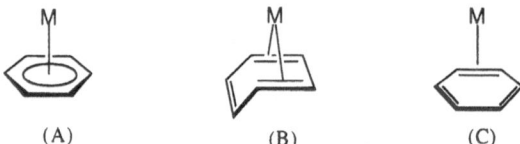

(A)   (B)   (C)

In contrast, the η^4-coordination mode (occupying 2 CUS) results in partial loss of the aromatic character and in an actual bending of the molecule, with the unbound C=C bond pointing away from the metal. Both these effects translate into a notable activation of the arene ring toward hydride attack. The η^2-coordination mode (requiring 1 CUS) only occurs in organometallic chemistry seldom, and for very electron-rich d^{10} metal complexes, and it does not seem to be involved in hydrogenation catalysis. Therefore it appears that the most favorable situation for promoting arene hydrogenation would be a 14-electron fragment in a complex, or a surface metal site with two coordinative unsaturations, allowing arene chemisorption as in (B). In such a site, chelate binding in the diene mode would be favored over coordination (chemisorption) of an olefin, so the activity for hydrogenation of aromatics should not be greatly affected by the presence of olefins. On the contrary, singly unsaturated sites allowing only η^2-coordination as in (C) are particularly well adapted for olefin binding and thus such coordination of aromatics, although possible, would be disfavored if olefins are competing for the same site. Such considerations could explain why poisoning effects are different for olefin and aromatic hydrogenation reactions on HDS catalysts [1-2] (See also Chapter 1). This would be in line with the idea that those two types of reactions occur on different active sites, one important difference being the number of coordinative unsaturations.

It is tempting to establish a parallel between the way in which the unsaturations are created in the hydrogenation catalysts (through ligand loss or modification of the bonding) and the fascinating ideas advanced by Whitehurst on how one single geometrical configuration in the active Co-Mo-S site can perform two different functions ("the two site dilemma", see Chapter 1) [2]. There are no contradictions in these views if one considers the possibility of a surface sulfur readily changing its bonding from e.g. a triple bridge to a double bridge *when needed*, as has also been discussed by Curtis using the concept of *latent vacancies* in CpCoMoS clusters capable of desulfurizing thiols and thiophene in solution under mild conditions [63] (see Section 1.2.4.2 and Chapter 4). Also, Byskov *et al.*, based on DFT calculations on MoS_2 surfaces, have found *surface reconstruction* to take place when sulfur is removed from a catalytic site and spontaneous migration from the bulk takes place as to partially fill the "vacancy" [64].

Therefore, it is possible to envisage the existence of two closely related but yet different cobalt active sites on a Co-Mo-S HDS surface. They could easily interconvert by low energy metal-sulfur bond breaking and rearrangement processes, depending on the reaction conditions employed and the relative concentrations and binding energies of the various components present in the mixtures being passed over the catalyst.

3.3. Homogeneous Hydrogenation of Heteroaromatic Hydrocarbons

3.3.1. SULFUR CONTAINING HETEROAROMATICS

As discussed in Chapter 1, two major pathways have been advanced in the literature for HDS of thiophenes, depicted in Figs. 1.7-1.10. Based on reactor studies, on the observation of intermediate products and/or on the kinetic analyses of the reactions some authors believe that prehydrogenation of the sulfur containing rings is possible -or even necessary- before C-S bond cleavage occurs; hydrogenative routes involving the benzene rings also appear to become more competitive for larger S-containing molecules like benzonaphthothiophene, and for highly refractory substituted molecules such as 4-methyldibenzothiophene and 4,6-dimethyldibenzothiophene. Other researchers defend the idea that direct sulfur extrusion takes place predominantly (or in parallel) with the hydrogenation route [1] (for details, see Chapter 1); it seems that both pathways are possible and the predominance of either one depends strongly on the nature of the substrate and the catalyst, and on the reaction conditions. In order to help elucidate this interesting question, a number of groups have studied the metal complex catalyzed homogeneous hydrogenation of thiophenic molecules and established the major reaction pathways in solution. Apart from this mechanistic modeling interest, the development of highly active catalysts for the hydrogenation of thiophenes would be a very welcome addition to HDS technology, since in general cyclic thioethers are desulfurized more readily by conventional catalysts than the corresponding thiophenes; thus, a two-stage process could be envisaged incorporating the selective prehydrogenation of the sulfur-containing ring by e.g. a liquid biphasic or a solid-supported metal complex catalyst, followed by mild HDS on conventional heterogeneous catalysts.

3.3.1.1. Thiophene hydrogenation.

Apart from the early report by Wender on the hydrogenation of thiophenes by cobalt carbonyl under forcing conditions (180-190 °C, 230-265 atm $2H_2/1CO$, 2-3 turnovers in 4-5 h) [65], the only example of a well defined system for the reduction of thiophene is the work of Bianchini and Sánchez-Delgado [66] who have demonstrated its conversion to tetrahydrothiophene by reaction of $[Ir(H)_2(\eta^1S\text{-}T)_2(PPh_3)_2]PF_6$ [67] with H_2 under extremely mild conditions (1 atm and 80° C). Although in a formal sense the reaction is catalytic, the complex is strongly poisoned by the product after only two cycles, due to the fact that tetrahydrothiophene binds much more strongly to iridium than thiophene itself to form the very stable $[Ir(H)_2(\eta^1S\text{-}THT)_2(PPh_3)_2]PF_6$ which was characterized by X-ray diffraction [67]; this is in line with the binding preferences of thiophenes to metals already described in Chapter 2, Section 2.2). Although of no practical use for this reason, this system provided a unique opportunity to study the detailed mechanism of the hydrogenation of thiophene, which is shown in Fig. 3.5. Some of the most interesting aspects of this cycle are:

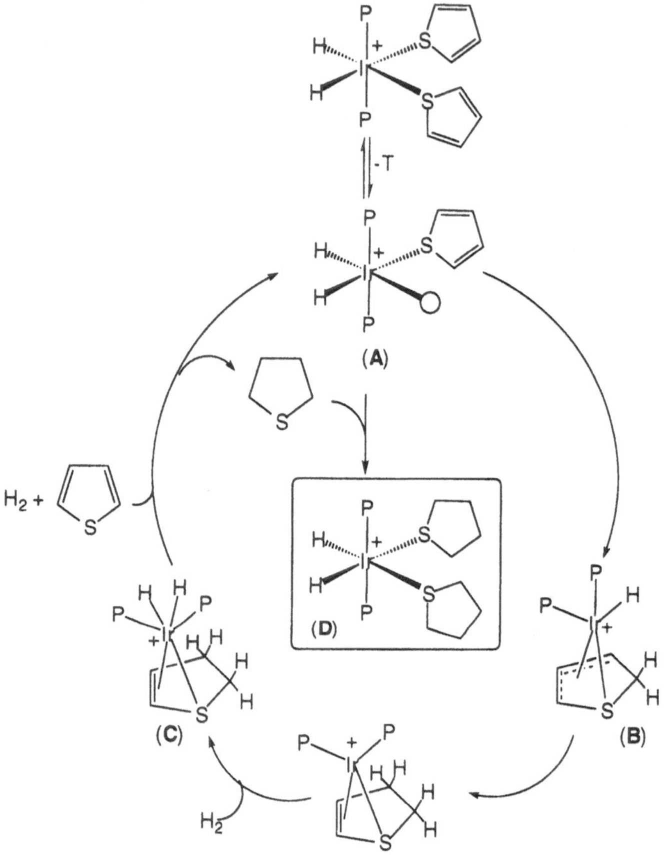

Fig. 3.5. The mechanism of thiophene hydrogenation by $[Ir(H)_2(\eta^1\text{-}SC_4H_4)_2(PPh_3)_2]PF_6$ (Adapted from ref. 66).

(1) The fact that one single metal center in intermediate (**A**) -derived from a well characterized bis(thiophene)Ir complex- is enough to activate both thiophene and hydrogen. Although this is highly unlikely to take place in a real HDS working catalyst containing many neighboring metal atoms, it provides an intriguing example of the high reactivity that an isolated metal site can display if the electronic and steric characteristics are adequate. This activation of hydrogen and thiophene in a single metal site has been invoked by Whitehurst in his model for Co-Mo-S sites [2].

(2) Intramolecular stepwise transfer of two hydrides to the bound thiophene leads to a thioallyl intermediate (**B**), characterized by NMR spectroscopy, in which the sulfur-containing ligand itself has come to formally occupy *three coordination sites*; this is of interest if a similar mechanism were to be invoked for a heterogeneous catalyst, since then a highly unsaturated site (3 CUS) would be required, assuming that hydrogen activation takes place elsewhere, or has taken place previously at the same site.

(3) Dissociation of one thiophene ligand takes place before the first hydride is actually transferred; this would allow the remaining thiophene to bind in any one of four possible modes, *viz.* L-type η^1S, $\eta^2(C=C)$, L_2 type $\eta^3(S,C=C)$, or η^4-diene like, and/or to interconvert between them. Precedents for $\eta^1 \leftrightarrow \eta^2$ isomerization and its involvement in BT hydrogenation reactions [68, 69], as well as for the $\eta^1 \leftrightarrow \eta^4 \leftrightarrow$ thioallyl transformations in the hydride attack to coordinated thiophenes are available [70, 71]; the latter requires at least a doubly unsaturated (14e) site. Also, the $\eta^3(S,C=C)$ bonding mode has been proposed in heterogeneous catalysis as part of a mechanism proceeding via prehydrogenation of the ring. Therefore, again it is seen that the capability of generating 1, 2, or 3 CUS would be related to the possible occurrence of thiophene binding and activation toward C=C bond hydrogenation and eventual hydrodesulfurization.

(4) It was demonstrated that 2,3-dihydrothiophene (and not 2,5-dihydrothiophene) is the actual intermediate toward ring saturation in (**C**), in agreement with some proposals made for the heterogeneous hydrogenation of thiophene in the context of HDS reactions which involve "multi point" $\eta^3(S,C=C)$ adsorption [69-71]. Methyl substituted thiophenes react in an analogous manner as long as there is at least one unsubstituted carbon atom next to sulfur, but *e.g.* 2,5-Me$_2$T was unreactive [72].

In a related set of stoichiometric reactions depicted in Eq. 3.3, Angelici has shown [68, 69] that η^5-coordinated thiophene is reduced to 2,3-dihydrothiophene by nucleophilic attack of an external hydride source on $(\eta^5\text{-}SC_4H_4)Mn(CO)_3^+$ to yield a neutral $(\eta^3,\eta^1S\text{-thioallyl})Mn(CO)_3$ derivative analogous to **B**. Subsequent protonation with HCl produced a stable complex containing the hemireduced 2,3-dihydrothiophene

(3.5)

bonded to Mn through the sulfur atom and the remaining C=C bond (analogous to **C**). Interestingly, a similar reaction of $[CpRu(\eta^5\text{-}SC_4H_4)]^+$ with hydrides, also reported by Angelici [73], evolves by C-S bond rupture rather than hydrogenation (Eq. 3.4), presumably via the analogous thioallyl complex $(\eta^3,\eta^1S\text{-thioallyl})RuCp$ as an unstable intermediate.

(3.6)

Since both starting complexes are isoelectronic 18-electron species the explanation for them following such different behavior must be found in the greater effective electron density on the metal in the fragment CpRu(II) as compared with the $(CO)_3Mn(I)$ moiety being better suited for C-S bond breaking. An intermediate electronic situation is provided by $Re(H)_7(PPh_3)_2$ which also hydrogenates thiophene to tetrahydrothiophene stoichiometrically but hydrogenolyzes it to the corresponding thiolates photochemically (see Section 4.2.1)

(5) Prolonged contact of the tetrahydrothiophene iridium complex **D** (Fig. 3.5) with hydrogen even under more forcing conditions (up to 120 °C and 50 atm H_2) did not lead to C-S bond activation, showing that hydrogenation sites of this kind are not adequate for hydrogenolysis and/or desulfurization reactions. It would seem that the right combination of a sufficiently high degree of coordinative unsaturation and electron density on the metal center must be achieved in order to produce C-S bond rupture, which is not the case in Ir(III) complexes like the PPh_3 derivative $[Ir(H)_2(\eta^1S\text{-}THT)_2(PPh_3)_2]PF_6$.

3.3.1.2. Benzothiophene hydrogenation.
Hydrogenation of benzothiophene (BT) to 2,3-dihydrobenzothiophene (DHBT) takes place much more readily than reduction of thiophene, since the C(2)=C(3) bond in this molecule has more of a localized double bond character, instead of the pseudoaromatic nature of an isolated thiophene ring or of a heterocycle in a fused-ring system like dibenzothiophene and its derivatives. As shown by the data in Table 3.2, several Ru, Os, Rh, and Ir complexes are known to be very efficient catalysts for the reduction of BT to DHBT under moderate reaction conditions [74-86].

TABLE 3.2. Homogeneous catalysts for the hydrogenation of benzothiophene

Catalyst	T (°C)	P(H_2) (atm)	Ref.
$RuCl_2(PPh_3)_3$	85	21	77
$RuCl_2(PPh_3)_3$	170	110	78
$RuHCl(CO)(PPh_3)_3$	170	110	78
$RuCl_3.3H_2O$ + TPPMS + base	100	30	79, 80
$[(triphos)Ru(NCMe)_3][BF_4]_2$	25	1-30	81
$OsHCl(CO)(PPh_3)_3$	170	110	78
$RhCl(PPh_3)_3$	85	21	77
$RhCl(PPh_3)_3$	170	110	78
$[Cp*Rh(NCMe)_3][BF_4]_2$	40	30	82
$[Rh(COD)(PPh_3)_2]PF_6$	170	110	78
$[Rh(COD)(PPh_3)_2]PF_6$	125	1	83-85
$[Rh(COD)(PPh_3)_2]PF_6$	40	1	86
$(triphos)Rh[\eta^3\text{-}S(C_6H_4)(CH=CH_2)]$	36	1-30	87
$[Ir(COD)(PPh_3)_2]PF_6$	170	110	78
$[Ir(COD)(PPh_3)_2]PF_6$	40	1	86
$(sulphos)Rh(COD)$	200	30	88
$(sulphos)Ru(NCMe)_3]SO_3CF_3$	100	30	88, 89

TPPMS = $Ph_2P(C_6H_4)SO_3Na$
triphos = $MeC(CH_2PPh_2)_3$; sulphos = $NaO_3S(C_6H_4)CH_2C(CH_2PPh_2)_3$

Detailed mechanisms for the reduction of benzothiophene have been discussed for several active systems and the kinetics of these and other hydrogenation processes were recently reviewed by Sánchez-Delgado and Rosales [90]. In pioneering studies based on deuterium labeling experiments and *in situ* NMR studies, Fish and coworkers elucidated the cycle shown in Fig. 3.6 for $[Cp*Rh(NCMe)_3]^{2+}$ [91] which starts with a $[\eta^2(C=C)-BT]Rh$ polyhydrido species. Stepwise intramolecular transfer of cordinated hydrides to C(2) and C(3) of BT results in the formation of DHBT which remains in the coordination sphere of Rh through η^6-bonding to the benzene ring of the product; DHBT is finally displaced by a new molecule of the substrate to re-start the cycle; the intermediacy of a η^6-arene complex preceding the last step is suggested by the selective incorporation of deuterium in C(7) of the benzene ring of the product. Neither the oxidation state nor the stoichiometry of the intermediates could be clearly established.

More recently, Sánchez-Delgado and Bianchini [86] found that the same complex and its Ir analogue are extremely active for the hydrogenation of BT if 1,2-dichloroethane is used as the solvent, while the rate law for the Rh-catalyzed reaction was identical to that displayed in 2-methoxyethanol, indicating that the mechanisms in both solvents are essentially very similar. Additionally, moving from rhodium to the more stable iridium chemistry opened the possibility not only of obtaining highly active catalysts, but also of isolating and characterizing some of the key species thought to be involved in the catalysis, such as $[Ir(H)_2(\eta^1S-BT)_2(PPh_3)_2]PF_6$ -a rather stable 18-electron compound amenable to analytical and spectroscopic characterization- and $[Ir(H)_2(\eta^1S-DHBT)_2(PPh_3)_2]PF_6$, which was isolated from the catalytic runs and could be characterized by X-ray diffraction.

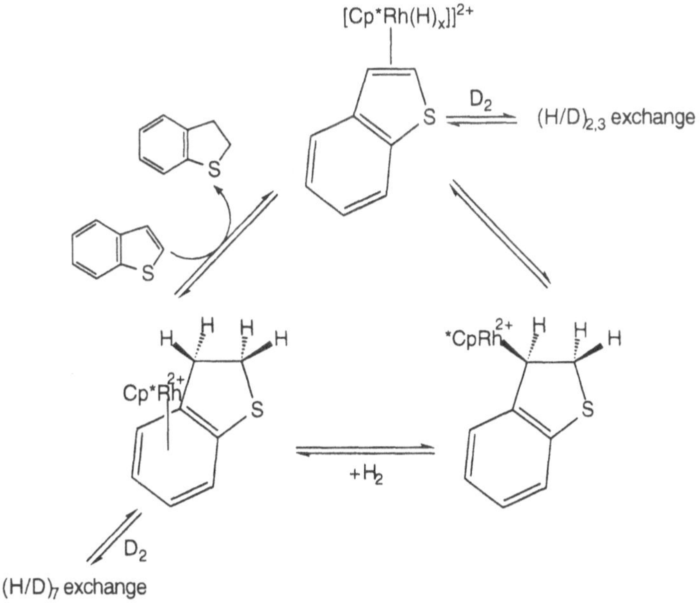

Fig. 3.6. The mechanism of benzothiophene hydrogenation catalyzed by $[Cp*Rh(H)_x]^{2+}$
(Adapted from ref. 91).

The kinetics and other experimental findings disclosed in these two papers led to a generic catalytic cycle for both systems, which is schematically represented in Fig. 3.7. Two BT molecules initially bind in a η^1S-mode to a M(III) dihydride precursor, producing $[M(H)_2(\eta^1S\text{-}BT)_2(PPh_3)_2]PF_6$ (M = Rh, Ir, isolated as a stable compound in the case of iridium); loss of one BT ligand yields the 16-electron S-bonded species $[M(H)_2(\eta^1S\text{-}BT)(PPh_3)_2]PF_6$, which is very likely in equilibrium with the $\eta^2(C=C)$-BT complex. This olefin-like intermediate undergoes selective hydrogenation of the C(2)=C(3) bond trough a hydrido-2-benzothienyl species formed by transfer of the first hydride to C(3), as supported by calculations, to finally yield $[M(\eta^1S\text{-}DHBT)(PPh_3)_2]^+$. Displacement of dihydrobenzothiophene by a new molecule of benzothiophene leads to $[M(\eta^1S\text{-}BT)(PPh_3)_2]^+$ which reacts with H_2 to form the dihydride and restart the cycle. Originally it was thought that the η^5-bonded BT complex $[Rh(\eta^5\text{-}BT)(PPh_3)_2]PF_6$ was involved in the hydrogenation cycle [83-85]; although some NMR evidence as well as theoretical calculations support the existence of such η^5-bonded BT derivatives, they have not been properly authenticated as yet, and their participation in the catalysis, although possible, is now believed to be minor (if any), and thus the predominant catalytic cycle proceeds in the way described by Fig. 3.7 for both Rh and Ir, in 1,2-dichloroethane or in 2-methoxyethanol [86].

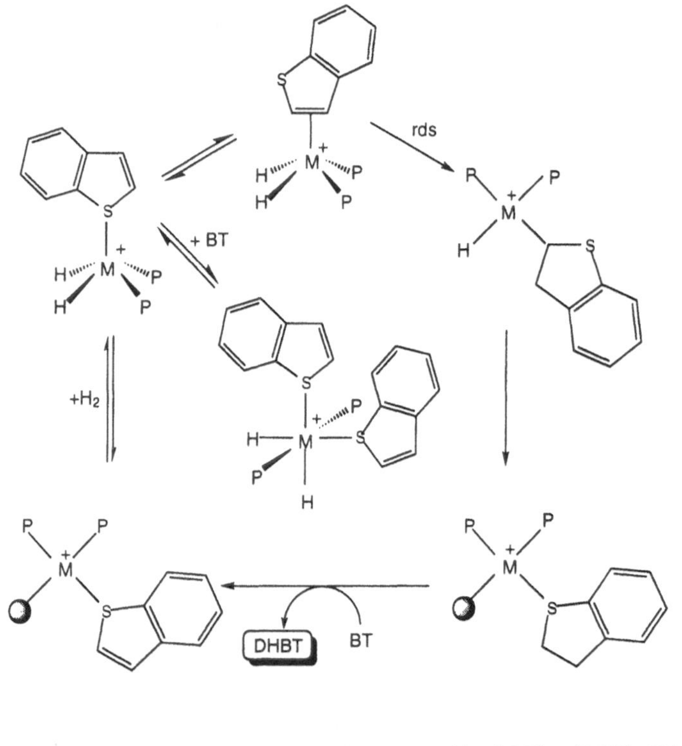

Fig. 3.7. The mechanism of benzothiophene hydrogenation catalyzed by $[M(H)_2(\eta^1S\text{-}BT)_2(PPh_3)_2]PF_6$ (M = Rh, Ir; P = PPh$_3$; adapted from ref. 86).

The very large solvent effect observed in these reactions (rates of the order of 10^6 times faster in dichloroethane than in 2-methoxyethanol!) has been interpreted in terms of solvation effects, which in the case of 2-methoxyethanol can prevent the binding of the substrate to a considerable extent. This would not be the case for the much more poorly coordinating 1,2-dichloroethane, once again pointing to the importance of vacant or readily available coordination sites for efficient catalysis to occur.

Still a much more active catalyst precursor for the hydrogenation of BT based on ruthenium was recently reported by Bianchini and Sánchez-Delgado [81] *viz.* [(triphos)Ru(NCMe)$_3$][BF$_4$]$_2$ [(triphos = MeC(CH$_2$PPh$_2$)$_3$] in THF solution. The extremely rapid rate of the hydrogenation reaction (500 turnovers per hour at 30 atm of H$_2$ and 100 °C, the fastest rates achieved so far for the homogeneous hydrogenation of BT) was found to be first order in catalyst, substrate and hydrogen concentrations. In addition, two stable intermediates, *viz.* [(triphos)RuH(NCMe)$_2$]$^+$ and [(triphos)RuH(DHBT)(NCMe)]$^+$, could be adequately characterized by their independent synthesis and/or NMR observations at low and high pressures of hydrogen. All these data led to the mechanism represented in Fig. 3.8. The fact that only two species, namely [(triphos)RuH(NCMe)$_2$]$^+$ and [(triphos)RuH(DHBT)(NCMe)]$^+$ were detected by NMR spectroscopy, together with the fact that no deuterium incorporation was observed in either unreacted BT or the arene ring of DHBT indicates that transfer of the hydrides to coordinated benzothiophene is fast and irreversible. This mechanistic scheme differs from the one deduced for BT hydrogenation catalyzed by Rh- and Ir-PPh$_3$ complexes in that in this case the hydride transfer to the coordinated substrate appears to be rate-limiting. Otherwise, the elementary steps that compose the proposed cycle for the (triphos)Ru complex are similar to the ones described in other BT hydrogenation mechanisms.

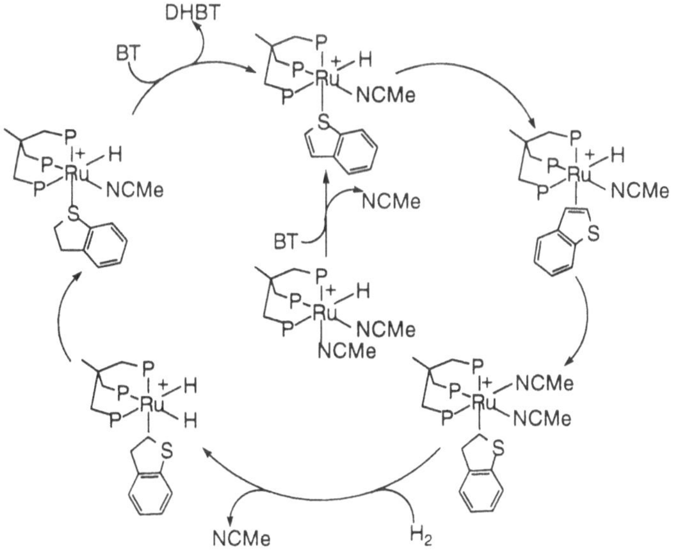

Fig. 3.8. The mechanism of benzothiophene hydrogenation catalyzed by [(triphos)RuH(NCMe)$_2$]BF$_4$ (Adapted from ref. 81).

Interestingly, Bianchini and his coworkers have reported [92] that the closely related anionic d^8 Ru(0) 16e complex [(triphos)RuH]⁻ is a very good catalyst precursor for the selective benzothiophene *hydrogenolysis* to 2-ethylbenzenethiol (for details, see Chapter 4); this is of great interest in relation to the question of whether identical or different sites are needed for hydrogenation and desulfurization reactions on surfaces, since both complexes contain reactive tetracoordinated Ru-monohydride fragments with the same geometry but differing only in the formal oxidation state, and consequently on the electron count. The cationic Ru(II) fragment [(triphos)RuH]⁺ is a highly electrophilic 14e (2 CUS) species, whereas the electron rich anion [(triphos)RuH]⁻ would be a 16e (1 CUS) intermediate, and therefore two extra electrons, rather than an extra ligand, occupy one coordination site:

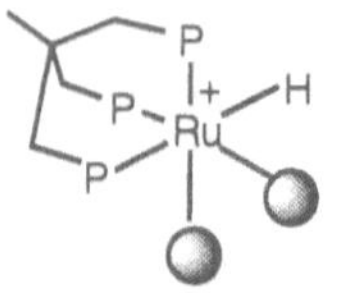

Ru(II), 14e, 2CUS 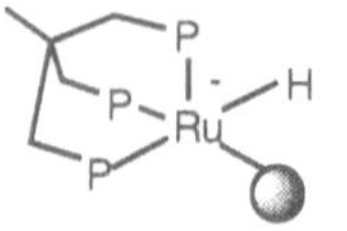   Ru(0), 16e, 1CUS

This may be again related to Whitehurst's "two site dilemma" on Co-Mo-S phases [2], since transfer of electrons through sulfur to the metal, which would be expected to be easy for metal sulfides, could change the nature of an active surface site (and thus its behavior), without much geometrical disturbance. It appears further that the higher electron density on the metal is what drives the reaction towards C-S bond breaking of the η^1S-bonded BT instead of hydride transfer to the coordinated C=C bond in the Ru(II) case, as will be discussed in greater detail in Chapter 4. This is similar to what was pointed out in p. 77 (Eqs. 3.5 and 3.6), where hydride transfer to a Mn-thiophene complex led to hydrogenation, while the analogous reaction on an electron richer Ru-thiophene complex produced C-S bond scission.

Other catalytic systems are known or presumed to operate through mechanisms involving similar features to those described in Figs. 3.6-3.8. For instance, a catalyst system patented by INTEVEP, the Venezuelan Petroleum Research Company [79, 80] (see Section 3.3.1.3. below) is prepared *in situ* by mixing $RuCl_3.3H_2O$ with a sulphonated phosphine and aniline (or another nitrogen base), to yield $RuHCl(PPh_3)_2(Aniline)_2$ as the active species entering a catalytic cycle whose major features are analogous to those described for the Ru(II)-triphos catalyst in Fig. 3.8, *i.e.* displacement of a labile aniline ligand by benzothiophene, hydride transfer to yield a 2-thienyl intermediate, oxidative addition of dihydrogen, and reductive elimination of the hydrogenated product, which regenerates the active catalyst and re-starts the cycle.

Some general key points, common to all the well understood catalysts, that merit further comment are the following:

(1) All homogeneous and hybrid (biphasic or supported) catalysts reported so far to be active in BT hydrogenation are based on "HDS promoter metals" (Ru, Os, Rh, Ir) [1, 2] and are also specific for the hydrogenation of the sulfur-containing ring of BT.

Saturation of the benzene ring has never been detected in solution when using such precursors. In the case of Cp*Rh(NCMe)$^{2+}$ some H-D exchange has been observed in the 7-position of dihydrobenzothiophene, and this has been taken as indicative that the hydrogenated product remains bonded to the metal in a η^6-coordination mode. This latter point seems to be a characteristic of this particular complex and not a general trend that can be related to active sites in HDS catalysts.

(2) What may be of interest in relation to Co-Mo-S and related phases is that in all the mechanisms elucidated so far in solution studies, coordination of benzothiophene to metals occurs in a η^1S or a η^2(C=C) fashion prior to hydrogen transfer. Although the olefin-type coordination has not been observed under catalytic conditions, Angelici has clearly demonstrated the occurrence of the η^1S $\leftrightarrow$ η^2(C=C) equilibrium in metal benzothiophene complexes [68, 69]. Both modes of binding require only a single coordinative unsaturation, and thus, if a parallel is drawn with solid catalysts, *benzothiophene* hydrogenation can take place on the same active sites as olefin hydrogenation, which are not necessarily the same sites required for arene or thiophene hydrogenation that seem to require two or three vacancies. Also, the desulfurization sites probably require not only a higher degree of unsaturation than benzothiophene hydrogenation but also a higher electron density at the metal center (see Chapter 4); this is very likely related to the fact that none of the metal complexes that reduce benzothiophene to 2,3-dihydrobenzothiophene -Ru(II), Os(II), Rh(III), Ir(III)- promotes C-S bond scission to 2-ethylthiophenol, which is observed for Ru(0), Rh(I) or Ir(I). An exception for this is found in the (triphos)Rh(I) fragment which is in fact a good catalyst for the hydrogenolysis of benzothiophene, but also hydrogenates it -albeit slowly- to the saturated thioether [87].

(3) The first intramolecular hydride migration from the metal to BT may lead to either one of the two possible isomeric 2- or 3-benzothienyl intermediates, and the available experimental evidence does not allow a clear distinction between them. However, semi empirical (CNDO) [83-85] as well as *ab initio* [93] calculations indicate that the C(2) atom of free benzothiophene is more negatively charged than C(3) and thus it should be more susceptible to electrophilic attack by the metal; therefore a 2-benzothienyl intermediate is more likely to be involved in the catalytic cycle.

3.3.1.3. Benzothiophene hydrogenation as a pretreatment for HDS.

Besides the mechanistic considerations presented in the preceding sections, an interesting possibility for a practical application of metal complex catalyzed hydrogenation reactions in HDS of refined fuels or refinery cuts has emerged by use of water soluble catalysts for the biphasic reduction of sulfur-containing heterocycles, as represented in Fig. 3.9. As described in a series of Patents filed by INTEVEP, the Venezuelan Petroleum Research Company [82], the catalysts are generated *in situ* by *e.g.* reaction of RuCl$_3$.3H$_2$O with *m*-monosulfonated or trisulfonated triphenylphosphine (TPPMS, TPPTS) in the presence of a basic co-catalyst such as aniline or quinoline. The resulting mixtures can hydrogenate benzothiophene at reasonably fast rates under 30 atm H$_2$ at 120° C. This partial reduction of benzothiophene can be envisaged as a pretreatment for *e.g.* naphtha, aimed at saturating the thiophenic rings of BT; the hydrogenated naphtha

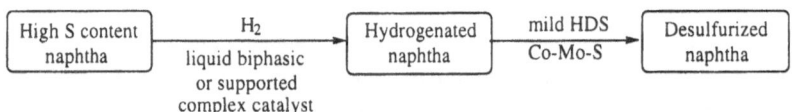

Fig. 3.9. A possible two stage process for improved HDS of fuels or refinery cuts.

containing the saturated cyclic thioethers would be subsequently subjected to mild HDS over conventional catalysts, thus improving sulfur removal without compromising the benzene rings which contribute to a high octane number. Some preliminary experiments involving this type of two-stage treatment of a high sulfur content naphtha indicate that prehydrogenation over the liquid biphasic Ru catalyst increases the degree of sulfur removal by a considerable amount.

Bianchini has extended the idea of biphasic BT hydrogenation catalysis by use of the water soluble complexes [(sulphos)Rh(COD)] and [(sulphos)Ru(NCMe)$_3$]SO$_3$CF$_3$ (sulphos = NaO$_3$S(C$_6$H$_4$)CH$_2$C(CH$_2$PPh$_2$)$_3$, the sulfonated analogue of triphos) [88, 89]; the latter complex is a much faster catalyst than those previously reported. Also, these Ru and Rh complexes containing the sulphos ligand have been supported on silica, and the heterogenized catalysts are able to reduce benzothiophene at rates superior to those observed for the homogeneous or the liquid biphasic systems [89]. Further research into biphasic hydrogenation, as a pretreatment for HDS of difficult thiophenic substrates would be welcome. This concept could be of particular use for the treatment of refractory fractions containing dibenzothiophenes, and most notably for the removal of 2,6-Me$_2$DBT contained in diesel fuels, if sufficiently good catalysts were to be found for the reduction of one or both of the arene rings in such molecules. Also, metal complexes supported on inorganic oxides or on supported metal catalysts, like the ones mentioned in this Chapter, also offer promise for possible applications of this type. However, there are no reports available to date on the metal complex-catalyzed hydrogenation of dibenzothiophenes, and thus this remain an exciting challenge for future work.

3.3.2. NITROGEN CONTAINING HETEROAROMATICS

The hydrogenation of nitrogen containing heteroaromatic compounds present in petroleum fractions (which are predominantly in the form of quinoline and indole-derivatives with short chain alkyl substituents) is no doubt of interest in relation to the hydrodenitrogenation (HDN) process, which is of great industrial and environmental importance, as described in Chapter 1, Section 1.3. Nitrogen removal by HDN takes place on the same or similar catalyst as HDS and usually concurrently with it during hydrotreatment. In contrast with HDS mechanisms -for which direct sulfur removal or prehydrogenation followed by C-S bond scission can be envisaged- there seems to be a general agreement in the literature in that in HDN processes, reversible and selective hydrogenation of the nitrogen-containing rings always takes place prior to the C-N bond breaking reactions leading to the actual nitrogen removal. This is exemplified for quinoline in Fig. 1.13 (p. 28), where it is shown that only saturated N-containing rings are hydrogenolyzed.

3.3.2.1. Hydrogenation of pyridines.

Pyridines are not easily reduced presumably due to their high aromatic character [94], but a few examples are available of metal complex-catalyzed reduction of the pyridine ring. Early studies involved the use of mixtures of $RhCl_3(py)_3$ and $NaBH_4$, which functioned in DMF as the solvent under hydrogen gas at atmospheric pressure [94]; the active species in this case is thought to be $RhCl_2(DMF)(py)_2(BH_4)$ but the reaction mechanism was not investigated in detail. Laine used $Rh_6(CO)_{16}$ under water gas shift conditions (CO + H_2O) to reduce pyridine, but the efficiency of the system was low and the mechanistic details were not elucidated [95]. On the other hand, Fish carried out some rather extensive work on the hydrogenation of N-heterocycles, including 2-methylpyridine to 2-methylpiperidine, by use of $[Cp*Rh(NCMe)_3]^{2+}$ as the catalyst precursor [91, 98]. Mechanistic studies, based predominantly on deuteration experiments and isolation of reaction intermediates, suggest that hydrogen addition to both C=N and C=C bonds is reversible, and that initial C=N bond reduction is the key step of the cycle, since it involves breaking the aromaticity of the ring. It was also demonstrated that partially hydrogenated pyridines are readily dehydrogenated back under the reaction conditions; this is important in that it confirms that re-aromatization is thermodynamically favored over C-N bond breaking during HDN.

3.3.2.2. Hydrogenation of polynuclear N-heterocycles.

Several homogeneous systems have been reported to catalyze the hydrogenation of polynuclear N-heterocycles efficiently; the regioselective reduction of the N-containing ring in quinoline, isoquinoline, indole, benzoquinolines, acridine and other related molecules can be achieved with relative ease and under moderate conditions of temperature and pressure. A number of metals are known to be active (Cr, Mo, W, Mn, Fe, Ru, Os, Co, Rh, Ir) but most of the published work has concentrated on ruthenium and rhodium precursors [95-100] as shown in Table 3.3.

The $RhCl_2(DMF)(py)_2(BH_4)$ system reported by Jardine and McQuillin [94] and described in the previous section, is also capable of hydrogenating quinoline under mild conditions. Surprisingly, this catalyst, discovered 30 years ago has not been further exploited or studied since then. Metal carbonyls were also found to display some hydrogenation activity for a variety of polynuclear N-heterocycles under H_2, H_2/CO or CO/H_2O [49, 50, 61]; the heterocyclic ring was regioselectively hydrogenated in all cases were activity was observed, but conditions were rather harsh and turnover frequencies relatively low. This makes this type of catalyst unattractive for either practical uses or mechanistic studies.

Other ligand environments (e.g. cyclopentadienyl or phosphine metal complexes) have proved more adequate for detailed investigations of the hydrogenation of N-heterocycles. The groups of Fish [91], Sánchez-Delgado [99] and Rosales [100] have provided extensive studies including kinetic data, deuteration experiments, detailed NMR studies, isolation of important intermediates, and some theoretical calculations.

TABLE 3.3. Homogeneous catalysts for the hydrogenation of N-heterocycles.

Catalyst	Substrates	T (°C)	P(H_2) (atm)	Ref.
$RhCl_2(DMF)py_2(BH_4)$	pyridine quinoline			94
$Rh_6(CO)_{16}$	pyridine		CO/H_2O	95
$Rh_6(CO)_{16}$	quinolines	150	30 60 (CO/H_2)	96
$Os_3(CO)_{12}/H_2Os_3(CO)_{10}$	quinoline			??
Metal carbonyls Fe, Co, Mn, Rh, Ru, W, Mo, Cr	polynuclear N-heteroaromatics	180-200	45 (CO) 90 (CO/H_2)	49, 50, 61
$Fe(CO)_5$	polynuclear N-heteroaromatics	150-300	30-50	51
$RhCl(PPh_3)_3$	polynuclear N-heteroaromatics	85	20	77
$RuCl_2(PPh_3)_3$	polynuclear N-heteroaromatics	85	20	97
$[Cp*Rh(NCMe)_3]^{2+}$	pyridine polynuclear N-heteroaromatics	40	30	91, 98
$RuCl_2(PPh_3)_3$	quinoline	150	30	78
$RuHCl(CO)(PPh_3)_3$	quinoline	150	30	78
$OsHCl(CO)(PPh_3)_3$	quinoline	150	30	78
$RhCl(PPh_3)_3$	quinoline	150	30	78
$[Rh(COD)(PPh_3)_2]PF_6$	quinoline	150	30	78
$[Rh(COD)(PPh_3)_2]PF_6$	quinoline	120	1	99
$[Ir(COD)(PPh_3)_2]PF_6$	quinoline	150	30	78
$[RuH(CO)(NCMe)_2(PPh_3)_2]BF_4$	quinoline	25	1-30	100
$RuCl_3.3H_2O$ + TPPMS + base[a]	quinoline	120	30	79, 80
$[M(COD)Cl]_2$ + Ktp'[b] (Rh, Ir)	quinoline	100	30	101
$RuCl_2(NCMe)_4$ + KTp'[b]	quinoline	100	30	101

[a]TPPMS = *m*-monosulfonated triphenylphosphine, base = aniline; [b]Tp' = hydrotrispyrazolylborates

From this rich information, sound mechanistic proposals for the hydrogenation of quinoline and related molecules using Ru and Rh complexes as catalyst precursors have been deduced. The first experimentally supported mechanism for the homogeneous hydrogenation of N-heterocycles was provided by Fish, using $[Cp*Rh(NCMe)_3]^{2+}$ as the catalyst precursor [91, 98]. Extensive studies with deuterium gas, and monitoring of the reaction intermediates and products by GC-MS and *in situ* high pressure NMR data, led to a rather complete hydrogenation cycle (shown in Fig. 3.10) for the transformation of Q into THQ in CH_2Cl_2 at 40° C under H_2 (*ca.* 35 atm). The most interesting points of this mechanism are: (1) the importance of η^1-N binding of Q (as in a) to initiate the cycle, in line with heterogeneous proposals involving vertical adsorption of the heterocycle onto catalytic surfaces. (2) The possibility of η^2(C=N) bonded intermediates (b) being involved in the hydrogenation step. (3) The reversible hydrogenation of the C=N bond, followed by migration of the Rh fragment to the C(3)=C(4) bond (as in c), which is also reversibly hydrogenated; this is also in agreement with the reversibility of hydrogenation during HDN of quinoline over solid catalysts. (4) Binding of the product THQ through the carbocyclic ring in a η^6-mode A (as in d), and displacement of the product by an incoming Q molecule, to restart the cycle.

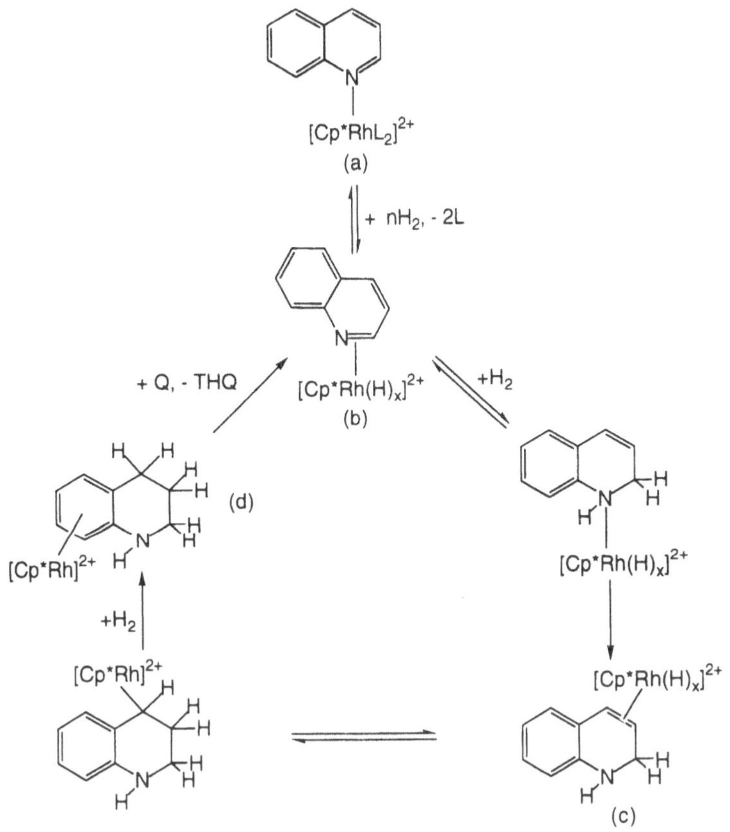

Fig. 3.10. The mechanism of quinoline hydrogenation by use of [Cp*Rh(NCMe)₃]²⁺ (Adapted from ref. 91).

Some points of this mechanism, however, are not entirely clear, such as the number of hydride ligands attached to rhodium, or the actual oxidation state of the metal, that is inferred to be higher in this case than the most usual situations for this metal in organometallic complexes. One interesting mechanistic alternative, not envisaged by these authors, would be to consider the heterolytic activation of hydrogen assisted by the substrate itself as the acting base (Eq. 3.7). Such a reduction of the C=N bond would then be occurring through a usual Rh(III) state via common singly charged monohydride intermediates:

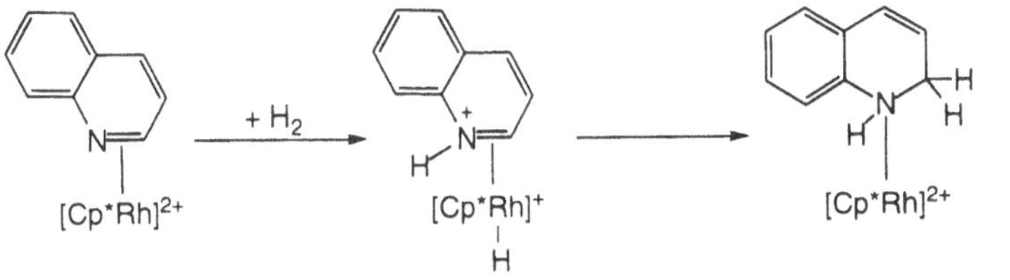

The mechanism of Fig. 3.10 is easily adapted to explain the hydrogenation of pyridine mentioned in Section 3.3.2.1, as well as that of other polynuclear N-heterocycles, that are reduced by this catalyst at rates that decrease with increasing basicity and steric hindrance at the nitrogen atom:

Acridine > quinoline > 5,6-benzoquinoline > 2-Me-quinoline > 2-Me-pyridine

Recently, Alvarado *et al.* reported a related set of active catalysts for quinoline hydrogenation, formed *in situ* by addition of pyrazolylborate ligands Tp, hydrotris(pyrazolyl)borate and Tp*, hydrotris(3,5-dimethylpyrazolyl)borate –which are isolectronic with cyclopentadienyl ligands- to $[M(COD)Cl]_2$ (M = Rh, Ir, COD = 1,5-cyclooctadiene), $[Ir(COE)Cl]_2$ (COE = cyclooctene), and $RuCl_2(NCMe)_4$. Again, rhodium turned out to be the most efficient metal, and the activity was related with the ease of formation of Tp or Tp* complexes. No mechanistic details were provided [101].

Transition metal phosphine complexes provide another important class of hydrogenation catalysts. Wilkinson's complexes, $RhCl(PPh_3)_3$ and $RuHCl(PPh_3)_3$, well known for their olefin hydrogenation activity, were shown by Fish [77, 97] to be also good precursors for the reduction of polyaromatic substrates under mild reaction conditions (85° C, *ca.* 20 atm H_2), following a general activity trend consistent with a combination of electronic and steric factors:

Phenanthridine > acridine > quinoline > 5,6-benzoquinoline > 7,8-benzoquinoline

Regioselectivity for the reduction of the heterocycle was high in all cases except for acridine, which was hydrogenated to a mixture of 9,10-dihydroacridine and 1,2,3,4-tetrahydroacridine. Competition experiments showed that quinoline hydrogenation was inhibited by the presence of pyridine or 1,2,3,4-tetrahydroquinoline, presumably due to stronger binding of these molecules to the metal center than quinoline itself. On the other hand, the presence of pyrrole, indole or carbazole, as well as that of thiophene, benzothiophene or dibenzothiophene had a beneficial effect on the rate of quinoline hydrogenation. The latter effects are interesting but were not clearly explained. Deuterium labeling experiments, together with other relevant data led to the conclusion that, similarly to what was described above for the Cp*Rh catalysts, the hydrogenation of the C=N bond is reversible and initiates the cycle, whereas reduction of the C(3)=C(4) bond is in this case irreversible (and *cis* stereoselective); also, activation of the C(8)-H bond was deduced, and explained through a cyclometalation process.

Similar metal-phosphine derivatives previously shown to be good catalysts for C=C and C=O bond reduction, were reported by Sánchez-Delgado and González [78] to be efficient also for quinoline hydrogenation at 150° C and 30 atm H_2. The order of activity observed was:

$$[Rh(COD)(PPh_3)_2]PF_6 > RhCl(PPh_3)_3 > RuCl_2(PPh_3)_3 > RuHCl(CO)(PPh_3)_3 >$$
$$[Ir(COD)(PPh_3)_2]PF_6 > OsHCl(PPh_3)_3$$

Turnover numbers as high as 200 (mol product)(mol cat)$^{-1}$h^{-1} were achieved with the cationic rhodium complex, whereas third row metals were much less active. No mechanistic proposals were provided in these early studies. Nevertheless, [Rh(COD)(PPh$_3$)$_2$]PF$_6$ was later studied in detail by the same group, using a combination of chemical, spectroscopic and kinetic methods [99]. Quinoline was regioselectively reduced to THQ at *ca.* 100° C and 1 atm H$_2$, according to a third order rate law $-d[Q]/dt = k_{cat}[Rh][H_2]^2$, where k_{cat} (298 K) = 1.7 ± 0.5 M^{-2}s^{-1}, and the following thermodynamic parameters: $\Delta H^\neq = 9 \pm 1$ kcal mol^{-1}, $\Delta S^\neq = -27.0 \pm 0.3$ eu, and $\Delta G^\neq = 19.1 \pm 0.3$ kcal mol^{-1}. The active species was shown to be a phosphine-free complex, formed by displacement of PPh$_3$ by Q, *viz.* [Rh(COD)(Q)$_2$]PF$_6$, that operates through the cycle depicted in Fig. 3.11. This mechanism operates through conventional Rh(I)-Rh(III) couples and dihydride intermediates. The reduction of the C=N bond is reversible, and it is followed by the irreversible hydrogenation of the C(3)=C(4) bond in dihydroquinoline, in accord with the rate law and with the fact that preformed dihydroquinoline was found to be hydrogenated at the same rate as quinoline.

A similar mechanism was deduced by Rosales [100] for the hydrogenation of quinoline, using the cationic Ru(II) complex [RuH(CO)(NCMe)$_2$(PPh$_3$)$_2$]BF$_4$ as the catalyst precursor at 125° C. The experimentally determined rate law is $-d[Q]/dt = k_{cat}[Rh][H_2]$, where k_{cat} (298 K) = 0.27 M^{-2}s^{-1}, and the thermodynamic parameters were found to be $\Delta H^\neq = 42 \pm 6$ kJ mol^{-1}, $\Delta S^\neq = -115 \pm 2$ JK^{-1}mol^{-1}, and $\Delta G^\neq = 92 \pm 8$ kJ mol^{-1}.

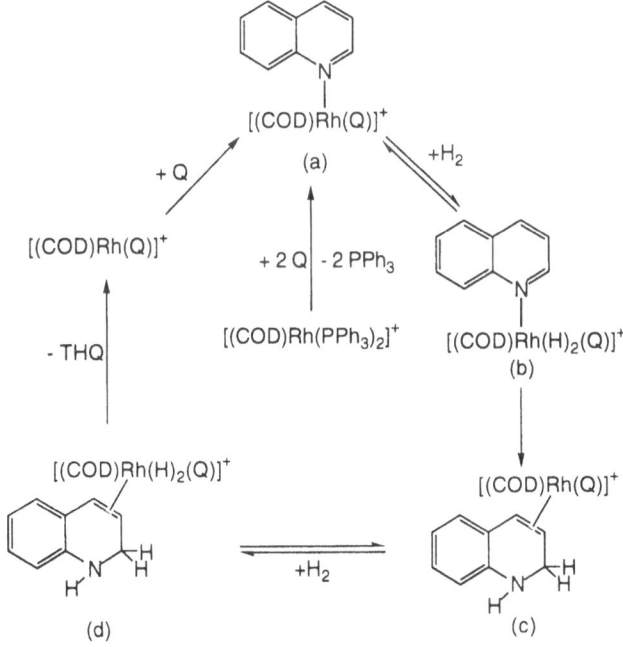

Fig. 3.11. The mechanism of quinoline hydrogenation by use of [Rh(COD)(PPh$_3$)$_2$]PF$_6$. (Adapted from ref. 99.)

Kinetic, chemical and theoretical data indicate that the mechanism of hydrogenation can be described by the set of reactions depicted in Eqs. 3.8-3.11, consistent with a rate law $-d[Q]/dt = \{(k_2K_1)/(1 + K_1[H_2])\}[Rh][H_2]^2$, which can be approximated at low H_2 pressures to $-d[Q]/dt = k_2K_1[Rh][H_2]^2$, identical to the experimental law if $k_{cat} = k_2K_1$. This mechanism again implies a rapid and reversible hydrogenation of a η^1-bonded quinoline to dihydroquinoline, followed by an irreversible C=C reduction to yield 1,2,3,4-tetrahydroquinoline.

$$[RuH]^+ \; \underset{+ L}{\overset{+ Q}{\rightleftharpoons}} \; [RuH(\eta^1\text{-}Q)]^+ \qquad\qquad (3.8)$$

$$[RuH(\eta^1\text{-}Q)]^+ + H_2 \; \overset{K_1}{\rightleftharpoons} \; [RuH(\eta^2\text{-}DHQ)]^+ \qquad\qquad (3.9)$$

$$[RuH(\eta^2\text{-}DHQ)]^+ + H_2 \; \overset{k_2}{\longrightarrow} \; [RuH(\eta^1\text{-}THQ)]^+ \qquad\qquad (3.10)$$

$$[RuH(\eta^1\text{-}THQ)]^+ + Q \; \overset{K_3}{\rightleftharpoons} \; [RuH(\eta^1\text{-}Q)]^+ + THQ \qquad\qquad (3.11)$$

The hydrogenation of other polynuclear N-heterocycles was also reported by Rosales' group, by use of the same Ru(II) catalyst precursor operating at 125° C and 3 atm H_2. The reactivity order acridine > quinoline >> 5,6-benzoquinoline > 7,8-benzoquinoline > indole > isoquinoline was in line with other trends previously observed, reflecting steric and electronic effects. A kinetic analysis of the hydrogenation of acridine reveals a different, second order, rate law $-d[A]/dt = k_{cat}[Rh][H_2]$, consistent with the hydrogenation of acridine to 9,10-dihydroacridine as the rate –determining step. In summary, some mechanistic similarities as well as some differences emerge from one complex to another, and a number of common features can be identified in the various proposed reaction pathways for quinoline hydrogenation.

The most relevant elementary steps involved in all the important quinoline hydrogenation catalysts are represented in Fig. 3.12 in the form of a generic catalytic cycle for both dihydride and monohydride catalysts that may be easily applied to other N-heterocyclic substrates. A number of points concerning these mechanisms are noteworthy:

(1) Similarly to the case of benzothiophene (Section 3.3.1), the homogeneous reduction of the heterocyclic ring of quinoline and other nitrogen heteroaromatic compounds is achieved with high regioselectivity, no saturation of the benzene rings or C-N bond breaking reactions being observed with these catalysts. In fact, very few metal complexes are capable of activating C-N bonds [103, 104].

(2) The coordination requirements for these catalytic cycles are essentially limited to single vacancies, since the key intermediates only seem to involve η^1-N and η^2(C=C) bonding modes. This reinforces the idea that this type of hydrogenation on

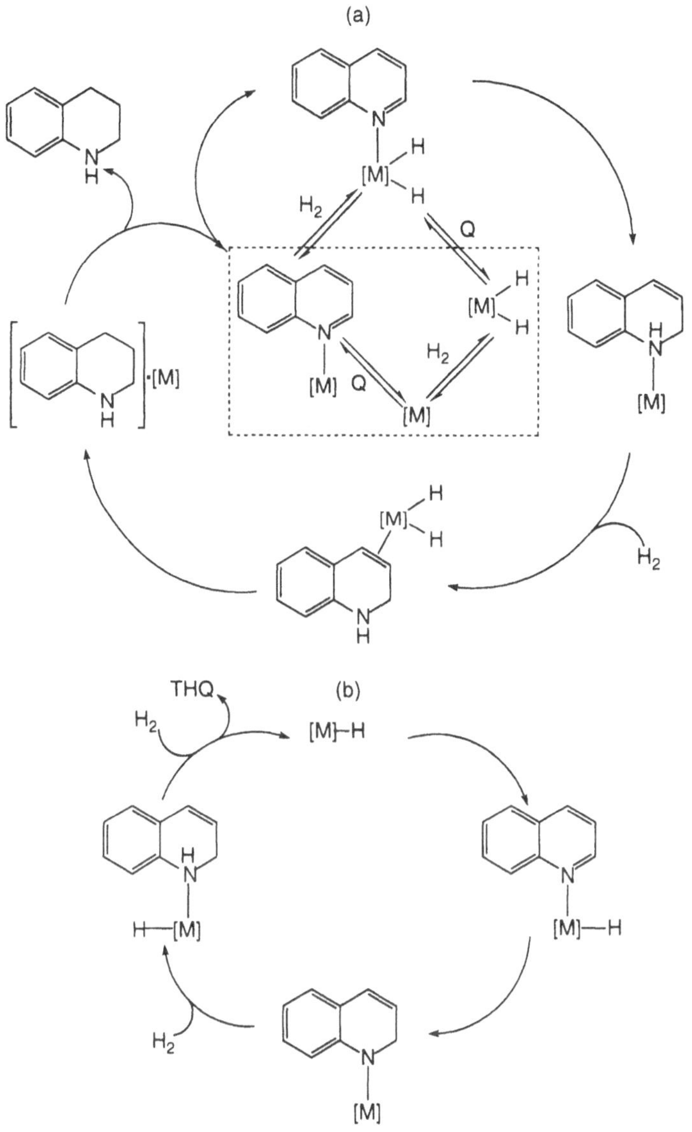

Fig. 3.12. The mechanism of quinoline hydrogenation by (a) dihydride and (b) monohydride catalysts.

HDS surfaces preferentially occurs on fairly saturated sites, conceivably the same as for olefin hydrogenation but not necessarily the same that are active for C-S and C-N bond scission. In fact, nitrogen-containing molecules are known to poison olefin hydrogenation sites in Co-Mo-S and related catalysts under certain reaction conditions [99], and also to inhibit HDS in most cases [1].

(3) The first hydrogenation of quinoline to 2,3-dihydroquinoline in solution has been shown to be generally reversible, in line with previous knowledge on heterogeneous hydrogenation prior to HDN [105]. Besides quinoline, the hydrogenation of isoquinoline and indole with ruthenium catalysts has been studied by Rosales [100]. Isoquinoline is only reduced with difficulty, and this has been associated with its high basicity. Indole is also difficult to hydrogenate, and very few catalysts have been reported for this reaction; $RuHCl(PPh_3)_3$ and $[RuH(CO)(NCMe)_2(PPh_3)_2]BF_4$ are capable of effecting the reduction to indoline, but with very low turnover frequencies. This may be related to the fact that, as described in Chapter 1, Section 1.3.1, indole tends to bind to metals through the $C(2)=C(3)$ bond, rather than η^1-N, and thus the activation of the C=N bond is not very marked. Better results were obtained by use of $[(triphos)Rh(DMAD)]PF_6$ in the presence of acid at 60° C and 30 atm H_2, and it was demosntrated that indoline was actually formed by reduction of the protonated form, the 3H-indolium cation which posseses a localized C=N bond.

The water soluble catalysts patented by PDVSA-INTEVEP (see p. 83), composed of $RuCl_3.3H_2O$, *m*-monosulfonated or trisulfonated triphenylphosphine (TPPMS, TPPTS), and a basic co-catalyst such as aniline or quinoline, that are active for the biphasic reduction of sulfur-containing heterocycles, can also be employed for the hydrogenation of N-heterocycles [82]. The active species is thought to be $[RuHCl(P)_2(N)_2]$ (P = TPPMS, TPPTS; N = Q, THQ, aniline), and the mechanism is the typical one for monohydride catalysts (Fig. 3.12b). Bianchini's water-soluble complexes containing sulfonated ligands, *viz.* $[(DPPDS)Rh(H_2O)_2]Na$ (Na_2-PPPDS = $[\{NaO_3S(C_6H_4)CH_2\}_2C(CH_2PPh_2)_2]$), $[(sulphos)Rh(COD)]$, and $[(sulphos)Ru(NCMe)_3]$-SO_3CF_3 (sulphos = $[NaO_3S(C_6H_4)CH_2C(CH_2PPh_2)_3]$) [88, 89, 106] are also very efficient in catalyzing the regioselective hydrogenation of Q under reasonable reaction conditions. These compounds have been also supported on inorganic solids [107, 108] and on organic polymers [109], and in both cases high catalytic activities for the hydrogenation of quinoline have been found.

No reports are available on the homogeneous hydrogenation of carbazoles, which would be interesting substrates analogous to dibenzothiophenes; again this would amount to finding much needed good metal complex based arene hydrogenation catalysts.

References

1.   H. Topsøe, B. S. Clausen, and F. E. Massoth, in J. R. Anderson and M. Boudart, eds.: *Catalysis Science and Technology*, Springer-Verlag, New York, Vol. 11 (1966).
2.   D. D. Whitehurst, T. Isoda, I. Mochida: *Adv. Catal.* **42**, 345 (1988).
3.   V. Lamure-Meille, E. Schulz, M. Lemaire and M. Vrinat: *Appl. Catal. A* **131**, 143 (1995).
4.   V. Meille, E. Schulz, M. Lemaire and M. Vrinat: *J. Catal.* **170**, 9 (1997).
5.   V. Meille, E. Schulz, M. Lemaire and M. Vrinat: *Appl. Catal. A* **187**, 179 (1999).
6.   M. Macaud, A. Milenkovic, E. Schultz, M. Lemaire and M. Vrinat: *J. Catal.* **193**, 255 (2000).
7.   P. Sabatier, J. B. Senderens: *Bull. Soc. Chim. Fr.* [3] **31**, 101 (1904).
8.   P. Sabatier, L. Espil: *Bull. Soc. Chim. Fr.* [4] **15**, 228 (1914).
9.   P. N. Rylander: *Catalytic Hydrogenation over Platinum Metals*, Academic Press, New York, N. Y. (1967).
10.   R. L. Agustine: *Catalytic Hydrogenation*, Marcel Dekker, New York, N. Y. (1971).
11.   I. Wender, R. Levine, M. Orchin: *J. Am. Chem. Soc.* **72**, 4375, (1950).
12.   S. Friedman, S. Metlin, A. Svedi, I. Wender: *J. Org. Chem.* **24**, 1287 (1959).

13. E. L. Muetterties, F. J. Hirsekorn: *J. Am. Chem. Soc.* **96**, 4063 (1974).
14. E. L. Muetterties, F. J. Hirsekorn: *J. Am. Chem. Soc.* **96**, 7920 (1974).
15. E. L. Muetterties, M. C. Rakowski, F. J. Hirsekorn, W. D. Larson, V. J. Basus, F. A. L. Anet: *J. Am. Chem. Soc.* **97**, 1266 (1975).
16. M. C. Rakowski, F. J. Hirsekorn, L. S. Stuhl, E. L. Muetterties: *Inorg. Chem.* **15**, 2379 (1976).
17. L. S. Stuhl, M. Rakowski Du Bois, F. J. Hirsekorn, J. R. Bleeke, A. E. Stevens, E. L. Muetterties: *J. Am. Chem. Soc.* **100**, 2405 (1978).
18. E. L. Muetterties, J. R. Bleeke: *Acc. Chem. Res.* **12**, 324 (1979).
19. J. R. Bleeke, E. L. Muetterties: *J. Am. Chem. Soc.* **103**, 556 (1981).
20. N. Murgesan, S. Sarkas: *Indian J. Chem.* **14A**, 107 (1976).
21. K. J. Klabunde, B. B. Anderson, M. Bader, L. J. Radonovich: *J. Am. Chem. Soc.* **100**, 1313 (1978).
22. S. J. Laporte, W. Schuett: *J. Org. Chem.* **28**, 1947 (1963).
23. S. J. Laporte: *Ann. N. Y. Acad. Sci.* **158**, 510 (1969).
24. Y. Chauvin, J. Gaillard, J. Leonard, P. Booifay, J. W. Andrews: *Hydrocarbon Processing* 110 (1982).
25. A. Alvanipour, L- D. Kispert: *J. Mol. Catal.* **48**, 277 (1988).
26. F. K. Shmidt, Yu. S. Levkovskii, N. M. Ryutina, T. I. Bakunina: *Kinet. Katal.* **23**, 299 (1982).
27. K. Jonas: *Angew. Chem. Int. Ed. Engl.* **24**, 295 (1985).
28. V. A. Avilov, Yu. G. Borod'ko, V. B. Panov, M. L. Khidekel', P. S. Chekrii: *Kinet. Katal. (Engl. Transl.)* **9**, 582 (1968).
29. O. N. Efimov, O. N. Eremenko, A. G. Ovcharenko, M. L. Khidekel', P. S. Chekrii: *Bull. Acad. Sci. USSR, Div. Chem. Sci.* 778 (1969).
30. O. N. Efimov, M. L. Khidekel', V. A. Avilov, P. S. Chekrii, O. N. Eremenko, A. G. Ovcharenko: *J. Gen. Chem. USSR* **38**, 2581 (1968).
31. M. J. Russel, C. White, P. M. Maitlis: *J. Chem. Soc. Chem. Commun.* 427 (1977).
32. P. M. Maitlis: *Acc. Chem. Res.* **11**, 301 (1978).
33. A. J. Sivak, E. L. Muetterties: *J. Am. Chem. Soc.* **101**, 4878 (1979).
34. D. Pieta, A. M. Trzeciak, J. J, Ziólkowski: *J. Mol. Catal.* **18**, 193 (1983).
35. A. M. Trzeciak, T. Glowiak, J. J. Ziolkowksi: *J. Organomet. Chem.* **552**, 159 (1998).
36. C. R. Landis, J. Halpern: *Organometallics* **2**, 840 (1983).
37. K. R. Januszklewicz, H. Alper: *Organometallics* **2**, 1055 (1983).
38. H. Gao, R. J. Angelici: *J. Am. Chem. Soc.* **119**, 6937 (1997).
39. H. Gao, R. J. Angelici: *Organometallics* **18**, 898 (1999).
40. J. W. Johnson, E. L. Muetterties: *J. Am. Chem. Soc.* **99**, 7395 (1977).
41. M. Y. Darensbourg, E. L. Muetterties: *J. Am. Chem. Soc.* **100**, 7425 (1978).
42. M. A. Bennet, T. -N. Huang, T. W. Turney: *J. Chem. Soc. Chem. Commun.* 312 (1979).
43. R. A. Grey, G. P. Pez, A. Wallo: *J. Am. Chem. Soc.* **102**, 5949 (1980).
44. R. Wilczynski, W. A. Fordyce, J. Halpern: *J. Am. Chem. Soc.* **105**, 2066 (1983).
45. D. E. Linn, J. Halpern: *J. Am. Chem. Soc.* **109**, 2969 (1987).
46. L. Plasseraud, G. Süss-Fink: *J. Organomet. Chem.* **539**, 163 (1997).
47. E. Garcia Fidalgo, L. Plasseraud, G. Süss-Fink: *J. Mol. Catal. A: Chemical* **132**, 5 (1998).
48. P. J. Dyson, D. J. Ellis, D. G. Parker, T. Welton: *J. Chem. Soc. Chem. Commun.* 25 (1999).
49. R. H. Fish, A. D. Thormodsen, G. A. Cremer: *J. Am. Chem. Soc.* **104**, 5234 (1982).
50. R. H. Fish: *Ann. N. Y. Acad. Sci.* **415**, 292 (1983).
51. T. J. Lynch, M. Banah, H. D. Kaesz, C. R. Porter: *J. Org. Chem.* **49**, 1266 (1984).
52. J. S. Yu, B. C. Ankianiec, M. T. Nguyen, I. P. Rothwell: *J. Am. Chem. Soc.* **114**, 1927 (1992).
53. M. C. Poyten, I. P. Rothwell: *J. Chem. Soc. Chem. Commun.* 849 (1995).
54. I. P. Rothwell: *J. Chem. Soc. Chem. Commun.* 1331 (1997).
55. M. S. Eisen, T. J. Marks: *J. Am. Chem. Soc.* **114**, 10358 (1992).
56. R. D. Profilet, A. P. Rothwell, I. P. Rothwell: *J. Chem. Soc. Chem. Commun.* 42 (1993).
57. J. C. Cheng, J. Maiorello, J. W. Larsen: *Energy & Fuels* **3**, 321 (1989).
58. B. R. James: *Homogeneous Hydrogenation*, Wiley, New York (1973).
59. B. R. James, in G. Wilkinson, F. G. A. Stone and E. W Abel, eds.: *Comprehensive Organometallic Chemistry*, Pergamon, Oxford (1982). Vol. 8, Ch. 51.
60. P. A. Chaloner: *Handbook of Coordination Catalysis in Organic Chemistry*, Butterworths, London (1986). See also P. A. Chaloner, M. A. Esteruelas, F. Jóo and L. A. Oro: *Homogeneous Hydrogenation*, Kluwer Academic Publishers: Dordrecht, Netherlands, 1993.
61. R. H. Fish in R. Ugo, ed: *Aspects of Homogeneous Catalysis*, Kluwer Academic Publishers: Dordrecht, Netherlands, 1990. Vol. 7, pp. 65-83.
62. B. Cornils, W. A. Herrman, eds.: *Aqueous-Phase Organometallic Catalysis-Concepts and Applications*, VCH, Weinheim (1998).
63. K. E. Dungey, M. D. Curtis: *J. Am. Chem. Soc.* **119**, 842 (1997).
64. L. S. Byskov, B. Hammer, J. K. Nørskov, B. S. Clausen, H. Topsøe: *Catal. Letters* **47**, 177 (1997).
65. H. Greenfield, S. Metlin, M. Orchin, I. Wender: *J. Org. Chem.* **23**, 1054 (1958).
66. C. Bianchini, A. Meli, M. Peruzzini, F. Vizza, V. Herrera, R. A. Sánchez-Delgado: *Organometallics* **13**, 721 (1994).
67. R. A. Sánchez-Delgado, V. Herrera, C. Bianchini, D. Masi, C. Mealli: *Inorg. Chem.* **32**, 3766 (1993).
68. M. -G. Choi, M. J. Robertson, R. J. Angelici: *J. Am. Chem. Soc.* **113**, 4005 (1991).

69. M. -G. Choi, R. J. Angelici: *Organometallics* **11**, 3328 (1992).
70. G. N. Glavee, L. M. Daniels, R. J. Angelici: *Organometallics* **8**, 1856 (1989).
71. N. N. Sauer, R. J. Angelici: *Inorg. Chem.* **26**, 2160 (1987).
72. H. Kwart, G. C. A. Schuit, G. C. A. Gates: *J. Catal.* **61**, 128 (1980).
73. E. J. Markel, G. L. Schrader, N. N. Sauer, R. J. Angelici: *J. Catal.* **116**, 11 (1989).
74. R. J. Angelici: *Acc. Chem. Res.* **21**, 387 (1988).
75. V. Herrera, M. Rosales, R. A. Sánchez-Delgado: unpublished results.
76. J. W. Hachgenei, R. J. Angelici: *Angew. Chem. Int. Ed. Engl.* **26**, 909 (1987).
77. R. H. Fish, J. L. Tan, A. D. Thormodsen: *J. Org. Chem.* **49**, 4500 (1984).
78. R. A. Sánchez-Delgado, E. González: *Polyhedron* **8**, 1431 (1989).
79. D. E. Paez, A. Andriollo, R. A. Sánchez-Delgado, N. Valencia, F. Lopez-Linares and R. Galiasso: *U.S. Patents 5,753,584 5* (19 May 1998), *5,958,223* (28 Sept. 1999)and *5,981,421* (Nov. 1999) to INTEVEP.
80. D. E. Paez, A. Andriollo, F. Lopez-Linares, R. E. Galiasso, J. A. Revete, R. A. Sanchez-Delgado, A. Fuentes: *Am. Chem. Soc. Div. Fuel. Chem. Symp. Prepr.* **84**, 563 (1998).
81. C. Bianchini, A. Meli, S. Moneti, W. Oberhauser, F. Vizza, V. Herrera, A. Fuentes R. A. Sánchez-Delgado: *J. Am. Chem. Soc.* 121, 7071 (1999).
82. R. H. Fish, E. Baralt, S. J. Smith: *Organometallics* **10**, 54 (1991).
83. R. A. Sánchez-Delgado, V. Herrera, L. Rincón, A. Andriollo, G. Martín: *Organometallics* **13**, 553 (1994).
84. R. A. Sánchez-Delgado, in M. Graziani, C. N. R. Rao (eds.): Advances in Catalyst Design, World Scientific Publishing Co., Singapore, p. 214 (1991).
85. R. A. Sánchez-Delgado, in M. Graziani, C. N. R. Rao (eds.): Advances in Catalyst Design, World Scientific Publishing Co., Singapore, Vol. 2, p. 95 (1993).
86. V. Herrera, A. Fuentes, M. Rosales, R. A. Sánchez-Delgado, C. Bianchini, A. Meli and F. Vizza: *Organometallics* **16**, 2465 (1997).
87. C. Bianchini, V. Herrera, M. V. Jiménez, A. Meli, R. A. Sánchez-Delgado, F. Vizza: *J. Am. Chem. Soc.* **117**, 8567 (1995).
88. C. Bianchini, A. Meli, V. Patinec, V. Sernau, F. Vizza: *J. Am. Chem. Soc.* **119**, 4945 (1997); C. Bianchini, A. Meli: *Acc. Chem. Res.* **31**, 109 (1998).
89. C. Bianchini, A. Meli and F. Vizza: *Eur. J. Inorg. Chem.* 43 (2001).
90. R. A. Sánchez-Delgado, M. Rosales: *Coord. Chem. Rev.* **196**, 249 (2000).
91. E. Baralt, S. J. Smith, I. Hurwitz, I. T. Horváth, R. H. Fish: *J. Am. Chem. Soc.* **114**, 5187 (1992).
92. C. Bianchini, A. Meli, S. Moneti, F. Vizza: *Organometallics* **17**, 2636 (1998).
93. A. Hinchliffe, H. J. Soscún-Machado: *J. Mol. Struct. (Theochem)* **334**, 235 (1995).
94. P. Abley, I. Jardine, F. J. McQuillin: *J. Chem. Soc. C* 840 (1971).
95. R. M. Laine, D. W. Thomas, L. W. Cary: *J. Org. Chem.* **44**, 4964 (1979).
96. S.-I. Murahashi; Y. Imada, H. Irai: *Tetrahedron Lett.* **28**, 77 (1987).
97. R. H. Fish, J. L. Tan, A. D. Thormodsen: *Organometallics* **4**, 1743 (1985).
98. R. H. Fish, T.-J. Kim, J. E. Babin, R. D. Adams: *Organometallics* **5**, 2193 (1986).
99. R. A. Sánchez-Delgado, D. Rondón, A. Andriollo, V. Herrera, G. Martín, B. Chaudret: *Organometallics* **12**, 4291 (1993).
100. M. Rosales, Y. Alvarado, M. Boves, R. Rubio, H. Soscún, R. A. Sánchez-Delgado: *Transition Met. Chem.* **20**, 246 (1995).
101. Y. Alvarado, M. Busolo, F. López-Linares: *J. Mol. Catal. A: Chemical* **142**, 163 (1999).
102. J. Miciukiewicz, W. Zmierczak, F. E. Massoth: *Proc. 8th Int. Congr. Catal.*, Verlag Chemie, Weinheim, Vol. 2, p. 671 (1984).
103. P. A. Fox, M. A. Bruck, S. D. Gray, N. E. Gruhn, C. Grittini and D. E. Wigley: *Organometallics* **17**, 2729 (1998).
104. T. S. Klecley, J. L. Bennet, P. T. Wolczansky and E. B. Lobkovsky: *J. Am. Chem. Soc.* **119**, 247 (1997).
105. J. F. Cocchetto, C. N. Satterfield: *Ind. Eng. Chem. Proc. Res. Dev.* **20**, 49 (1981).
106. C. Bianchini, A. Meli and F. Vizza: *PCT/EP97/06493* (1999) and *IT FI96A000272* to CNR.
107. C. Bianchini, D. G. Burnaby, J. Evans, P. Frediani, A. Meli, W. Oberhauser, R. Psaro, L. Sordelli and F. Vizza: *J. Am. Chem. Soc.* 121, 5961 (1999).
108. C. Bianchini, V. Dal Santo, A. Meli, W. Oberhauser, R. Psaro and F. Vizza: *Organometallics* 19, 2433 (2000).
109. C. Bianchini, M. Frediani, G. Mantovani and F. Vizza: *Organometallics* **20**, 2660 (2001).

CHAPTER 4
RING OPENING, HYDROGENOLYSIS AND DESULFURIZATION
OF THIOPHENES BY METAL COMPLEXES.

4.1. Introduction

In Chapter 2 the various ways in which thiophenes can be activated by coordination to transition metals in molecular complexes have been analyzed, in connection with their reactivity in solution and the possible relevance of this chemistry to HDS-related reactions. In Chapter 3, the homogeneous transition metal complex-catalyzed hydrogenation of aromatic hydrocarbons, as well as of sulfur- and nitrogen-heteroaromatic rings, was described in some detail as a model of an important part of most of the HDS and HDN mechanisms normally envisaged in heterogeneous catalysis. Nevertheless, it must be remembered that the breaking of C-S bonds is the family of reactions ultimately responsible for the actual desulfurization of thiophenes and other organosulfur compounds. Thus the understanding of the pathways through which such reactions take place, is of prime concern in dealing with modeling HDS mechanisms and eventually with new catalyst design. Therefore, a great proportion of the published work in the field of modeling hydrodesulfurization with transition metal complexes has been directed toward achieving such C-S bond rupture reactions under relatively mild reaction conditions. Particular emphasis has been placed on the unraveling of the various mechanisms involved and on the determination of the different thermodynamic and kinetic factors that influence their course.

As will hopefully become evident throughout this Chapter, C-S bond scission induced by transition metal complexes in solution -which 15-20 years earlier was a rarity- has become a fairly common event for a variety of compounds of different metals and varied stoichiometries and structures. More importantly, the intimate mechanisms involved in C-S activation by metal ions can now be considered very well understood. Also interestingly, some homogeneous *catalytic* systems for the hydrogenolysis and hydrodesulfurization of thiophenes have recently begun to emerge in the specialized literature. Whether such *homogeneous* systems will be further developed into practical applications in the refining industry still remains to be seen. However, their modification toward biphasic (aqueous or ionic liquids), or solid supported catalysts –all of which would be much easier to recover and recycle- appears as a very promising alternative, specially for the treatment of refined cuts or intermediate refinery products with relatively low sulfur contents. In any case, the extensive mechanistic knowledge gained thus far from organometallic models represents a significant contribution to the understanding of the nature of the active sites and of reaction networks operating in heterogeneous catalysts.

4.2. Stoichiometric ring opening, hydrogenolysis, and desulfurization of thiophenes

In dealing with C-S bond breaking reactions by transition metal complexes, it is important to note that C-S bond energies are of the order of 62 Kcal mol[-1] (in aliphatic thiols and thioethers), while C=S bonds of thiophenes and related molecules are of around 114 Kcal mol[-1]. These values are comparable to the range of energies encountered for C-H bonds (*ca.* 100 Kcal mol[-1]) [1]. Interestingly, the activation of C-H bonds by transition metal complexes is now a well established phenomenon and a wide variety of metal complexes have been found to promote such reactions through several mechanisms which are currently very well documented and understood [2-4]. Thus the search for metal complexes capable of breaking the C-S bonds of thiophenes and other related compounds has profited from the knowledge previously gained in C-H activation studies. Although less extensively developed, the activation of C-S bonds by metal complexes in solution has experienced an explosive growth within the last few years, to the point that it can no longer be considered a curiosity [5-14].

Ring opening of thiophenes through what can be formally viewed as an oxidative addition of the C-S bond across the metal center in a complex has been achieved in a variety of ways for a considerable number of elements. The reaction pathways most frequently discussed in the literature are represented in Fig. 4.1:

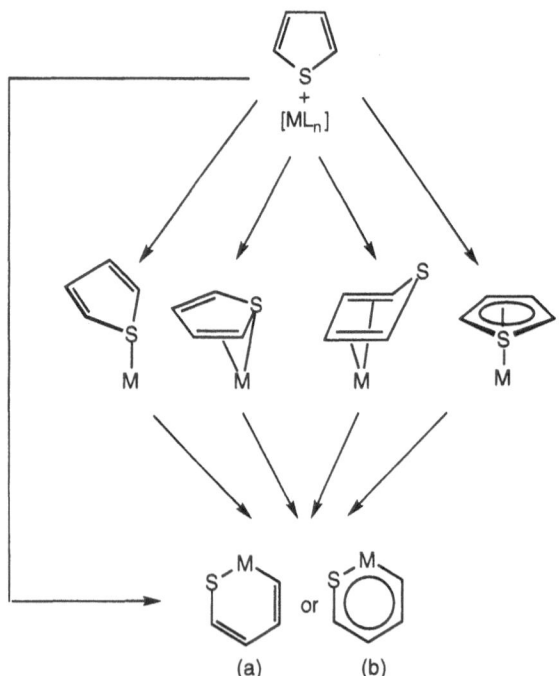

Fig. 4.1. Ring-opening reactions leading to η^2(C,S) thiophene-inserted metal complexes.

This C-S activation can be considered as a concerted process taking place as the thiophene approaches the metal ion, or as a two-step reaction involving intermediates that contain the intact thiophene molecule coordinated to the metal, in any one of the common binding modes described in Chapter 2 (see Fig. 2.1, p. 36). The ring-opened η^2(C,S) products thus formed have been well characterized, mainly by use of NMR spectroscopy, but crystallographic methods have also been employed in a good number of cases, and some thorough theoretical studies have nicely complemented the experimental data.

These metallacycles have been described in some cases as containing a thiabutadiene fragment with alternating localized C=C bonds -as in structure (a) in Fig. 4.1- while in others, a delocalized thiametallabenzene ring has been demonstrated or suggested to better represent the structure -as in (b) in Fig. 4.1. The delocalization of the bonding is thought to be related, on the one hand, to the degree of electron deficiency at the metal center, and on the other, to the particular orbital structure associated with specific coordination geometries. Also, X-ray diffraction studies have shown that the metallacyclic ring is very commonly planar, but important deviations from planarity have also been observed in some instances [5-14]. Molecular mechanics calculations provided by Harris and Jones indicate that the geometric details of individual complexes, and specifically the planarity or non-planarity of the metallacycle, are dictated mainly by steric rather than by electronic effects [15].

4.2.1. REACTIONS OF THIOPHENES ON MONONUCLEAR COMPLEXES

A good number of mononuclear metal complexes are known to activate C-S bonds, and it is clear that a high electron density around the metal atom generally favors the oxidative addition reaction. One commonly encountered case is the interaction of electron rich, coordinatively unsaturated (16-electron) metal fragments with thiophenes which results in the rupture of one of the C-S bonds with the consequent formation of the corresponding saturated 18-electron thiametallacycles.

4.2.1.1. Complexes with cyclopentadienyl and related ligands.
Pioneering mechanistic studies of this type of C-S bond activation reaction were provided by Jones and his coworkers, using the very reactive d^8 ML_4 16-electron fragment [Cp*Rh(PMe$_3$)] (Cp* = C$_5$Me$_5$); this very reactive species can be produced by thermolysis of the hydrido-phenyl precursor [Cp*Rh(PMe$_3$)(H)(Ph)], or by photolysis of the dihydride [Cp*Rh(PMe$_3$)(H)$_2$] (see Eq. 4.1) [16, 17]. Upon thermal reaction with thiophene (T), benzothiophene (BT) or dibenzothiophene (DBT), the unsaturated Rh fragment yielded the corresponding thiametallacycles Cp*Rh(PMe$_3$)[η^2(S,C)-Th] (Th = T, BT, DBT). Interestingly, the low temperature photochemical experiment yielded the same thiametallacycles, but mixed with the corresponding 2-thienyl hydride complexes Cp*Rh(PMe$_3$)(H)(η^1C-Th) resulting from activation of the thiophene C-H bond carbon α to sulfur; the latter slowly converted to the thermodynamically more stable ring-opened product in solution. This demonstrates that both C-S and C-H bond activation pose similar energy requirements on the metal fragment, and that both the thermodynamic and the kinetic factors influencing the course of the reaction must be

taken into account in order to understand the product distribution and the mechanisms involved.

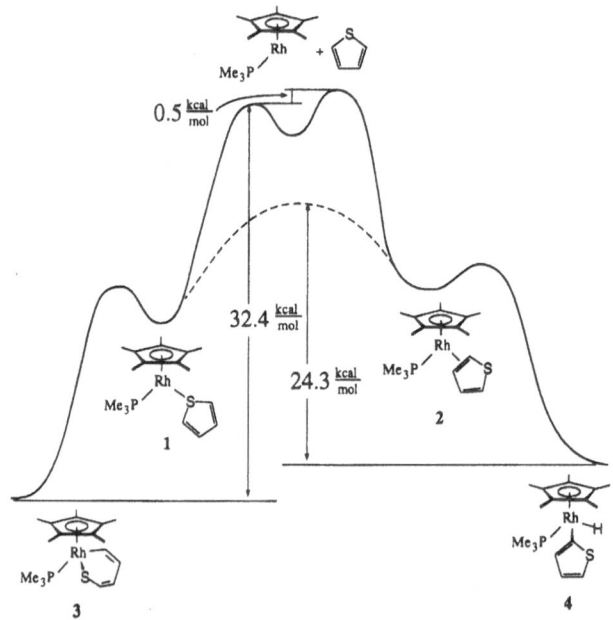

(4.1)

The high selectivity observed for the ring-opened product in this case was explained by Jones in terms of a slight preference (of the order of 0.5 Kcal mol^{-1}) for η^1S-coordination of thiophene to Rh -which was claimed to lead to C-S bond scission- over η^2(C=C) bonding which was thought in turn to be responsible for C-H bond activation, as represented in Fig. 4.2.

Fig. 4.2. C-S vs. C-H activation by Cp*Rh(PMe₃) (Reproduced with permission from Ref. 18).

Recent theoretical (DFT) calculations performed by Sargent and Titus [18] on the same system indicate that the C-S bond-breaking pathway proposed by Jones, going through the η^1S-thiophene isomer, is indeed a very reasonable alternative. Nevertheless, these calculations also demonstrate that C-H bond activation is *not* restricted to the η^2(C=C) intermediate as initially thought, since the S-bonded isomer can provide an additional low energy pathway toward C-H activation by considering some sensible intramolecular distortions, without any need for a previous dissociative isomerization to the olefin-like intermediate, as proposed by Jones. Me$_x$T as well as BT and DBT react in a similar manner to thiophene, with the peculiarity that activation invariably takes place at the unsubstituted C-S bond. This latter regioselectivity of the C-S insertion is believed to be directed mainly by steric hindrance [19, 20]. In a similar set of reactions, η^1S-coordination of thiophene and benzothiophene followed by ring opening has been observed by Selnau and Merola for (PMe$_3$)$_3$Ir(Th)Cl [21].

It is also interesting to note that Carmona's group reported the reaction of the complex Tp*Rh(PMe$_3$)(C$_2$H$_4$) [Tp* = hydrotris(3,5-dimethyl-1-pyrazolyl)borate] with thiophene -either thermally or photochemically- to give mixtures of the corresponding C-H and C-S activation products, but with the former being the predominant species regardless of the reaction conditions [22]. Polypyrazolylborate ligands have often been claimed to be electronically very similar to Cp, and therefore the fragment Tp*Rh(PMe$_3$) -presumably formed in this reaction- should be analogous to Jones' intermediate Cp*Rh(PMe$_3$), and yet the selectivity is inverted. This shows that relatively small variations in the electronic and steric environment around the active metal center can lead to important changes in the resulting reactivity toward C-H or C-S bonds of thiophenes, even if the formal oxidation state and coordination geometries of the sites remain essentially unchanged.

Despite the fact that cobalt-thiophene chemistry is obviously of particular relevance in relation to Co-Mo-S catalysts, this area of HDS modeling is very little developed. In a rare example involving a cobalt derivative, Jones has reported the reaction of Cp*Co(C$_2$H$_4$)$_2$ with T, BT, and DBT leading to insertion of the metal into the C-S bond, presumably through an undetected η^1S-bonded intermediate [23, 24]; since both ethylene ligands are readily displaced from the metal, the resulting electron deficient thiametallacycle Cp*Co(η^2C,S-T) achieves electronic saturation by binding to a second Cp*Co moiety, as shown in Eq. 4.2.

$$(4.2)$$

Other particularly important recent examples of thiophene ring opening by cyclopentadienyl metal complexes involve Parkin's finding [25] that the fragment [*ansa*-Cp$_2$''Mo] [Cp$_2$'' = Me$_2$Si(C$_5$Me$_4$)$_2$] -generated by photolysis of the dihydride precursor [*ansa*-Cp$_2$''Mo(H)$_2$]- readily inserts into the C-S bond of T and BT to produce the corresponding ring-opened products, as exemplified for T in Eq. 4.3:

$$(4.3)$$

No intermediates containing coordinated intact T or BT could be observed, although they are probably involved in this reaction. In agreement with this possibility, DBT coordinates to Mo to yield the stable η^1S-adduct Cp$_2$''Mo(η^1S-DBT) which could be characterized by X-ray diffraction (see Section 2.2, p. 43). This S-bonded derivative, however, does not react further to open the ring; in fact, the thermal reaction (80 °C) of this η^1S-DBT adduct with T leads to displacement of the DBT ligand and production of the ring opened product of thiophene. The difference in reactivity observed is clearly related with the much greater difficulty generally associated with the activation of the C-S bonds of dibenzothiophene in comparison with thiophene or benzothiophene (see Chapter 1). This set of reactions described by Parkin and his coworkers are remarkable in that they are the only examples available to date of homogeneous thiophene C-S bond breaking reactions induced by *molybdenum* complexes, a very important achievement indeed, in connection with modeling MoS$_2$-based catalysts.

Interestingly, Jones had reported earlier an analogous photochemical reaction of thiophene with the unsubstituted-Cp analogue Cp$_2$MoH$_2$ which led only to C-H activation [26]; this clearly indicates that the presence of the *ansa* bridge and the methyl ring-substituents in Parkin's complexes are playing an important role in this chemistry. In fact, the failure to activate the C-S bond of dibenzothiophene, as well as the small tilt angle observed in Cp$_2$''Mo(η^1S-DBT) can be attributed to steric effects of the Cp'' ligand. On going to the tungsten analogue Cp$_2$WH$_2$, Jones observed a transient product of thiophene ring opening that rearranged to the thermodynamically stable C-H activation product [26]. This again shows that metal complexes can in principle be fine-tuned by relatively minor variations of the metal-ligand environment to promote either C-S or C-H activation.

Other interesting Cp systems which have been used in HDS-related chemistry include Angelici's Cp*Ir(η^4-T), which is formed by reduction of the corresponding η^5-coordinated precursor; this η^4-T derivative isomerizes to the ring opened isomer by the action of basic alumina or triethylamine, or alternatively, by UV irradiation. Also, attack on cationic [CpRu(η^5-T)]$^+$ by H$^-$ or other nucleophiles promotes ring opening [8] (see Section 2.5).

Boncella has very recently reported [81] the use of the –non Cp- complex [W(NPh){o-(Me$_3$SiN)$_2$C$_6$H$_4$}(C$_5$H$_5$N)$_2$], to open the rings of thiophene, 2-methylthiophene, and benzothiophene at 65° C in toluene solution. This is another particularly important case representing the only example of C-S activation induced by W(IV) , and the only stable complexes containing ring-opened thiophenes on W, one of the most frequently used metals in heterogeneous HDS:

4.2.1.2. Complexes with phosphine ligands.

A different set of thoroughly studied cases involves a variety of phosphine ligands. Bianchini and Sánchez-Delgado [5, 7 9, 10, 14] have extensively used the fragments [(triphos)Ir]$^+$ or [(triphos)MH] (M = Rh, Ir), thermally generated from the stable precursors [(triphos)Ir(η^4-C$_6$H$_6$)]$^+$, [(triphos)Ir(H)$_2$(C$_2$H$_5$)], or [(triphos)Rh(H)$_3$], [triphos = MeC(CH$_2$PPh$_2$)$_3$]. These 14- or 16-electron fragments exhibit a very rich chemistry related to HDS, and in some cases the compounds obtained actually *catalyze* the homogeneous hydrogenolysis or hydrodesulfurization of some thiophenes (for details, see Section 4.3 below). The unsaturated Rh and Ir phosphine fragments readily undergo oxidative addition across the C-S bond of T [27], BT [28], DBT [29], and some of their methylated derivatives [30], leading to the corresponding thiametallacycles. The propensity of these Rh(I) and Ir(I) species to activate the C-S bond is very probably related to the fact that the tripodal ligand is fairly rigid and thus it is constrained to bind to the metal in a way which imposes a very different geometry from the preferred square planar disposition of ligands for d^8 Rh(I) or Ir(I) systems. The net result is a high energy ground state which provides the driving force for the oxidative addition of the C-S bond; such a reaction leads the metal ion into a "normal" 18-electron octahedral (or 16-electron square based pyramidal) M(III) product, containing the triphos ligand coordinated in a *fac* arrangement. It is also important to note, as discussed by Bianchini [30], that "due to the availability of filled metal d-π orbitals of appropriate energy and symmetry to interact with the C(2)-S π* orbital, the cleavage of the C(2)-S bond of thiophenes by 16e metal fragments of the type [(triphos)MH] (Rh, Ir) is a low energy process indeed.

In particular, it is lower in energy than the thermal processes which are necessary to generate the 16e fragments from their 18e precursors".

Because of these special characteristics the complex [(triphos)Ir(η⁴-benzene)]BF₄ has been successfully used as a precursor in modeling a complete stepwise hydrodesulfurization reaction in solution by the sequential addition of thiophene, H⁻ and H⁺ as shown in Fig. 4.3 [27].

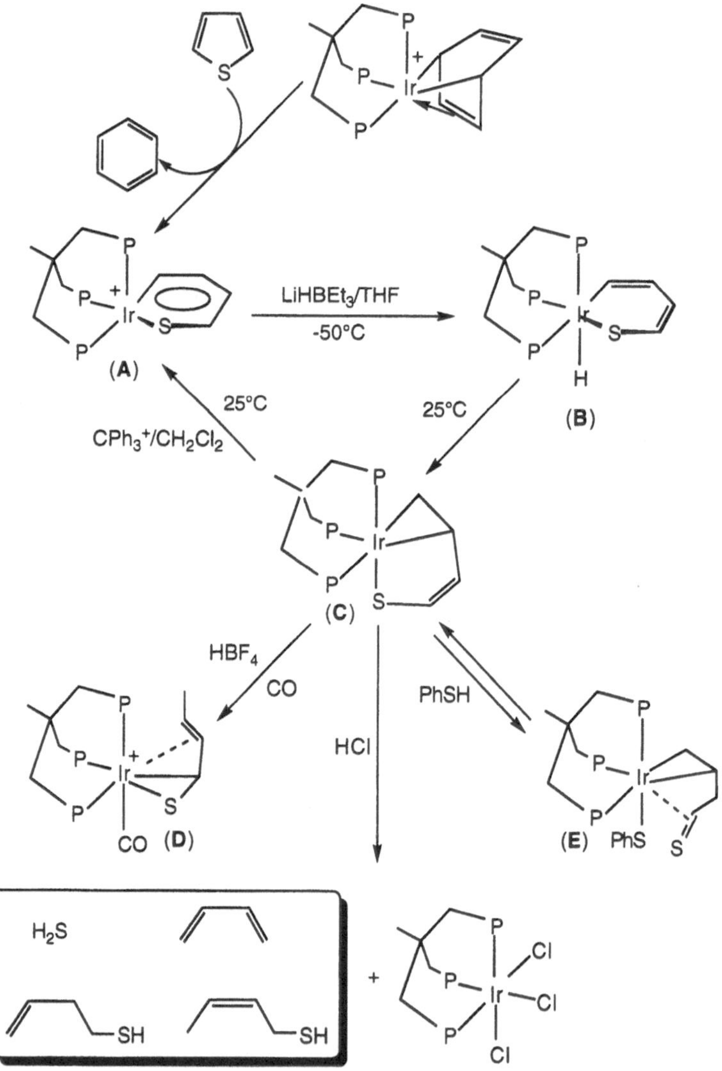

Fig. 4.3. Stepwise HDS of thiophene by [(triphos)Ir(η⁴-benzene)]BF₄.

After displacement of benzene (and possibly coordination of thiophene through the sulfur and one C=C bond, as has been observed for benzothiophene (see Eq. 4.4 below), oxidative addition across a C-S bond of thiophene takes place readily to yield the thiametallacycle **A**, which seems to be delocalized, on the basis on NMR data. Although this stable ring-opened product fails to react with molecular hydrogen, treatment with an external hydride source such as LiHBEt$_3$ yields a neutral hexacoordinated Ir(III) hydride **B** which could be detected by NMR spectroscopy at -50° C. Upon slow warming up to 50° C the H$^-$ ligand in **B** can be seen to migrate regiospecifically to C(5) of the metallacycle, yielding the very stable butadienethiolate complex **C** which was well characterized by NMR spectroscopy. This illustrates once more that electron rich systems do indeed promote ring-opening of thiophene but also importantly, that addition of hydrogen to an activated thiophene can take place via metal-hydride intermediates, a reaction path that has been invoked in some heterogeneous mechanisms. Subsequent protonation with HBF$_4$ also takes place at C(5) to yield **D** -which was unstable but could be isolated as a CO adduct amenable to characterization- while with the weaker acid PhSH (more similar to protons in the -SH groups present in metal sulfide surfaces) protonation is rapid and exclusively directed to C(3) forming the thioaldehyde complex **E**. This step is believed to occur through initial protonation of the S-atom of the ring-opened thiophene ligand, followed by migration to C(3), another reaction path that has been presented in some heterogeneous mechanisms [31]. If a 5-fold excess of gaseous HCl is added to the butadienethiolate intermediate **C**, the hydrodesulfurization is completed with the generation of 1,3-butadiene + H$_2$S, as well as small amounts of the corresponding intermediate butenethiols, all of which are indeed known to be primary products in the heterogeneous HDS of thiophene. The iridium is quantitatively recovered in the form of the known complex [(triphos)IrCl$_3$] which is not capable of activating C-S bonds or of easily regenerating the chloride-free fragment and therefore the HDS reaction cannot be rendered catalytic. Nevertheless, this work represents an excellent organometallic model in which various steps involved in the degradation of thiophene into butadiene plus H$_2$S could be authenticated and understood.

All these elementary steps authenticated in this model system find a parallel in surface chemistry, since the presence of metal hydrides and of protons of different degrees of acidity have been frequently mentioned in the heterogeneous literature. Thus a reaction scheme similar to the one described in Fig. 4.3 may be envisaged on *e.g.* a Co-Mo-S site. Also it is interesting to note that the mild HDS in this case does not take place under gaseous H$_2$, which might indicate that this complex is not adequate for activating hydrogen under the reaction conditions used. The fact that it has been achieved by using H$^-$/H$^+$ couples suggests that heterolytic splitting of H$_2$ may be what is needed; this type of activation is nowadays considered to be an important step in the activation of dihydrogen on solid catalysts [31].

An analogous set of reactions has been described for benzothiophene with the important difference that in this case a stable intermediate toward ring-opening, containing an intact BT molecule, and coordinated in a unique η^3(S,C=C) fashion (see Chapter 2, Section 2.4, p. 46) was isolated and characterized by multinuclear NMR spectroscopy (Eq. 4.4). This intermediate is stable in THF solution at room temperature but it begins to convert to the corresponding ring-opened isomer at temperatures above 40° C. The latter was fully characterized by spectroscopic as well as X-ray diffraction data, all of which are consistent with the thiametallacycle in this complex being delocalized and planar.

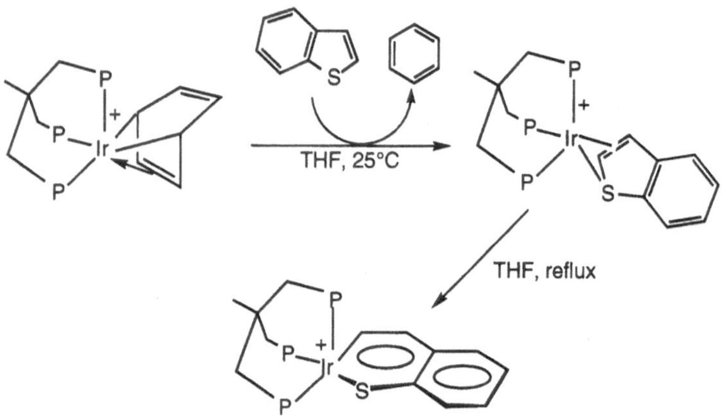

$$(4.4)$$

This $\eta^3(S,C=C)$ as well as other possible bonding modes available to d^8 ML_3 fragments have been studied theoretically by Rincón using *ab initio* and DFT methods on the model compound $[(PH_3)_3Rh(T)]^+$ [32] and the closer analogue $[(triphos*)Ir(T)]^+$ (triphos* is a simplified version of triphos in which the phenyl substituents on the phosphorus atoms have been replaced by methyl groups) [33]. The results of these calculations indicate that the thioallylic coordination identified experimentally is indeed a particularly stable bonding situation for these d^8 14e fragments. They also show that its easy conversion into the metallacycle is well justified, while the η^1S-bonded isomer, also considered in the calculations, does not seem to be a reasonable intermediate toward ring-opening in this case. It is important to point out that adsorption of thiophene through the sulfur atom plus a double bond on a MoS_2 surface appeared in one of the earliest proposals for HDS of thiophenes on metal sulfide catalysts, the so called "multi point adsorption" mechanism (see Sections 1.2.6 and 2.4).

    Also importantly, in the case of benzothiophene, the reactions with H_2 or with H^-/H^+ couples lead only as far as the formation of ethylbenzenethiolate, in contrast to the complete desulfurization of thiophene. Thus this system adequately models the hydrogenolysis of BT to the corresponding thiol, but not its complete hydrodesulfurization to ethylbenzene plus H_2S [28]. The benzothiophene-derived metallacycle in fact reacts with H_2 at 5 atm and 25°C (which is not the case for thiophene) yielding products containing S-bonded ethylbenzenethiolate ligands but curiously, the reaction follows different courses for different counteranions as shown in Eqs. 4.5 and 4.6. If the BPh_4 anion is used, decomposition into BPh_3 and benzene is observed, together with the formation of a stable mononuclear dihydrido-thiolate product (Eq. 4.5), while the analogous complex containing PF_6^- as the anion leads to a thiolate-bridged dinulear species containing only one hydride per iridium atom. This implies the presence of protons in the reaction medium, which in turn suggests again the intervention of a heterolytic mechanism in the activation of the H_2 molecule by the Ir(III) complex, and thus this reaction was also modeled by means of a stepwise addition of H^-/H^+, followed by H_2.

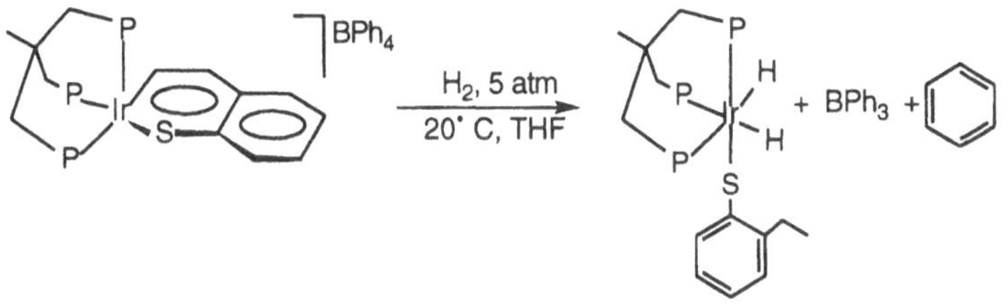

$$(4.5)$$

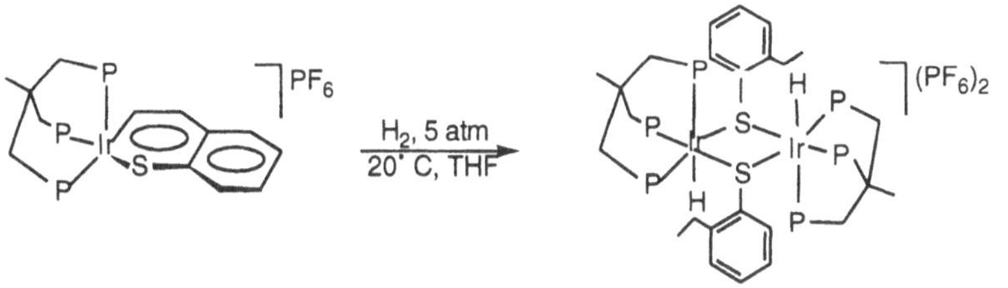

$$(4.6)$$

The most relevant of these reactions are summarized in Fig. 4.4. Addition of H^- takes place at the Ir center at low temperature, as demonstrated by *in situ* NMR measurements, but upon warming to room temperature the hydride is seen to migrate to the carbon atom already attached to the metal. Subsequent protonation with HBF_4 is also directed to the same carbon atom. The intermediate thus formed can be further hydrogenated (1-5 atm H_2) to yield a thiolate-bridged dimeric complex; the latter can also be obtained by direct hydrogenation of the BT ring-opened metallacycle (see Eq. 4.6); alternatively, a second sequence of H^-/H^+ addition yields the same dimer. Finally, this thiolate-bridged dimer can also be formed by initial reaction with H_2 to yield a mononuclear Ir(dihydrido)ethylbenzenethiolate (*cf.* Eq. 4.5), followed by protonation with HBF_4, which causes the elimination of hydrogen.

A probable overall mechanism for the formation of the thiolates is shown in Fig. 4.5. It is not clear why the second C-S bond is not broken in the benzothiophene case to complete HDS as was observed for thiophene; presumably, some delicate electronic differences are responsible for the different behavior in these two apparently very similar systems.

This extensive [(triphos)Ir]$^+$ model chemistry demonstrates that all the necessary steps for the complete HDS of thiophene or for the hydrogenolysis of benzothiophene can be achieved without any need to previously saturate the heterocyclic ring, in agreement with one of the commonly invoked mechanisms in HDS catalysis –"the desulfurization route"- as discussed in Chapter 1.

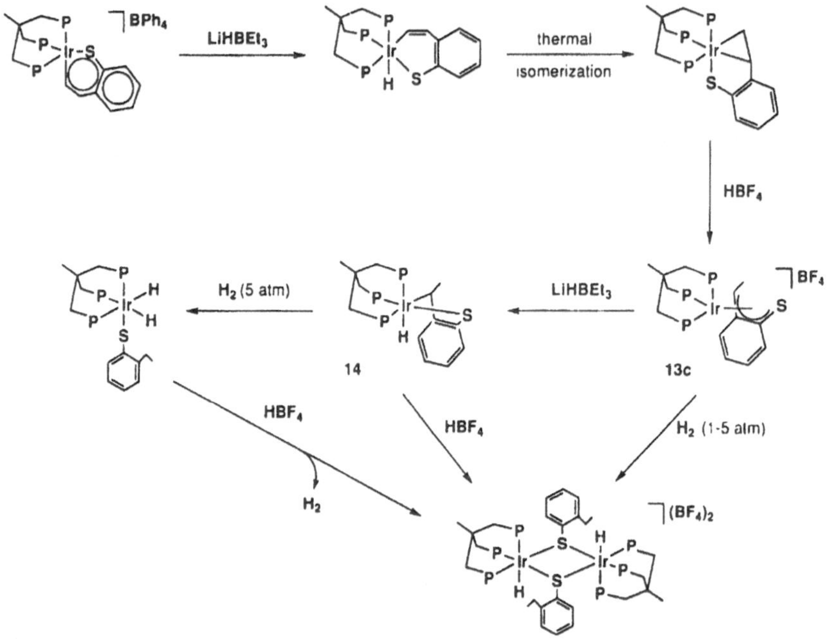

Fig. 4.4. Reactions of [(triphos)Ir(η^2-C,S-BT)]$^+$ with H$^+$, H$^-$, H$_2$ (Reproduced with permission from Ref. 28).

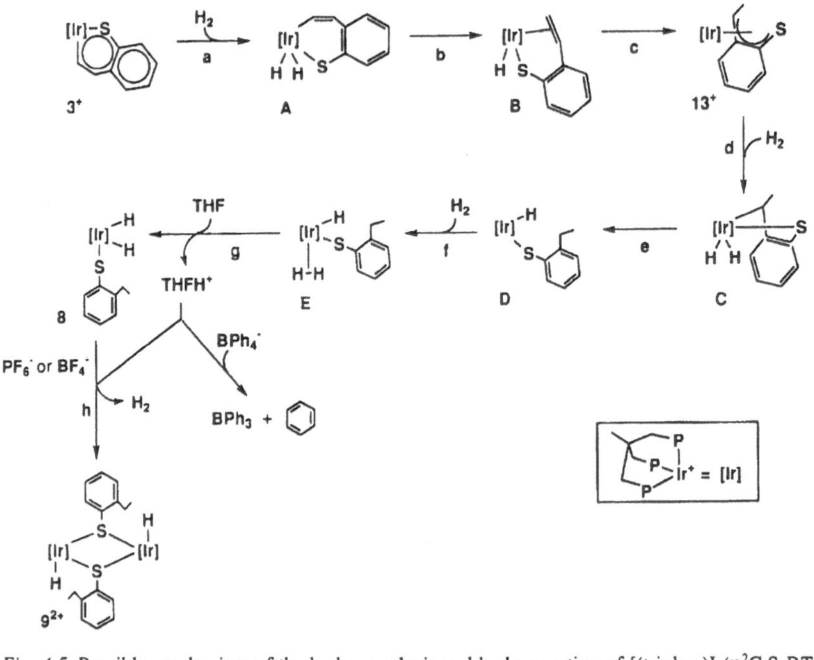

Fig. 4.5. Possible mechanism of the hydrogenolysis and hydrogenation of [(triphos)Ir(η^2C,S-BT)]$^+$ (Reproduced with permission from Ref. 28).

Furthermore, we noted in Chapter 3 that the systems that are very efficient for the catalytic hydrogenation of BT to DHBT (*e.g.* d^8 ML_2 $Ir(PPh_3)_2^+$ or d^6 ML_4 $Ir(H)_2(PPh_3)_2^+$ fragments) are generally incapable of subsequently breaking the C-S bonds of the saturated cyclic thioethers, consistent with the idea that BT (or olefin) hydrogenation sites on metal sulfides could be quite different from desulfurization sites both in being poorer in electron density and of a more flexible geometry; the existence of such different sites for different functions has been claimed many times in heterogeneous catalysis (see Chapter 1).

It is important to note, however, that Bianchini has recently reported the first example of the hydrogenolysis of a *saturated* cyclic thioether in solution [34]; by reacting the d^6 ML_5 16e fragment $[(triphos)Ir(H)_2]^+$ with 2,3-dihydrobenzothiophene in the presence of a strong base it was possible to produce ethylbenzenethiol plus $[(triphos)Ir(H)_3]$ through the reaction sequence depicted in Fig. 4.6, which was well established by *in situ* high pressure NMR experiments combined with the isolation of important intermediates.

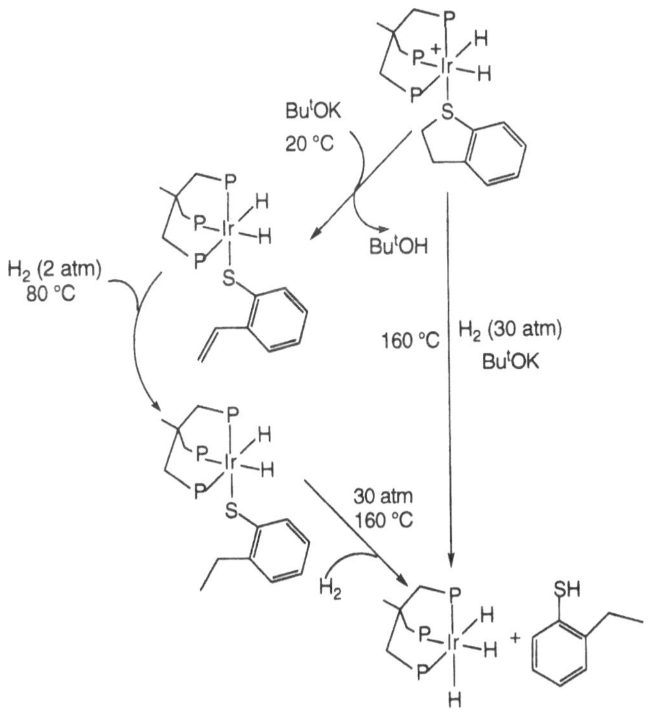

Fig. 4.6. Hydrogenolysis of 2,3-dihydrobenzothiophene by $[(triphos)Ir(H_2)]^+$ (Adapted from Ref. 34).

This is an interesting addition to the catalog of available organometallic HDS-model reactions, in that it demonstrates that even though presaturation of the thiophenic rings is not required for desulfurization, the C-S bonds of saturated cyclic thioether substrates can indeed be broken by use of (triphos)Ir(III) fragments which are geometrically very similar to -but less electron rich than- the ones which promote ring opening of thiophenes, which correspond to Ir(I). This chemistry may thus be taken as suggestive that the sites that are responsible for the cleavage of C-S bonds of thiophenes on solid catalysts are closely related to the ones which cleave the C-S bonds of pre-hydrogenated thiophenes. It is likely, though, that these two types of sites may be easily interconverted by addition of hydrogen, or by other simple 2-electron redox processes. Furthermore, these two kinds of sites may be in turn substantially different in their geometry and in their electronic properties from the surface sites required to effect the catalytic hydrogenation of the heterocycles to the corresponding saturated cyclic thioethers.

Also surprisingly, when the studies with triphos-iridium complexes were extended to the more refractory substrate dibenzothiophene, an unprecedented *homogeneous catalytic hydrodesulfurization* to produce biphenyl + H_2S was observed [29]; this reaction is described in detail in Section 4.3. Bianchini has also shown that the (triphos)MH (M = Rh, Ir) fragments are able to promote C-H activation in parallel with C-S bond scission at moderate temperatures even though the latter invariably predominates; furthermore, the thienylhydride products resulting from C-H activation isomerize to the thermodynamically more stable C-S activated thiametallacycles through mechanisms which are dependent on the electronic properties of the different substrates used [35, 36].

An example of rhenium-phosphine chemistry in the hydrogenolysis of thiophene to the corresponding thiol was provided by Rosini and Jones by use of the polyhydride $ReH_7(PPh_3)_2$ [37]; when employed in conjunction with the hydrogen acceptor 3,3-dimethyl-1-butene, this complex generated a stable thioallyl intermediate through an intramolecular endo attack of a coordinated hydride on a presumably η^4-bonded thiophene ligand, as depicted in Fig. 4.7. If this thioallylic species is subsequently photolyzed at low temperature in the presence of an excess of PMe_3, hydrogenolysis of one of the C-S bonds of the cyclic moiety takes place to yield a mixture of isomers containing coordinated 1-butenethiolates. Upon further hydrogenation the saturated *n*-butanethiolate group attached to Re is generated. In contrast, if the thioallyl intermediate is treated thermally, complete hydrogenation of the cyclic sulfur ligand to tetrahydrothiophene is observed instead. Presumably this reduction takes place by a mechanism analogous to the one elucidated for similar thioallyl species which have been observed in the hydrogenation of thiophene by Ir-PPh₃ complexes, as described in Section 3.3.1.

García and Maitlis have reported that some d^{10} ML_2 14-electron platinum phosphine fragments $[PtL_n]$ (L = PEt_3, PMe_3, L_2 = $Ph_2PCH_2CH_2PPh_2$ dppe) *reversibly* insert into the C-S bond of thiophenes -T, BT and DBT as well as of some of their methylated derivatives- to yield the corresponding ring-opened products $[PtL_2(\eta^2C,S-Th)]$, as exemplified in Eq. 4.7 for DBT [38-44].

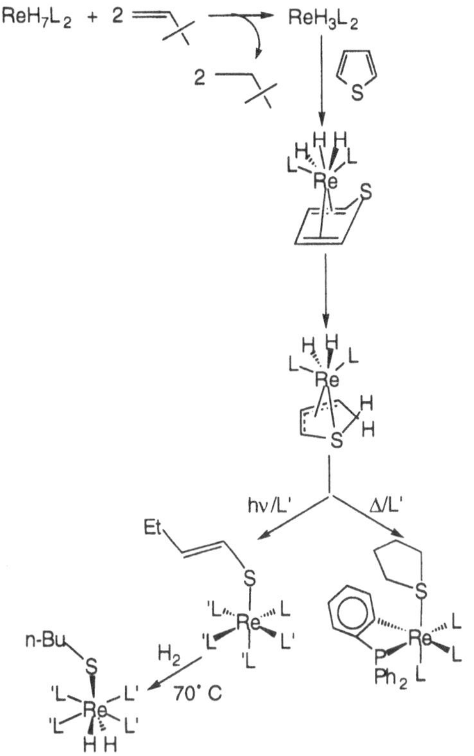

Fig. 4.7. Hydrogenolysis of thiophene by $ReH_7(PPh_3)_2$ ($L = PPh_3$, $L' = PMe_3$) (Adapted from Ref. 37; the formation of other isomers has been omitted for clarity).

In contrast with the triphos-iridium complexes of ring opened thiophenes described above, which are delocalized in nature, these Pt derivatives are better described as localized thiametallacycles, on the basis of NMR data and of the X-ray structure of the BT derivative. On heating these complexes in the presence of excess PEt_3 the reaction can be reversed; this allowed the determination of the corresponding equilibrium constants by use of NMR spectroscopy. The measured values indicate that the thiaplatinacycle derived from BT is about 10 times more stable than those derived from T or DBT, which are in turn, similar to each other.

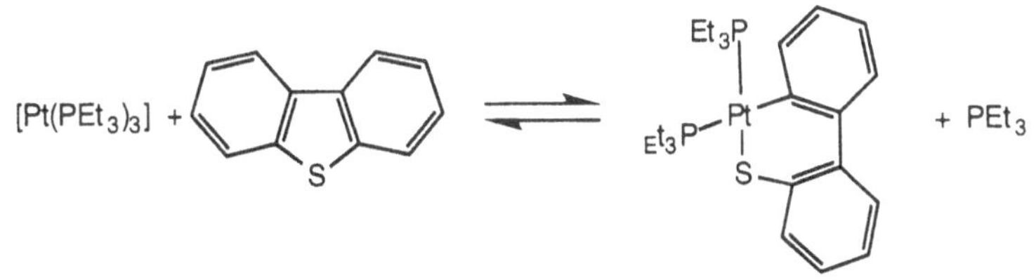

$$\qquad\qquad\qquad\qquad\qquad\qquad\qquad\qquad\qquad\qquad (4.7)$$

As shown in Fig. 4.8, the reaction of the PEt$_3$ thiametallacycle derived from DBT with an excess of Et$_3$SiH (a hydride source) results in the cleavage of both the Pt-C and the C-S bonds causing the *complete* HDS to yield biphenyl plus the interesting hydrido-hydrosulfido derivative [Pt(PEt$_3$)$_2$(H)(SH)]; upon exposure to HCl gas, the latter generates H$_2$S plus Pt(PEt$_3$)$_2$Cl$_2$, together with small amounts of [Pt(PEt$_3$)$_2$(H)(Cl)] thereby nicely completing the HDS scheme [38, 39]. Similar reactions with hydrides were observed for the ring-opened products of T and BT, and the efficiency of the HDS process was found to follow the trend BT > DBT >> T. These examples involving Pt complexes show again that a highly electron-rich d^{10} metal center is required to oxidatively add the C-S bonds of thiophenes; the hydride-promoted cleavage of the Pt-C and C-S bonds very likely takes place through Pt(IV) hydrides, although this could not be unambiguously demonstrated. The readiness with which the thiaplatinacycle derived from T reverts to the free ligand is probably the reason for the low HDS yields observed for that substrate.

In contrast to the reactions with hydrides, the interaction of the DBT and BT analogues with HCl produces the corresponding free thiols plus PtCl$_2$(PEt$_3$)$_2$, presumably through protonolysis of the Pt-C and the Pt-S bonds (Fig. 4.8) [40]. In this case the DBT derivative displays a higher reactivity than the BT derivative, while the T analogue simply reverts to the free ligand without producing the corresponding thiol. If HBF$_4$ is used instead of HCl, dicationic dinuclear Pt complexes containing bridging thiolate ligands reminiscent of the triphos-Ir complexes described in Fig. 4.4, are obtained; they can be further cleaved by excess HCl to generate the corresponding free thiols.

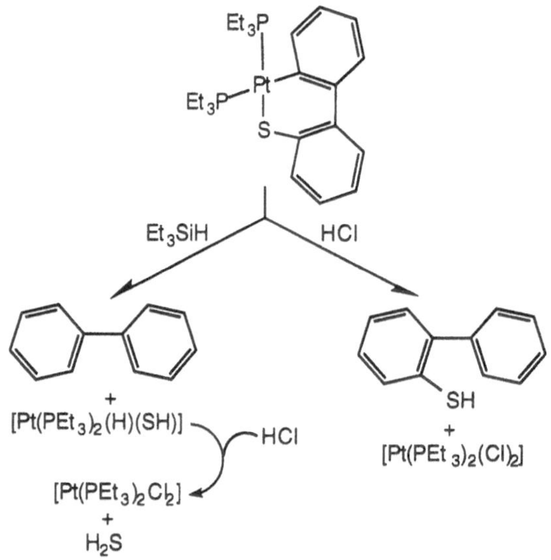

Fig. 4.8. Hydrogenolysis and desulfurization of DBT by Pt complexes.

This illustrates very well how the hydridic or protonic character of the hydrogen atoms being transferred to the ring-opened intermediates directs the reaction toward the Pt-C bond breaking process being accompanied by a second C-S bond rupture, or by a Pt-S bond scission, respectively. 2-MeBT and 3-MeBT also provide the corresponding thiaplatinacycles although the reactivity of the former is considerable lower due to the hindrance imposed by the methyl group next to sulfur; 4-MeDBT afforded the ring-opened product as a mixture of isomers [41]. Reaction of Pt(PEt$_3$)$_2$ with the highly refractory 4,6-Me$_2$DBT led to C-H activation only; nevertheless, the ring opened product of 4,6- Me$_2$DBT (DMDBT) could also be obtained by these authors, albeit by an indirect route involving the reaction of [PtCl$_2$(PEt$_3$)$_2$] with the substrate and metallic sodium under hydrogen at low pressure, as depicted in Eq. 4.8 [44]:

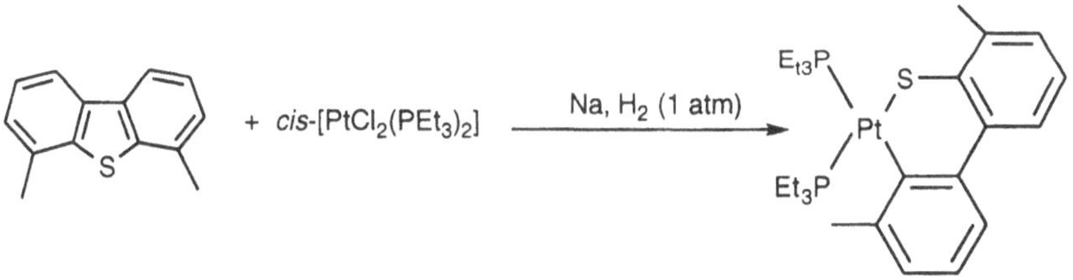

$$+ \; cis\text{-}[PtCl_2(PEt_3)_2] \xrightarrow{\quad Na, H_2 \; (1 \text{ atm}) \quad} \qquad\qquad\qquad\qquad (4.8)$$

If the PMe$_3$ and dppe (diphenylphosphino ethane) PtL$_2$ analogues are used, complete hydrodesulfurization of DBT and 4-MeDBT to biphenyl and 3-Me-biphenyl, respectively, takes place under 20 atm H$_2$ at 100 °C; interestingly, the presence of acid-washed alumina promoted the HDS reaction (see *e.g.* Eq. 4.9). Neither the fate of the sulfur nor the HDS mechanism is clear in this case [42, 43].

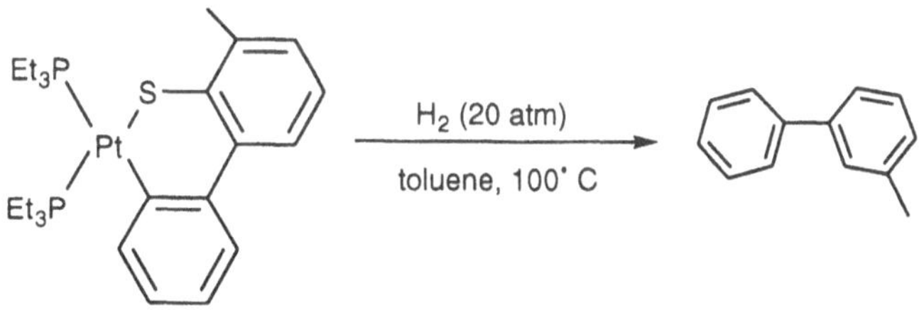

$$\xrightarrow[\text{toluene, 100° C}]{\quad H_2 \; (20 \text{ atm}) \quad} \qquad\qquad\qquad\qquad (4.9)$$

In related work, Vicic and Jones have achieved a remarkably mild HDS of DBT on nickel -a widely used promoter metal in industrial catalysts- according to the set of reactions summarized in Fig. 4.9 [45-48]. Although the starting material is dinuclear, most of the relevant reactions in this system occur on mononuclear fragments and thus the inclusion of this chemistry in this Section, rather than in the following one devoted to dinuclear and polynuclear complexes, seems justified.

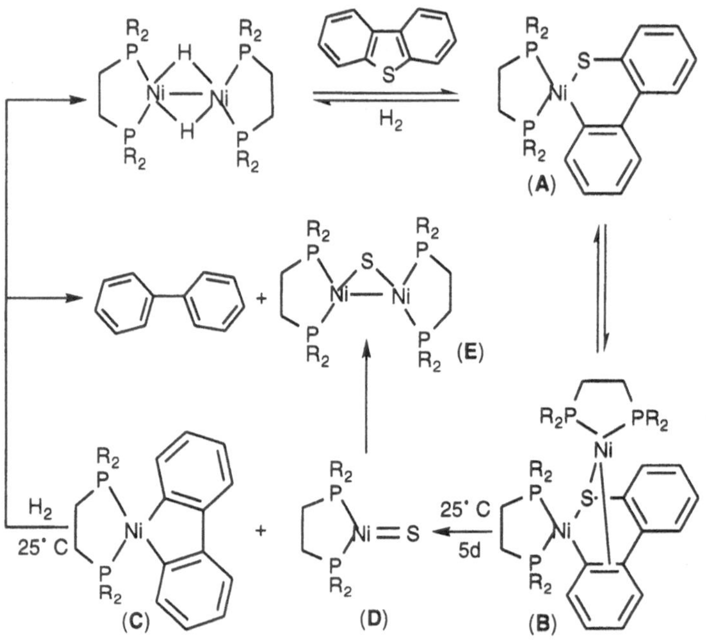

Fig. 4.9. Mild desulfurization of DBT on nickel; R = *i*-Pr.

The hydrido-bridged dimer [(dippe)NiH]$_2$ [dippe = (*i*-Pr$_2$PCH$_2$)$_2$] readily liberates dihydrogen to presumably form the electron-rich unsaturated (14-electron) intermediate [(dippe)Ni(0)], which then inserts quickly into the C-S bond of DBT to yield the corresponding Ni(II) thiametallacycle [(dippe)Ni(η^2-C,S-DBT)] (**A**). Over 5 days at room temperature in solution the complex evolves -via a dinuclear product (**B**) in which the thiophene is bonded in an analogous manner to that described for the Cp*Co system in Eq. 4.2- predominantly into [(dippe)Ni(2,2'-biphenyl)] (**C**) and a very interesting transient mononuclear intermediate (**D**) containing a terminal sulfido ligand; the latter could not be isolated since it spontaneously dimerizes into the very stable sulfido bridged complex [(dippe)$_2$Ni$_2$(μ-S)] (**E**) [47, 48]. Most remarkably, complex **C** reacts with H$_2$ at atmospheric pressure and room temperature to liberate biphenyl and regenerate the dimeric hydride. This demonstrates that the actual desulfurization step (**B** → **C** + **D**) can take place without any participation of hydrogen gas (or H$^+$/H$^-$ couples as described for the Ir-triphos chemistry). The presence of hydrogen seems to be required for this nickel system only in the subsequent release of the desulfurized product, biphenyl from **C** [48]. Further support for this interesting proposal came from the analogous reactions of T and BT with [(dippe)NiH]$_2$ which also led to the isolation of reasonably stable C-S insertion products; the latter are spontaneously and reversibly transformed in solution into dinuclear products analogous to **B**. On the other hand, 4-

MeDBT reacted with [(dippe)NiH]$_2$ at room temperature by insertion into the C-S bond distal to the methyl group, and strikingly, it reacted with an excess of the Ni hydride under hydrogen at atmospheric pressure at 90 °C to yield the hydrodesulfurization product 3-methylbiphenyl as the organic product, while the only metal-containing species identified was [(dippe)$_2$Ni] (see Eq. 4.10).

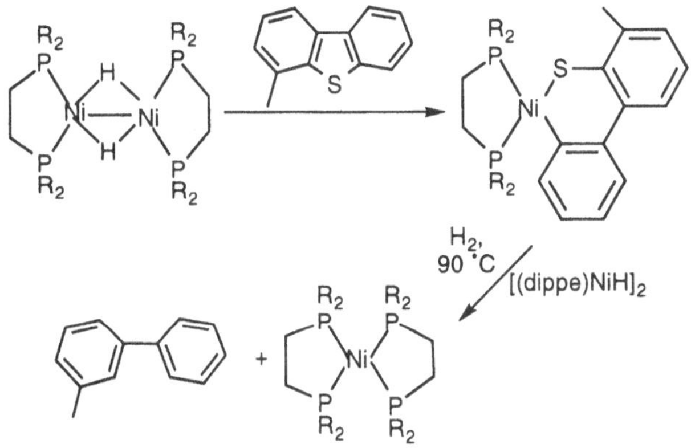

(4.10)

No nickel insertion products could be identified in similar reactions with 4,6-Me$_2$DBT but the reaction of the dimeric hydride with an excess of Me$_2$DBT under 1 atm H$_2$ at 90° C did indeed lead to the formation of the hydrodesulfurized product 3,3'-dimethylbiphenyl plus monomeric Ni(dippe)$_2$ as the only identifiable metal containing species [46].

In contrast, the reaction of the Pt analogue [(dippe)PtH]$_2$ with 4,6-Me$_2$DBT did provide the corresponding product of C-S activation, [(dippe)Pt(η^2-C,S-DMDBT)], a stable thiametallacycle resulting from the direct ring-opening of Me$_2$DBT by [(dippe)Pt(0)], as shown in Eq. 4.11.

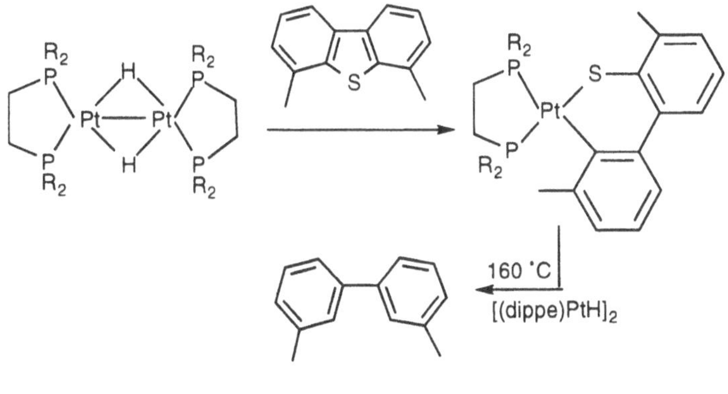

(4.11)

This compound reacted further with an excess of the Pt hydride at 160 °C to yield the HDS product 3,3'-dimethylbiphenyl and presumably a sulfido-bridged dimer [(dippe)$_2$Pt$_2$(μ-S)] which decomposed under the reaction conditions. This represents a unique example of the complete HDS of 4,6-Me$_2$DBT in solution. Analogous complexes originating from insertion of DBT and 4-MeDBT were also characterized for the Pt system, thus providing a nice family of complexes resulting from C-S activation of the dibenzothiophenes with varying degrees of methylation [46].

4.2.2. REACTIONS OF THIOPHENES ON DINUCLEAR AND POLYNUCLEAR COMPLEXES

As shown in the preceding sections, a variety of electron-rich mononuclear metal complexes containing different ligand environments are capable of readily inserting into C-S bonds of thiophenes, benzothiophenes, and dibenzothiophenes to produce the corresponding thiametallacycles. More importantly, the ring-opened products can undergo further reactions with hydrogen or hydrogen components under mild or moderate conditions leading to the transformation of the sulfur heterocycles into coordinated thiolate ligands, or free thiols. Additionally, some interesting examples of complete hydrodesulfurization of the organosulfur molecules to yield the resulting hydrocarbons in solution are available. Unfortunately, this latter type of sulfur extrusion reaction is still rather rare on transition metal complexes, indicating that breaking the second C-S bond of thiols or thiolates is not a facile process in most of those homogeneous systems. This is in clear contrast with heterogeneous catalysts, which are known to desulfurize thiols much more readily than thiophenes. Since it seems clear that multi-site surface reactions intervene in heterogeneous HDS, there is a growing agreement around the idea that more than one metal center may be needed also in solution in order to effect the complete removal of sulfur from the organic molecule. Therefore, binuclear or polynuclear metal complexes might actually be more suitable for either modeling or actually catalyzing HDS reactions in the liquid phase. This approach has been present in the literature for a number of years, including some particularly illustrative examples, some of which will be summarized in the following sections. However, a systematic use of homo- and hetero-polynuclear metal complexes in HDS modeling is by large still lacking, and thus further research along these lines is certainly to be expected and encouraged [6-11, 14].

A very instructive review on HDS modeling by transition metal clusters appeared recently [49].

4.2.2.1. Homopolynuclear complexes
The use of simple metal carbonyl clusters in HDS-related reactions was mentioned as early as 1960 by Stone and coworkers who reported that thiophenes interact with Fe$_3$(CO)$_{12}$ in refluxing benzene to produce mixtures of ferroles and thiaferroles [50]. No mechanistic details were given in that paper. Similar reactions were later studied in more detail by the groups of Weiss and Rauchfuss [51] for thiophenes and for benzothiophene, showing that the thiaferroles are actually the initial products arising from metal insertion into the less hindered C-S bond of the thiophene (see Fig. 4.10).

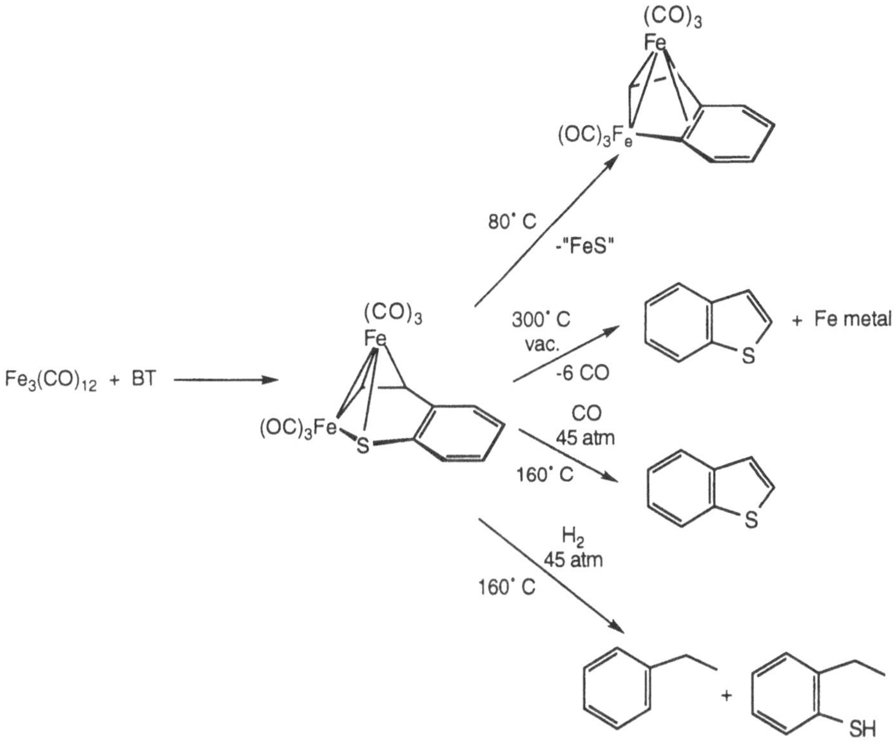

Fig. 4.10. Desulfurization of BT by $Fe_3(CO)_{12}$.

The thiaferrole derivatives are subsequently desulfurized to the corresponding ferroles in refluxing benzene, as illustrated for BT in Fig. 4.10. The benzothiaferrole can also revert back to the free BT upon thermolysis under vacuum (300 °C, 10^{-5} mm Hg), or via carbonylation at 160° C and *ca.* 45 bar CO. Perhaps more interestingly, the interaction of the benzothiaferrol with hydrogen gas at 160° C and *ca.* 45 bar yielded ethylbenzene and some related thiols. The fate of the sulfur being extracted was not clearly ascertained in any of these reactions. Dibenzothiophenes failed to undergo the analogous C-S bond activation under similar reaction conditions.

Arce *et al.* reported the reactions of $Ru_3(CO)_{12}$ with thiophenes and benzothiophene in solution [52], which give the desulfurized metallacyclic products analogous to the ferroles mentioned above, together with some C-H activation products. In the reaction of 2-MeT with the triruthenium cluster, an attractive tetranuclear compound was obtained (among other C-H activation and desulfurized products)   and fully characterized by X-ray diffraction. This new cluster contains the separated sulfur atom capping a triangle of Ru atoms, as well as the butadiene fragment that remains linked through two metal-carbon σ bonds to one of the Ru atoms in the triangle, and also η⁴ to a fourth "spiked" Ru center (see Eq. 4.12):

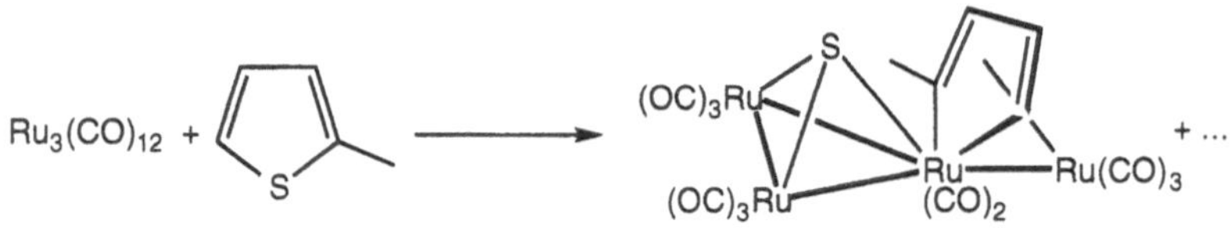

This remarkable molecule can be viewed as a "snapshot" of the desulfurization process taking place; unfortunately, the reaction of this compound with hydrogen was not reported. When the analogous Os clusters were used with the same substrates, only C-H activation products were observed for T and DBT, while for BT a minor product arising from a C-S bond cleavage reaction could also be detected.

Suzuki and coworkers demonstrated that the trinuclear ruthenium polyhydride $(Cp^*Ru)_3(\mu-H)_3(\mu_3-H)_2$ is capable of breaking the two C-S bonds of BT and DBT to yield ethylbenzene and biphenyl, respectively, according to the set of reactions depicted in Fig. 4.11 [53]. In the case of benzothiophene, the reaction pathway could be established as going through an intermediate thiaruthenacyclohexadiene complex formed in toluene at lower temperature (40° C) through insertion of one ruthenium atom across the vinylic C-S bond. This intermediate quantitatively evolves by a curious desulfurization process into a μ_3-alkylidyne-μ_3-sulfido-bis(μ_2-hydrido) cluster upon warming to 50° C for several hours; the latter cluster could be further hydrogenated in THF at 80° C and 7.2 atm H_2 for one week to produce the HDS product ethylbenzene and a new μ_3-sulfido complex $(Cp^*Ru)_3(\mu-H)_3(\mu_3-S)$.

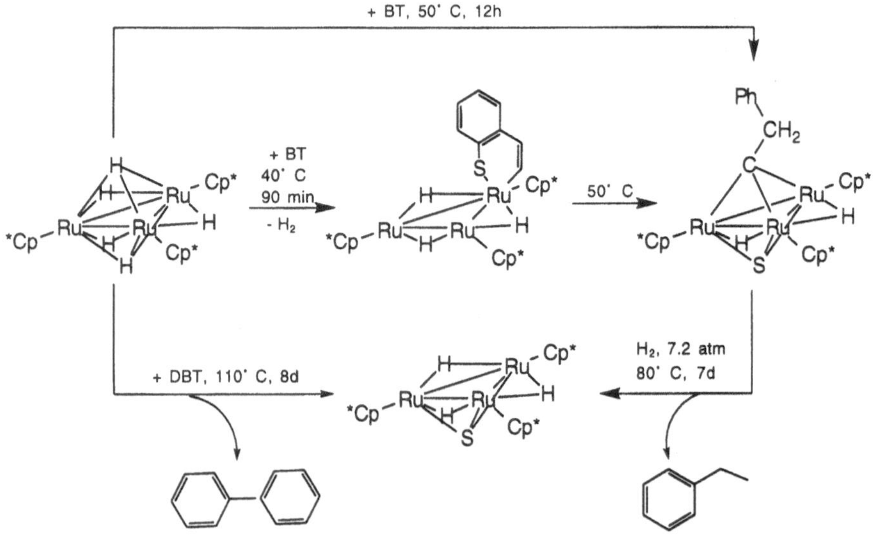

Fig. 4. 11. Desulfurization of BT and DBT by $(Cp^*Ru)_3(\mu-H)_3(\mu_3-H)_2$.

Dibenzothiophene is also cleaved by the same triruthenium cluster (Cp*Ru)$_3$(μ-H)$_3$(μ$_3$-H)$_2$ in toluene at 110° C over a period of 8 days to yield the fully desulfurized product biphenyl, plus the same μ^3-sulfido complex; the mechanism in this case is less clear.

Adams has published a series of papers describing C-S bond breaking reactions of *saturated* cyclic thioethers on metal clusters; C-S activation is clearly favored when the sulfur atom bridges two metal centers [54]. An example of this chemistry involving the thermolysis of Os$_3$(CO)$_{10}$(THT)$_2$ to yield an intermediate containing a coordinated butenethiolate ligand formed by ring-opening of THT via a Os$_3$(CO)$_{10}$(μ^2-THT) is shown in Eq. 4.13:

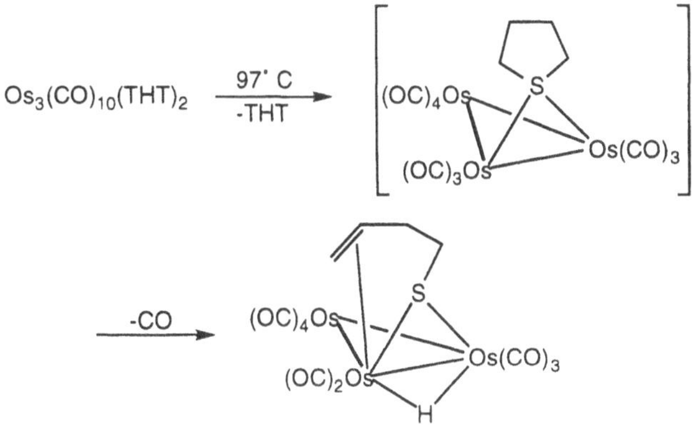

(4.13)

Metal carbonyl clusters are often unstable and their reactions frequently yield complicated mixtures of products in low yields. Possibly as a result of this, and despite the fact that a number of cluster containing HDS-related ligands are known, neither the reactivity of these derivatives resulting, nor the mechanisms of the desulfurization steps have been studied in any detail.

Recent efforts have concentrated on dinuclear complexes that appear to be more amenable to detailed investigations. They all have in common a bridging sulfur atom between the two metal centers, a situation that seems to be particularly favorable for achieving complete desulfurization. One interesting example of C-S bond activation on two metal atoms involves the sulfur-bridged dimer Cl$_3$W(μ-THT)$_3$WCl$_3$ (THT = tetrahydrothiophene) which undergoes heterolytic cleavage of the coordinated thioether upon nucleophilic attack by *e.g.* SR⁻ or H⁻, as shown in Fig. 4.12 [55].

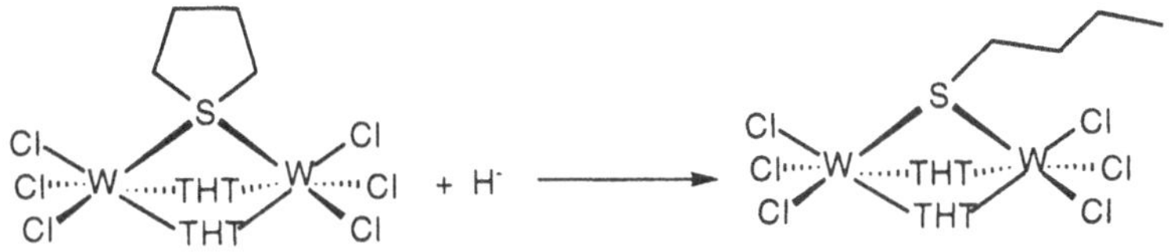

Fig. 4.12. Ring opening of THT by nucleophilic attack to Cl$_3$W(μ-THT)WCl$_3$.

This rare case of homogeneous C-S bond scission by a *tungsten* complex, shown in Fig. 4.12 has been extended to other nucleophiles and to other bridging thioethers (*e.g.* Et$_2$S). Its main relevance lies in the fact that it is based on one of the commonly used metals in practical HDS. Apparently, the acute W-S-W angle (*ca.* 62.5°) arising from a very short W-W bond may actually contribute to the weakening of the C-S bonds.

Jones has reported the reaction of the dinuclear Cp* iridium polyhydrido complex [Cp*IrH$_3$]$_2$ with thiophene in the presence of *tert*-butylethylene (as a hydrogen acceptor). Similarly to what happens with Arce's cluster mentioned above, this reaction yields a product containing the separated sulfur atom and the butadiene fragment both bridging the same two metal atoms (Eq. 4.14), in another remarkable "photograph" of a desulfurization process taking place. The product is presumably formed by the transfer of two hydrogen atoms from the metal to the α carbons of thiophene. The coordinated butadiene can be readily displaced from the complex by action of CO with simultaneous generation of [Cp*Ir(CO)]$_2$(μ-S); more interestingly, hydrogenation of Cp*Ir(μ-S)(μ-C$_4$H$_6$)IrCp* leads to the formation of butane, although in that case the fate of the sulfur or of the metal complex could not be clearly established [56].

butane + ?

2-MeT reacted in a similar manner with [Cp*IrH$_3$]$_2$, but 2,5-Me$_2$T, BT, and DBT failed to react [57]. In a related case also reported by Jones and coworkers, the chlorohydrido dimer [Cp*IrHCl]$_2$ reacted with T or BT at 90 °C under hydrogen to give products containing bridging thiolate ligands, [Cp*IrCl]$_2$(μ-H)(μ-SC$_4$H$_9$)] and [Cp*IrCl]$_2$(μ-H)(μ-S(C$_6$H$_4$)CH$_2$CH$_3$], respectively. The arise from hydrogenolysis of the thiophenic substrates, and both of them could be completely desulfurized to yield butane or ethylbenzene, by further reaction with hydrogen, as exemplified for BT in Eq. 4.15 [58].

Angelici and coworkers recently disclosed an interesting photochemical reaction of Re$_2$(CO)$_{10}$ with BT (Eq. 4.16) that produces a C-S bond cleavage product containing a bridging ring-opened BT, of similar structure to those described in Eq. 4.2 for Co and in Fig. 4.9 for Ni. Upon addition of excess PMe$_3$ two different phosphine substitution products are formed in one of which both the C-S and the Re-Re bonds have been cleaved [59].

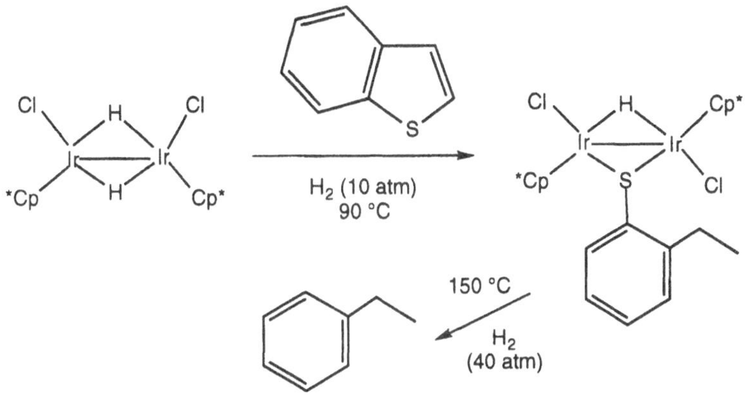

(4.15)

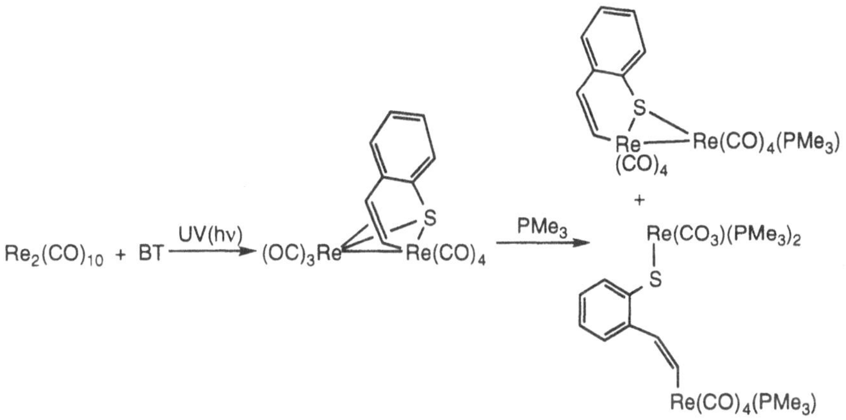

(4.16)

4.2.2.2. Heteropolynuclear complexes.

The best practical HDS catalysts are those composed of sulfides of two different metals (*e.g.* Co-Mo, Ni-W, see Chapter 1) and therefore heterobimetallic complexes are to be considered as better model complexes for HDS-related reactions. Despite this obvious consideration, such chemistry is still rather underdeveloped, possibly due to the increased difficulty that is generally encountered when dealing with heterobimetallic or heteropolymetallic systems as compared to mononuclear complexes; nevertheless, a few interesting examples of this type of cooperative chemistry have indeed appeared in the literature.

In an unusual approach, Rauchfuss and coworkers were able to show that $Fe_3(CO)_{12}$ reacts not only with free thiophenes -as described in the preceding section- but also with the η^4-activated thiophene in the rhodium complex $Cp*Rh(\eta^4\text{-}T)$, to yield predominantly a bimetallic product containing a ferrole unit π-bonded to Rh. The reaction was shown to proceed via a stable intermediate containing the T ligand simultaneously bonded η^4 to Cp*Rh and η^1 to a $Fe(CO)_4$ fragment; both complexes could be fully characterized by X-ray diffraction methods and their structures are shown in Fig. 4.13(a) and (b), respectively [60].

Similarly, Angelici has demonstrated that thiophenes which have been previously C-S activated on a single metal center, as for instance in the ring-opened derivative $Cp*Ir(\eta^2-C,S-Th)$ (Th = 2,5-Me$_2$T), can further react with e.g. iron carbonyls through cleavage of both C-S bonds as shown in Eq. 4.17 to produce again a compound containing the separated sulfur atom and organic fragment, both bonded to the metal framework [6, 8].

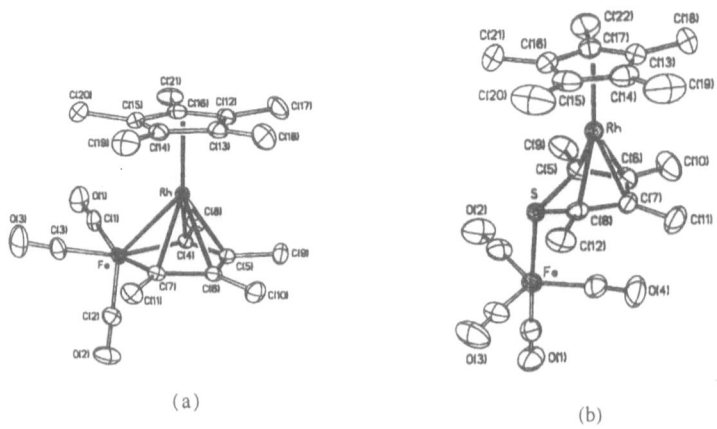

(4.17)

These studies indicate once more the importance of multimetallic cooperative effects for desulfurization to be accomplished. In another clear case of bimetallic cooperative C-S bond activation, Bianchini and coworkers have provided very pertinent information involving a complex containing a Group 8/Group 6 metal couple (Rh/W). The reaction of BT with [(triphos)RhH] leads to ring-opening followed by hydride transfer as shown in Eq. 4.18:

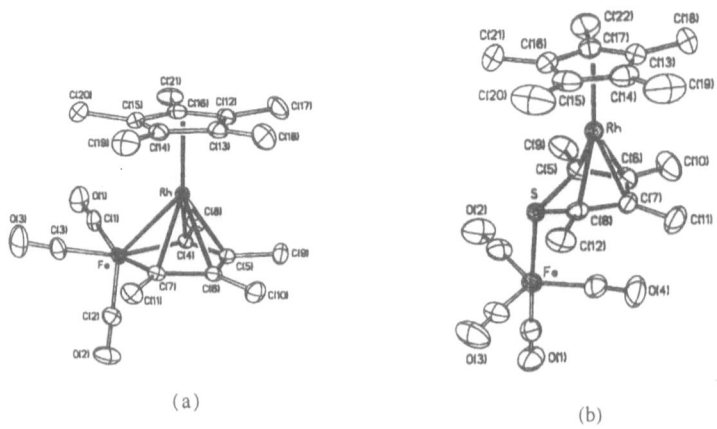

(a)   (b)

Fig. 4.13. The structure of (a) $Cp*Rh\{(\eta^5-C_4Me_4Fe(CO)_3\}$ and (b) $Cp*Rh\{(\eta^4:\eta^1-Me_4T)Fe(CO)_4\}$ (Reproduced with permission from Ref. 60).

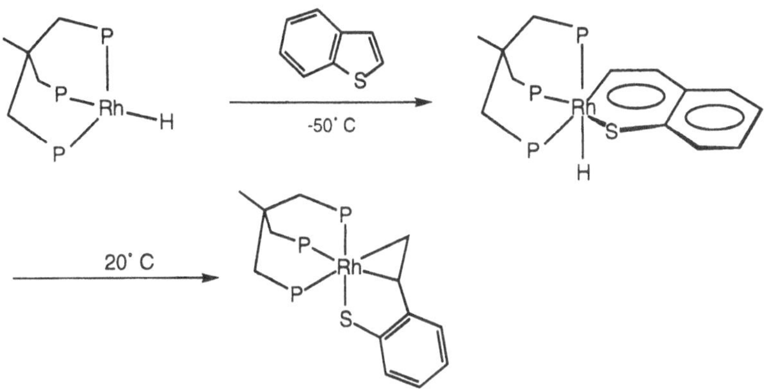

(4.18)

The intermediate thus produced is stable enough to be isolated and stored for long periods of time, but at the same time it is capable of *catalytically* transforming BT into ethylbenzenethiol under 30 atm of H_2 (see Section 4.3 below). More importantly here, the sulfur atom of the ring-opened BT, which remains coordinated to Rh, is particularly prone to add to an incoming electron deficient $W(CO)_5$ fragment yielding the stable new heterobimetallic sulfur-bridged species shown in Eq. 4.19.

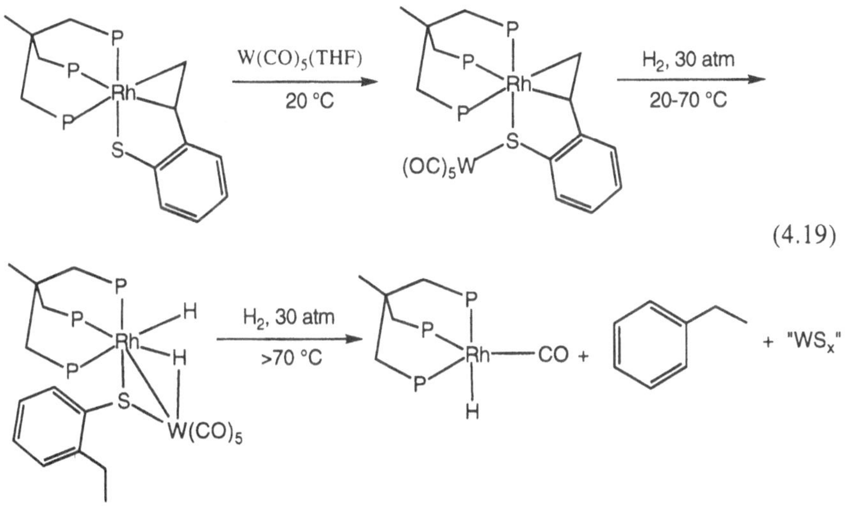

(4.19)

Upon thermolysis of this heterobimetallic species under hydrogen (30 atm), HDS of BT was accomplished yielding free *ethylbenzene*, plus [(triphos)RhH(CO)] together with an insoluble "W-S material". This is a particularly interesting example in that it shows that the bimetallic Rh-W couple -obviously related to the Co-Mo or Co-W heterogeneous catalysts- introduces the possibility of the sulfur atom bridging the two metal atoms and thereby switches the reactivity from hydrogenolysis -on Rh alone- to hydrodesulfurization [61]. It is tempting to speculate whether this well-characterized molecular case indicates that the role of "promoter" HDS metals in solid catalysts (late transition metals, Co, Ni) is actually the initial C-S activation of the thiophenes through ring opening reactions whereas the "catalytic" metal (Group 6, Mo, W) would be responsible for the transfer of hydrogen to the activated substrate in order to complete the HDS scheme. Also it seems that whenever the S-atom of the intermediate thiolate serves as a bridge between two metal centers, the remaining C-S bond of the thiolate ligand is weakened to an important extent and thus desulfurization can be achieved. This is a very likely pathway to be envisaged on metal sulfide surfaces where sulfides or thiolates bonded to two or more metal atoms are plentiful. However, this type of interaction is obviously an unavailable situation in mononuclear systems, and this could explain why ring opening is frequent but complete desulfurization is rare on such mononuclear derivatives.

Another type of activation of thiophenes by two metal centers worth mentioning in this section is the original approach recently disclosed by Sweigart and coworkers [62-66] in which π-coordination of a $Mn(CO)_3^+$ fragment to thiophenes facilitates the subsequent C-S bond scission by a second metal center under reducing conditions (e.g. by use of cobaltocene), as shown in Eq. 4.20; this reaction proceeds even for the sterically hindered 2,5-Me_2T which is unusual. Hydrogenation of the ring-opened intermediate led to the formation of dienethiolate ligands bridging the two Mn centers in a way in which each Mn atom is linked to one C=C bond and to the sulfur atom.

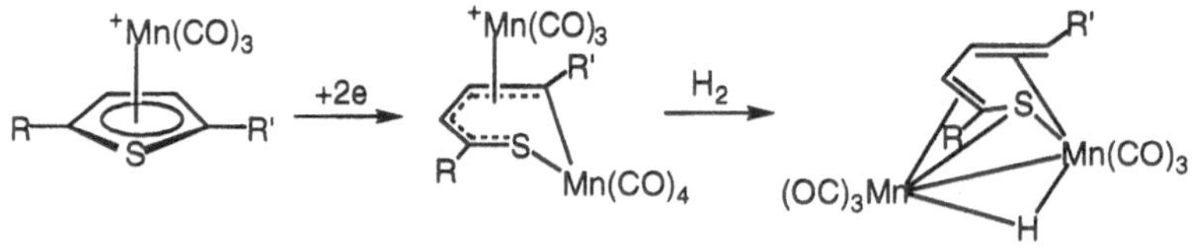

$$(4.20)$$

Also interestingly, the analogous reaction with BT starts with π-coordination of $Mn(CO)_3^+$ to the benzene ring, which induces ring opening of the thiophenic moiety by a second metal-containing unit, e.g. a second Mn-carbonyl fragment as exemplified for BT in Eq. 4.21. It is important to note that in BT ring opening reactions the metal most frequently inserts into the vinylic C-S bond, presumably due to selective weakening of that bond via previous η^1S-coordination; on the other hand, η^6-coordination to the benzene ring selectively activates the C(aryl)-S bond as the kinetically favored pathway, although isomerization to the thermodynamically stable C(vinyl)-S insertion products subsequently takes place at room temperature.

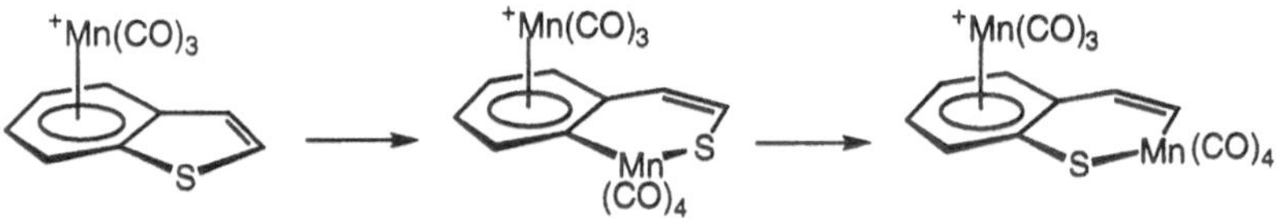

$$(4.21)$$

A variety of metal-containing fragments can produce this type of activation through η^6-bonding, and the second metal inserting into the C-S bond needs not be identical to the one π-bonding the arene ring. For instance, the weakly nucleophilic fragment $Pt(PPh_3)_2$ -which does not react with free BT- readily inserts into the C(vinyl)-S bond of $(\eta^6$-BT)ML_n complexes, the ease of insertion following the order $ML_n = Ru(C_6Me_6)^{2+}$, $Mn(CO)_3^+ > FeCp^+$, $RuCp^+ >> Cr(CO)_3$. Sweigart's group has very recently applied this chemistry to the difficult issue of "deep HDS" modeling. They were able to open the heterocyclic rings of a series of alkylated benzothiophenes and dibenzothiophenes by use of $Mn(CO)_3^+$ as the activating fragment on the carbocyclic ring, and $Pt(PPh_3)_2(C_2H_4)$ as the nucleophile [66]. With BT complexes bearing no substituents at the 2- or 3-position, coordination of Pt to the C=C bond in the heterocyclic ring takes place prior to insertion into the C(vinyl)-S bond. However, when an alkyl substituent is present at the olefinic bond, formation of the η^2(C=C) intermediate is blocked, the reaction rate slows, and insertion into the C(aryl)-S bond becomes possible or even preferred. In such cases, ring-opening is possibly preceded by a η^1S-bonded intermediate. Alkylated dibenzothiophenes, including 4,6-Me$_2$DBT, also underwent rapid ring opening by Pt through the "remote activation" by Mn, at the C-S bond closer to the Mn-coordinated ring, and η^1S-bonding before C-S bond breaking was also suggested. These results are very important in that they imply that η^6-coordination of benzo- and dibenzo-thiophenes on metal sulfide catalysts may be an important activation pathway for HDS, particularly of the highly refractory alkylated substrates, in agreement with some recent mechanistic proposals discussed in Chapter 1.

Using a similar reaction, Angelici and coworkers recently found that $Co_4(CO)_{12}$ reacts with DBT to yield the η^6-bonded derivative $Co_4(CO)_{12}(\eta^6$-DBT), which undergoes a peculiar further reaction with $Cr(CO)_3(NCMe)_3$ to produce $Co_4(CO)_9(\eta^6$-benzene) [67]; although the remaining products could not be identified, the formation of coordinated benzene implies that DBT underwent C-C and C-S bond breaking reactions. In the case of the reaction of BT with the cluster $Co_4(CO)_{12}$, the same degradation product $Co_4(CO)_9(\eta^6$-benzene) was directly obtained without intervention of a second molecule. No mechanistic information on these curious reactions involving Co - one of the most common HDS metals - is available at present.

4.2.2.3. Curtis' Co-Mo-S clusters.

Curtis has provided some homogeneous work that represents without doubt the closest model available for real Co-Mo-S catalysts. Heterobimetallic clusters, which contain the right combination of metals (Co-Mo) linked through sulfur bridges, were used, thereby beautifully mimicking a Co-Mo-S site [68, 69]. Even though this complex could be considered as just another example of a heterobimetallic cluster, its unique relation to real HDS catalysts seemed to justify its description in some detail in a separate section of its own.

The starting material is the "butterfly" shaped tetranuclear cluster $[Cp'_2Mo_2Co_2S_3(CO)_4]$ (Cp' = MeCp'), which reacts with thiophene in solution under hydrogen (150 °C, 15 atm) to produce the free C_4 hydrocarbons corresponding to HDS (together with small amounts of cracked products) and a new cubane cluster $[Cp'_2Mo_2Co_2S_4(CO)_2]$ in which the extruded sulfur has been incorporated (Eq. 4.22).

(4.22)

This cubane cluster can be converted back to the original one by reaction with CO (30 atm) to form COS, thus completing a cyclic reaction. Although this pair of reactions indicates the possibility of constructing a catalytic cycle for the HDS of thiophene, the presence of free CO unfortunately inhibits the forward desulfurization reaction and therefore the process cannot be rendered catalytic [68]. The exact mechanism of thiophene HDS by use of these clusters in not clearly understood, mainly because the reaction is fast and no intermediates have been isolated or identified. Nevertheless, Curtis has reported extensive related mechanistic studies on the desulfurization of a variety of thiols and sulfides using the same metal cluster $Cp'_2Mo_2Co_2S_3(CO)_4$ and as a result of that, the main reaction pathways for desulfurization of these simpler substrates are now well understood [69].

Table 4.1 shows a variety of compounds that have been desulfurized by this cluster. The rates of reaction of alkanethiols are similar to those found for arylthiols, despite the large differences in C-S bond energies between these two types of compounds. This is suggestive that C-S bond breaking is *not* the rate determining step of the overall process. The products observed are only those of direct desulfurization, not involving any skeletal rearrangements; this implies that the sulfur is not removed as the SH anion, as otherwise cations would be produced and the final products would be expected to be alkenes or rearranged alkanes. The kinetics of the desulfurization reactions follow a second order rate law, *e.g.* $r = k[ArSH][cat]$, and ΔS^* -5 eu to −22 eu and ΔH^* +27 kcal/mol to +19 kcal/mol. These results can be interpreted in terms of the rate determining step being actually the attachment of the thiol to the cluster, while the subsequent desulfurization was fast, not allowing the detection of intermediate products. Interestingly, since the ΔH^* of the desulfurization reaction has an upper limit of 27 kcal/mol and the C-S bond dissociation energy of aromatic thiols is *ca.* 80 kcal/mol, coordination to the cluster causes a reduction of this energy by a factor of 3-4, which is remarkable.

TABLE 4.1. Desulfurization by $Cp_2Mo_2Co_2S_3(CO)_4$.

Substrate	Product
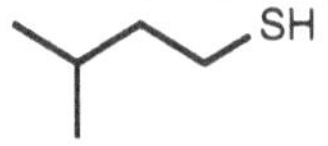 SH	
^tBuSH	tBuH
ArSH	ArH
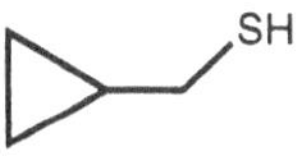 SH	
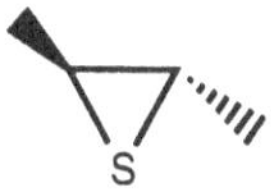 S	
 S	

The overall mechanism proposed by Curtis and coworkers for the Co-Mo-S cluster-catalyzed desulfurization of thiols is represented in Fig. 4.14. The reaction starts with the coordination of the thiol to the cluster to produce **A**, containing an intact terminal thiol bound to Co. This is believed to be the rate-determining step of the overall process, and it requires that a Co-S bond be broken, in order to create a vacant site to accommodate the incoming ligand. The coordinated thiol subsequently migrates so as to bridge the Co-Mo bond, thereby forming a μ^2-sulfide-μ^2-thiol intermediate **B**. Such a bridging coordination mode increases the acidity of the thiol proton, which may then be easily transferred to the μ^2-S ligand yielding the corresponding μ^2-hydrosulfido-μ^2-thiolate species **C**. The μ^2-thiolate in **C** subsequently becomes μ^3-bonded (to 2 Mo + 1 Co) as in **D**; this situation dramatically decreases the C-S bond dissociation energy, and the desulfurization can then be completed by a *homolytic* cleavage of the C-S bond, to generate the corresponding alkyl or aryl radicals. The latter pick up the S-H proton to yield the final organic products, while the cluster is oxidized by the sulfide radical with loss of two CO ligands.

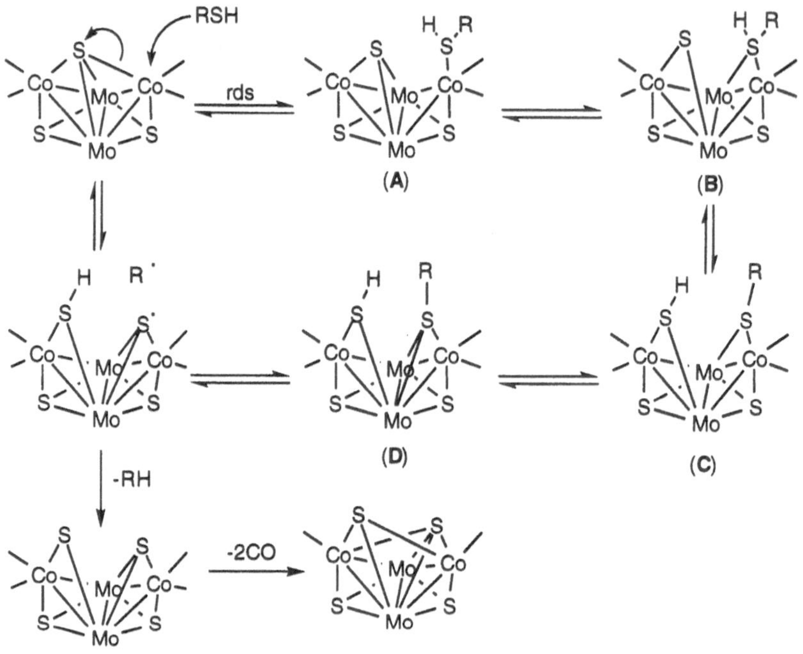

Fig. 4.14. The mechanism of HDS of thiols by Cp'$_2$Mo$_2$Co$_2$S$_3$(CO)$_4$ (other ligands ommited for clarity).

Harris has reported some theoretical work on this system showing that the reaction in which a sulfur atom adds to the cluster to form the cubane (Eq. 4.22) is accompanied by the transfer of two electrons from the Co centers to the added sulfur. It is not clear why the HDS proceeds readily in the butterfly cluster while in the other mononuclear or dinuclear complexes described above the reactions generally proceed through metal insertion into the thiophene ring to yield stable metallacyclic intermediates, which are most commonly incapable of breaking the second C-S bond. Still, the greatest value of this modeling chemistry is that if the mechanism deduced by Curtis is combined with the reaction networks which have been worked out for the ring-opening and hydrogenolysis of thiophenes by electron-rich late transition metal complexes, a very complete mechanistic model for heterogeneous HDS emerges. Also it is important to note that in this model, there are several low energy steps related to breaking and forming C-S bonds, and also that the proton is highly mobile. These features are also typical of HDS catalysts, which in essence are very dynamic metal sulfide surfaces acting as reservoirs of protons and electrons.

A further point of great interest that emerges from the work of Curtis and from the mechanism depicted in Fig. 4.14 is the concept of "latent vacancies and sulfur mobility". This can be related to some recent important considerations on the existence of real anionic vacancies on Co-Mo-S surfaces, a frequently encountered key feature of HDS mechanisms. The Co atom, which is known to be the primary site of attack of thiols in this case, is electronically saturated, but the empty site required for coordination of the incoming thiol is formed *only when needed* through an energetically little demanding rearrangement of a quadruply bridging sulfide (bonded to 2 Co + 2 Mo atoms) into a triply bridging mode (bonding 1 Co + 2 Mo atoms). This is consistent with the Co-S bond (promoter metal) being weaker than the Mo-S bond. At the end of the HDS reaction the cobalt loses the "extra ligand" and the quadruple bridge can be easily reconstructed. Within this framework, there is no need to invoke physical "vacancies", but rather a resting state of the cluster -or surface site on a solid catalyst- possessing *latent* vacancies. As mentioned in Section 1.2.4.2 this is a situation of lower energy than that corresponding to an "empty site" and intuitively more in line with nature's dislike for voids.

A similar situation arises in recently reported DFT calculations by Byskov *et al.* [70, 71] that show that if vacancies are introduced in a large cluster model of a MoS_2 surface by removing sulfur atoms, and relaxation is allowed taking into account several layers of the solid, a spontaneous surface reconstruction process takes place through migration of sulfide ions from the bulk to the surface, so as to "fill" the empty sites. Thus, although the distinction may be rather subtle, the idea of "latent" or "potential" vacancies in a "resting state" rather than actual physically empty sites around electron rich areas of space seems like an adequate and useful evolution of the classical concept of "anionic vacancies" in HDS catalysts.

4.3. Homogeneous Catalytic Hydrogenolysis and Hydrodesulfurization of Thiophenes.

Until fairly recent times, most of the reported homogeneous HDS-related reactions were limited to stoichiometric model transformations in which thiophenes and related molecules could be activated by coordination to the metal centers so as to undergo C-S bond scission and eventually hydrogenolysis to the corresponding free thiol, or in some cases even complete desulfurization to yield hydrocarbons.   This has represented a remarkable contribution to the understanding of many of the elementary steps involved in complex heterogeneous HDS schemes, but it had obvious limitations, and it offered little promise in terms of developing catalytic systems of any mechanistic or practical importance.

This situation has begun to change with the extension of some of the modeling chemistry developed by the groups of Bianchini and Sánchez-Delgado toward the *catalytic hydrogenolysis and hydrodesulfurization* of thiophenic substrates to produce their corresponding thiols or hydrocarbons in solution [5, 7, 9, 10, 14]. It is also interesting to note that, as described in previous Sections of this Chapter, numerous metal complexes are capable of cleaving C-S bonds of thiophenes and of adding hydrogen to the resulting intermediates. However, only very few of them have been found to do this in a catalytic manner and they are all characterized by a precise molecular architecture involving the rigid tripodal ligand triphos as well as other related P-donors. Also, the examples available to date all involve late transition metals (Ru, Rh, Ir) which can be related with "promoter" elements; this is also in correspondence with the trends showing that maximum HDS activities for pure metal sulfides lie around Groups 8 and 9.

These unprecedented homogeneous transformations have been achieved for a variety of substrates, viz. T [72], BT [72-74], DBT [29, 72], and dinaphthothiophene (DNT) [75], by use of the electron-rich 16-electron neutral or anionic fragments $[(triphos)MH]^n$ (M = Rh, Ir, n = 0; M = Ru, n = -1). The reactions have been typically performed at about 160 °C under 30 atm H_2, conditions that are well tolerated by the very stable triphos derivatives; the presence of strong bases such as NaOH or KO^{t}Bu dramatically increases the overall rate of the catalytic hydrogenolysis.

As an example, Table 4.2 contains the results of the homogeneous hydrogenolysis of benzothiophene catalyzed by the complex $[(triphos)Rh(\eta^3-S)(C_6H_4)CH=CH_2]$, a stable precursor resulting from the interaction of the 16-electron fragment [(triphos)RhH] with BT. From these data it is clear that the reaction is quite selective for the hydrogenolysis of BT to 2-ethylbenzenethiol, although a moderate activity for BT hydrogenation to 2,3-dihydrobenzothiophene was also observed. The very small amounts of ethylbenzene present at the end of the catalytic runs are probably due to hydrodesulfurization induced by traces of metallic rhodium formed by decomposition of the catalyst, which was found to be extensive above 200° C.

The homogeneity of the hydrogenolysis reaction –at temperatures up to 180° C- was demonstrated by the fact that metallic mercury did not affect the reaction rate. The hydrogen pressure did not seem to have a marked effect on the catalysis, within the range employed. Also, using THF as the solvent instead of acetone did not result in any important effect.

TABLE 4.2. Homogeneous catalytic hydrogenolysis of benzothiophene by use of $[(triphos)Rh(\eta^3-S(C_6H_4)CH=CH_2]^a$

| T(°C) | P(H$_2$) atm | T(h) | ETB | Product distribution (%)[b] | | | Rate (EtSH)[c] |
				ETSH	BT	DHBT	
160	30	2	0.2	25.5	73.2	1.1	12.7
160	30	4	0.2	39.8	57.4	2.6	9.9
160	30	8	0.3	45.1	51.8	2.8	5.6
160	30	12	0.4	51.0	44.6	4.0	4.2
160	30	16	0.4	57.4	37.6	4.6	3.6
160	30	16	-	57.6	37.9	4.5	3.6[d]
160	30	16	0.2	52.8	41.3	5.7	3.3[e]
160	15	16	0.3	55.3	40.2	4.2	3.5
160	60	16	0.4	60.2	34.3	5.1	3.8
120	30	4	-	2.0	97.6	0.4	0.5[e]
100	30	16	-	1.4	98.1	0.5	<0.1
180	30	16	0.9	64.2	29.0	5.9	4.0
220	30	16	3.5	43.3	45.8	6.9	2.7[e]
220	30	16	-	42.6	51.1	6.0	2.7[e]
160	30	16	0.3	61.2	34.0	4.5	1.9

[a]In acetone, [BT]:[Rh] = 100; [b]ETB (ethylbenzene), ETSH (2-ethylbenzenethiol), BT (benzothiophene), DHBT (2,3-dihydrobenzothiophene; [c]in mol ETSH per mol Rh per h; [d]in the presence of Hg; [e]in THF.

Kinetic studies, high pressure *in situ* multinuclear NMR measurements under actual working conditions, and the isolation and independent syntheses of key intermediates, together with the extensive knowledge previously gained from the modeling studies involving analogous (triphos)Ir complexes, have allowed the elucidation of the main reaction pathways conforming the catalytic cycles . The most important features of these experimentally supported mechanisms are exemplified in Fig. 4.15 for the hydrogenolysis of benzothiophene by use of [(triphos)RhH], a reaction that is accompanied by a low degree of competitive hydrogenation of the thiophenic ring of BT to produce DHBT.

The common intermediate for both reactions is the 18e S-bonded complex [(triphos)Rh(H)(η^1S-BT)] (**A**). In the hydrogenolysis cycle (Fig. 4.15, upper part), C-S insertion takes place to yield the corresponding thiametallacycle (**B**) which evolves through hydride migration to the carbon atom σ-bonded to Rh to yield the very stable intermediate (**C**); this compound is a close analogue of the Ir model complexes described in section 4.2.1.2, Figs. 4.3 and 4.4; moreover, it is stable enough to be stored for prolonged periods of time and thus it constitutes an excellent practical catalyst precursor. Reaction of **C** with hydrogen leads to a new complex (**D**) containing the corresponding thiolate attached to the Rh atom; this intermediate further reacts with a second hydrogen molecule to produce the corresponding dihydride (**E**). The thiol is then reductively eliminated to re-start the cycle in what is thought to be the rate-limiting step, regenerating the 16-electron d^8 Rh(I) monohydride [(triphos)RhH]. In the absence of hydrogen, complex **D** dimerizes to **F**, as demonstrated by *in situ* NMR studies. The minor DHBT product is formed through the sequence depicted in the hydrogenation cycle (Fig. 4.15, bottom part). This involves initial isomerization of the η^1S-intermediate **A** into the olefin bonded η^2(C=C) species (**G**), which undergoes C=C bond saturation by a mechanism similar to the ones described in Section 3.3.1.2. The promoting effect of strong bases on the rate of hydrogenolysis has been attributed to the rapid hydrolysis of intermediate **E** yielding the free thiolate salt plus water or an alcohol. This would provide a faster regeneration of the active species and thereby a higher overall rate of catalytic thiol production. It is interesting to remember at this point that the S-atom of intermediate **C** readily binds to the unsaturated $W(CO)_5$ fragment, and the resulting heterobimetallic intermediate reacts subsequently with H_2 to produce the complete desulfurization albeit not in a catalytic way.

In an similar manner, the anionic 16-electron d^8 Ru(0) fragment [(triphos)RuH] that is isoelectronic with the Rh complex described above, is also an active and selective catalyst for the conversion of BT to 2-ethylthiophenol above 60° C and 30 atm H_2 [77]. The catalytic cycle for this ruthenium system was shown experimentally to be in essence analogous to that discussed for Rh. Very interestingly, as noted in Section 3.3.1.2, the corresponding cationic 14-electron d^6 Ru(II) species [(triphos)RuH]$^+$, which has basically the same geometry but two electrons less than   [(triphos)RuH]$^-$, very efficiently saturates the thiophene ring of BT to form DHBT catalytically, but it is unable to break the C-S bonds [78]; this shows a remarkable switch in selectivity between hydrogenation and hydrogenolysis processes brought about simply by a 2e difference in the two catalysts, without any major geometrical change taking place, as illustrated in Fig. 4.16.

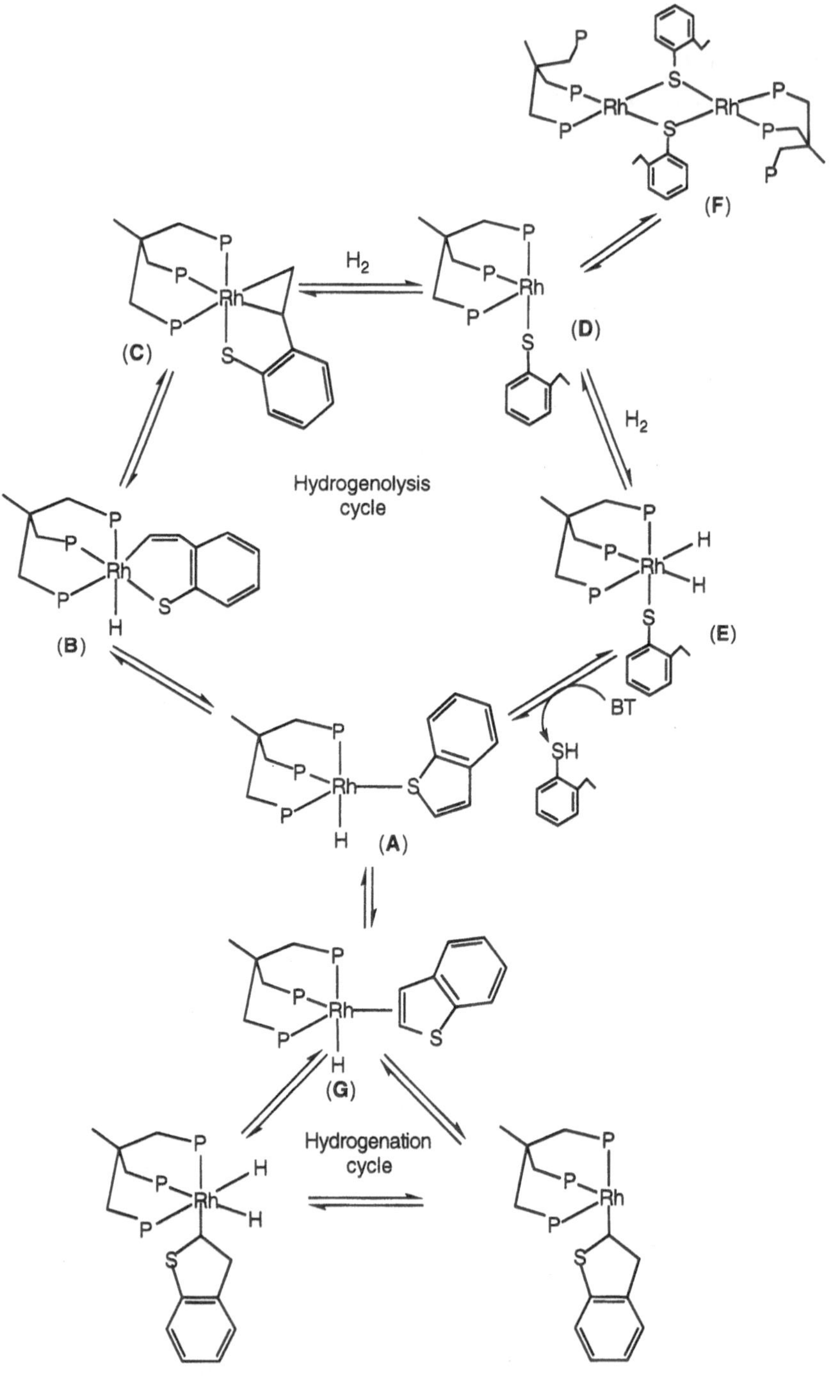

Fig. 4.15. Hydrogenolysis and hydrogenation of benzothiophene by [(triphos)RhH] (Adapted from ref. 73).

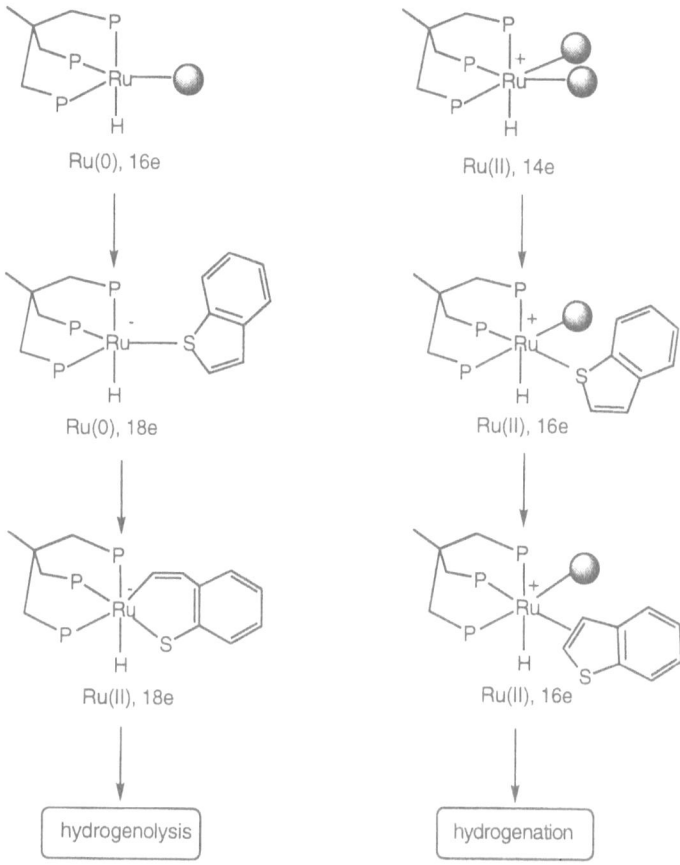

Fig. 4.16. Hydrogenation vs. hydrogenolysis of benzothiophene by [(triphos)RuH]^x.

This is consistent with the idea that electron-rich 16e fragments are particularly adapted for a metal $d\pi$ to ligand π^*(C-S) electron transfer and thereby a C-S bond scission. Whitehurst et al. have argued in what they called "the two site dilemma"(see Chapter 1, p. 10) [79] that in highly dispersed Co-Mo-S catalysts only one geometry is possible for the active sites, while more than one type of reaction is known to take place on such dispersed phases; the [(triphos)Ru]^x systems described above demonstrate that the chemical selectivity of a metal site can be fine-tuned by a simple 2e variation while maintaining essentially the same geometrical characteristics around the active metal. A Co-Mo-S surface is not only rich in electrons but also it allows them to migrate easily, mainly through the sulfur bridges which are easily formed, broken and reconstructed (as evidenced also in Curtis' elegant studies on Cp-Co-Mo-S carbonyl clusters, Section 4.2.2.3). Other low energy 2e redox pathways are available in metal sulfide surfaces, such as the interconversion of surface S_2^{2-}, S^{2-}, SH$^-$, etc. under HDS conditions; therefore, these organometallic systems may be providing important clues on how a single surface geometric configuration containing few metal atoms and several forms of sulfur ligands can give rise to several types of reactivity. The hydrogenolysis of T or DBT with the Rh-based catalyst is considerably less efficient than for BT and also produces considerable amounts of desulfurization products; a similar mechanism is thought to operate in these cases but the experimental data are less clear cut than for BT [72].

Besides these homogeneous catalysts, Bianchini and his coworkers have developed some very efficient liquid (aqueous) biphasic systems for the hydrogenolysis of benzothiophene. This involved the synthesis of a water-soluble –sulfonated-analogue of triphos, Na-sulphos (see Fig. 4.17), and its use in combination with Ru and Rh to generate active catalysts that are easily recovered and recycled through simple decantation [74]. The metal-sulphos derivatives have also been grafted onto silica surfaces through strong hydrogen bonds [76], or supported on organic polymers [82], generating the heterogenized version of the corresponding homogeneous systems. In both cases this resulted in very stable catalysts, capable of efficiently promoting the hydrogenolysis of benzothiophene to 2-ethylbenzethiol. The polymer-supported catalyst was also capable to promote, to a lesser extent, the complete hydrodesulfurization of benzothiophene to ethylbenzene. These novel hybrid catalysts could offer an interesting potential for practical applications.

Also of interest, Bianchini and Sánchez-Delgado have described the only well understood example so far available of a complete homogeneous *catalytic* hydrodesulfurization reaction of *dibenzothiophene* leading to biphenyl plus H_2S together with some hydrogenolysis products that were formed concurrently. The complex [(triphos)IrH], generated by thermolysis of [(triphos)IrH$_2$(Et)] in THF solution [29] only promotes C-H insertion reactions at temperatures up to 100 °C; at about 120° C metal insertion into a C-S bond is observed together with C-H activation, and at 170 °C the selectivity is switched entirely towards C-S bond activation. Under those conditions DBT is slowly degraded to the hydrogenolysis and hydrodesulfurization products.

The mechanism of this process was nicely determined from the isolation of most of the intermediates involved and through the independent study of the individual reactions conforming the catalytic cycle depicted in Fig. 4.18.

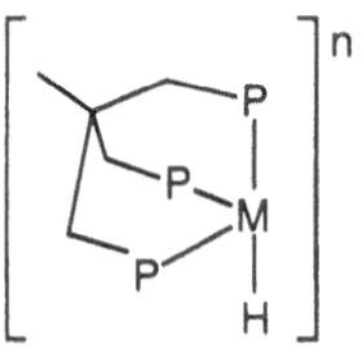

M = Rh, Ir; n = -1
M = Ru; N = -1

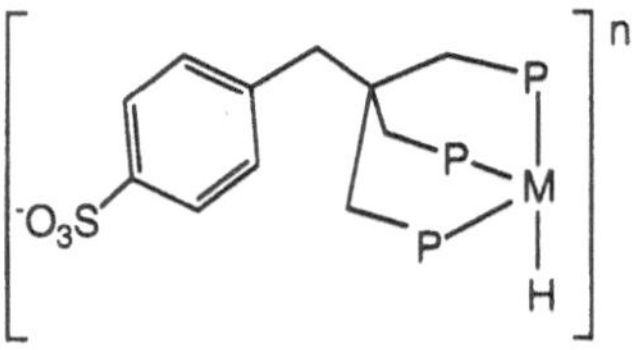

M = Rh; n = -1

Fig. 4.17. Some catalyst precursors used for the homogeneous
and liquid-biphasic hydrogenolysis of thiophenes

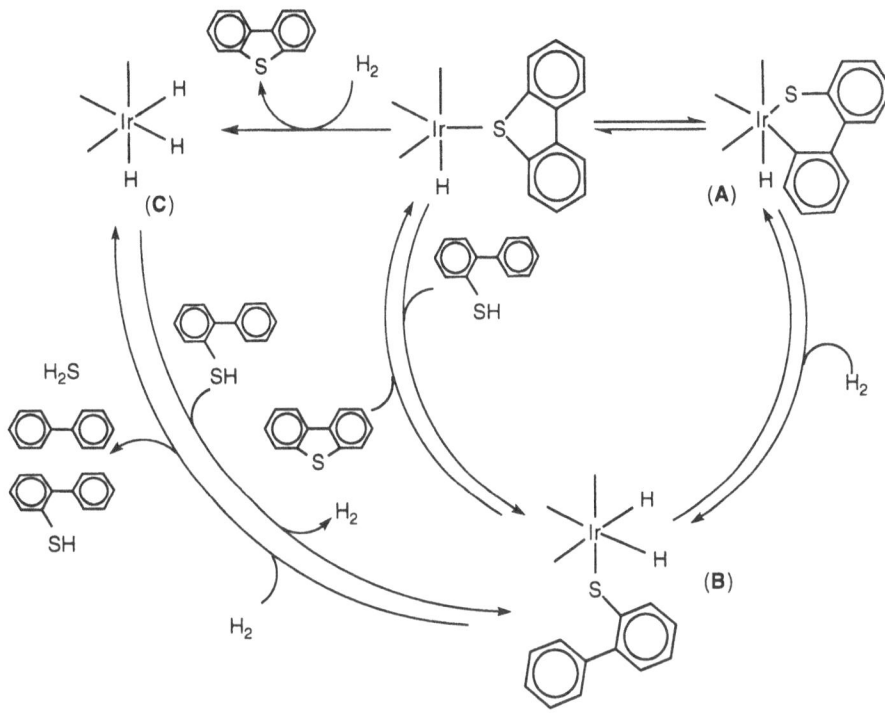

Fig. 4.18. Homogeneous HDS of DBT by [(triphos)Ir(H)₂(Et)](Adapted from ref. 29).

Hydrogenolysis of DBT takes place through the reaction of the C-S insertion product (**A**) with hydrogen to yield the stable (2-phenylthiolate)dihydride intermediate (**B**); reductive elimination of the free thiol from **B** may be promoted by either free DBT to regenerate the initial [(triphos)Ir(H)(η^1S-DBT)] or by H_2 which results also in the production of the known trihydride (**C**). This trihydride in turn slowly reacts with the free thiol produced during the catalysis to yield the dihydride **B** plus hydrogen. It was also demonstrated that the reaction of the 2-phenylthiolate dihydride intermediate **B** with H_2 results in the *evolution of H_2S plus biphenyl* with the concurrent formation of **C**. The details of the final C-S bond cleavage in the coordinated thiolate remain unclear, but some indirect evidence was obtained for the involvement of a [(triphos)Ir(H)₂(SH)] complex in the release of H_2S, analogous to the well characterized complex [(Et₃P)₂Pt(H)(SH)] reported by Garcia and Maitlis discussed in Section 4.2.1.2 (see Fig. 4.7). Although this triphos-Ir catalyst is far too slow for any practical application - possibly due to the sluggish "pendular" nature of the productive cycles - this piece of work remains as the only well understood example of the mechanism of a homogeneous catalytic HDS reaction of DBT.

Very recently, the groups of García and Maitlis reported the homogeneous HDS of a series of substituted thiophenes (2-MeOT, 3-MeOT, 3-AcT) by use of thiaplatinacycles derived from the reactions of [Pt(PEt$_3$)$_3$] with the corresponding thiophene as the catalyst precursors [83]; up to 81 turnovers were achieved at 100° C and ca. 20 bar of H$_2$ in THF. This is a unique example of HDS of thiophenes by metal complexes in solution; no mention is made of the analogous reaction with unsubstituted thiophene or with benzothiophenes or dibenzothiophenes. Presumably, mechanistic studies will be reported subsequently.

4.4. Further remarks on the relations between homogeneous and heterogeneous hydrogenolysis and HDS of thiophenes.

It is interesting at this point to try to establish some conceptual connections between the most striking features of homogeneous C-S bond breaking reactions and heterogeneous HDS catalysis. It has been shown throughout this Chapter that a rather wide variety of metal complexes are capable of stoichiometrically activating the C-S bonds of thiophenic substrates, including the highly refractory methylated dibenzothiophenes. Although this chemistry has been so far dominated by metals which are associated with the highest HDS activities in the form of pure sulfides (Ru, Rh, Ir, Ni), examples involving other common components of commercial heterogeneous catalysts (Mo, W, Co) have begun to appear. This obviously improves the quality of the modeling, and further examples of such chemistry will most probably be found and they will greatly contribute to complement the models already available.

There is an overwhelming evidence indicating that electron rich metal fragments (particularly 16-electron, 1 CUS species) are particularly prone to promote thiophene ring-opening; consequently the η^{1}S-bonding mode of thiophenes seems to be the most frequent form of activation toward C-S bond scission, and theoretical arguments support this idea. Only one example of a 14-electron fragment incorporating benzothiophene in a η^3(S,C=C)-bonding mode has been unambiguously identified, and it has been clearly associated with subsequent C-S bond breaking. This shows that the presence of two empty sites can aid desulfurization, but it is not required. Also, all the examples disclosed to date of homogeneous and biphasic catalytic hydrogenolysis and desulfurization contain the triphos ligand or a sulfonated derivative of it, that provide the right (rigid) geometry and favorable energetics for C-S bond scission plus hydrogen activation. They are all selective for hydrogenolysis vs. hydrogenation of the sulfur-containing ring, although in some cases minor amounts of hydrogenation products are observed.

The acummulated organometallic modeling data thus suggest that the most likely candidates to be the active sites in real catalysts are electron rich surface sites of strained geometry with only one CUS. In contrast, electron poorer centers (14-electron or less) are probably better suited for hydrogenation reactions of C=C bonds of olefins and S-heterocycles, particularly benzothiophenes. Olefinic substrates require a single coordinative unsaturation in order for hydrogenation to take place, and they are generally much better donor ligands than S-bonded thiophenes; therefore, their better binding capacity can account for the inhibition effects that olefins are known to exert on heterogeneous HDS reactions [80]. This can also be related to the fact that BT, in which both S-bonding and C=C-bonding can occur, generally prefers to be hydrogenated through a η^2(C=C) intermediate, in the presence of unstrained electron-poor Rh(III) and Ir(III) complexes containing monodentate phosphines. A major drawback in organometallic modeling concerns the lack of information concerning hydrogenation of aromatics that seems to require several coordinative unsaturations, and therefore electron deficient sites. In the absence of good models, it is difficult to try to establish possible characteristics linked to sites that are specifically active for arene hydrogenation, or to give some insight into whether such sites can be the same as the ones required for desulfurization. What does become clear from the examples presented above is that subtle electronic changes without major geometrical rearrangements can lead to drastic differences in reactivity, and it is interesting to extrapolate this concept to metal sulfide surfaces, where redox processes are facile and electron mobility is high.

In contrast to heterogeneous systems, breaking the C-S bonds of thiols or thiolates appears to be more difficult than breaking the first C-S bond of a thiophenic molecule in metal complexes –with the notorious exception of Curtis' clusters- and a good explanation for this is not yet available. The majority of examples known of complete desulfurization of thiophenes involves dinuclear species or clusters, and it appears that the presence of bridging sulfide ligands between two metal centers is a key feature required for sulfur extrusion. Further exciting results in the area of multimetallic and specially heterometallic complexes are to be expected, and this would contribute greatly to further understand HDS in solution and in the solid state.

References

1.   T. L. Cottrell: *The Strengths of Chemical Bonds*, 2nd. Ed., Butterworth, London (1958).
2.   J. A. Davies, P. L. Watson, A. Greenberg and J. F. Leibman, eds.: *Selective Hydrocarbon Oxidation and Functionalization*, VCH, New York (1990).
3.   R. H. Crabtree: *Chem. Rev.* **95**, 987 (1995).
4.   A. E. Shilov and G. B. Shul'pin: *Chem. Rev.* **97**, 2879 (1997).
5.   R. A. Sánchez-Delgado: *J. Mol. Catal.* **86**, 287 (1994).
6.   R. J. Angelici in R. B. King, ed.: *Encyclopedia of Inorganic Chemistry*, Wiley, New York (1994).
7.   C. Bianchini, A. Meli: *J. Chem. Soc. Dalton Trans.* 801 (1996).
8.   R. J. Angelici: *Polyhedron* **16**, 3073 (1997).
9.   C. Bianchini, A. Meli: *Acc. Chem. Res.* **31**, 109 (1998).
10. C. Bianchini, A. Meli: in T. Weber, R. Prins, and R. A. van Santen, eds., *Transition Metal Sulphides. Chemistry and Catalysis*, NATO ASI Series, Kluwer, Dordrecht, 1998, Vol. 60, pp. 129-154.
11. R. J Angelici, in T. Weber, R. Prins and R. A. van Santen, eds.: *Transition Metal Sulphides, Chemistry and Catalysis*, Vol. 3, John Wiley & Sons, New York, p. 1433 (1994).
12. S. Harris: *Polyhedron*, **16**, 3219 (1997).
13. M. S. Palmer, S. Rowe and S. Harris: *Organometallics* **17**, 3798 (1998).
14. R. A. Sánchez-Delgado, in I. T. Horváth, ed.: *Encyclopedia of Catalysis* John Wiley & Sons., New York (in press).
15. C. Blonski, A. W. Myers, M. Palmer, S. Harris and W. D. Jones: *Organometallics* **16**, 3819 (1997).
16. W. D. Jones and L. Dong: *J. Am. Chem. Soc.* **113**, 559 (1991).
17. L. Dong, S. B. Duckett, K. F. Ohman, W. D. Jones: *J. Am. Chem. Soc.* **114**, 151 (1992).
18. A. L. Sargent and E. P. Titus: *Organometallics* **17**, 65 (1998).
19. A. W. Myers, W. D. Jones and S. M. McClements: *J. Am. Chem. Soc.* **117**, 11704 (1995).
20. A. W. Myers and W. D. Jones: *Organometallics* **15**, 2905 (1996).
21. H. E. Selnau and J. S. Merola: *Organometallics* **12**, 1583 (1993).
22. M. Paneque, S. Taboada and E. Carmona: *Organometallics* **15**, 2678 (1996).
23. W. D. Jones and R. M. Chin: *Organometallics* **11**, 2698 (1992).
24   W. D. Jones and R. M. Chin: *J. Organomet. Chem.* **472**, 311 (1994).
25. D. G. Churchill, B. M. Bridgewater and G. Parkin: *J. Am. Chem. Soc.* **122**, 178 (2000).
26. W. D. Jones, R. M. Chin, T. W. Crane and D. M Baruch:*Organometallics* **13**, 4448 (1994).
27. C. Bianchini, A. Meli, M. Peruzzini, F. Vizza, P. Frediani, V. Herrera and R. A. Sánchez-Delgado: *J. Am. Chem. Soc.* **115**, 2731 (1993).

28.   C. Bianchini, A. Meli, M. Peruzzini, F. Vizza, S. Moneti, V. Herrera and R. A. Sánchez-Delgado, *J. Am. Chem. Soc.* **116**, 4371 (1994).
29.   C. Bianchini, V. Jimenez, A. Meli, F. Vizza, V. Herrera and R. A. Sánchez-Delgado, *Organometallics* **14**, 2342 (1995).
30.   C. Bianchini, A. Meli, S. Moneti and F. Vizza: *Organometallics* **17**, 2636 (1998).
31.   M. R. Blake, M. Eyre, R. B. Moyes, P. B. Wells: Proceedings of the 7th International Congress on Catalysis, T. Seiyama, K. Tanaba, Eds. Elsevier, Tokyo 1980, Vol. 7 Part A, p. 591.
32.   H. Diez y Riega, L. Rincón and R. Almeida: *J. Mol. Struct. (Theochem)* **493**, 259 (1999).
33.   L. Rincón, personal communication (2000).
34.   C. Bianchini, A. Meli, W. Oberhauser and F. Vizza,: *J. Chem. Soc. Chem. Commun.* 671 (1999).
35.   C. Bianchini, M. V. Jiménez, A. Meli, S. Moneti and F. Vizza,: *J. Organomet. Chem.* **504**, 27 (1995).
36.   C. Bianchini, J. A. Casares, D. Masi, A. Meli, W. Pohl and F. Vizza,: *J. Organomet. Chem.* **541**, 143 (1997).
37.   G. P. Rosini and W. D. Jones, *J. Am. Chem. Soc.* **114**, 10767 (1992).
38.   J. García and P. M. Maitlis: *J. Am. Chem. Soc.* **115**, 12200 (1993).
39.   J. Garcia, B. E. Mann, H. Adams, N. A. Bailey and P. M. Maitlis: *J. Am. Chem. Soc.* **117**, 2179 (1995).
40.   J. Garcia, A. Arevalo, V. Montiel, F. Del Rio, B. Quiroz, H. Adams and P. M. Maitlis: *Organometallics* **16**, 3216 (1997).
41.   J. Garcia, A. Arevalo, S. Capella, A. Chehata, M. Hernandez, V. Montiel, G. Picazo, F. Del Rio, R. A. Toscano, H. Adams and P. M. Maitlis: *Polyhedron* **16**, 3185 (1997).
42.   A. Iretskii, H. Adams, J. J. García, G. Picazo and P. M. Maitlis: *J. Chem. Soc. Chem. Commun.* 61 (1999).
43.   A. Iretskii, J. J. García, G. Picazo and P. M. Maitlis: *Catal. Letters* **51**, 129 (1998).
44.   A. Arévalo, S. Bernès, J. J. García, G. Picazo and P. M. Maitlis: *Organometallics* **18**, 1680 (1999).
45.   D. A. Vicic and W. D. Jones: *J. Am. Chem. Soc.* **119**, 10855 (1997).
46.   D. A. Vicic and W. D. Jones: *Organometallics* **17**, 3411 (1998).
47.   D. A. Vicic and W. D. Jones: *J. Am. Chem. Soc.* **121**, 4070 (1999).
48.   D. A. Vicic and W. D. Jones: *J. Am. Chem. Soc.* **121**, 7606 (1999).
49.   M. Brorson, J. D. King, K. Kiriakidou, F. Prestopino and E. Nordlander: in P. Braunstein, L. A. Oro and P. R. Raithby, eds., *Metal Clusters in Chemistry*, Vol. 2, Wiley-VCH, Weinheim, Ch. 2.6 (1999).
50.   H. D. Kaesz, R. B. King, T. A. Manuel, L. D. Nichols and F. G. A. Stone: *J. Am. Chem. Soc.* **82**, 4749 (1960).
51.   P. Hubener and E. Weiss: *J. Organomet. Chem.* **129**, 105 (1977). A. E. Ogilvy, M. Draganjac, T. B. Rauchfuss and S. R. Wilson: *Organometallics* **7**, 1171 (1988).
52.   A. J. Arce, P. Arrojo, A. J. Deeming and Y. De Sanctis: *J. Chem. Soc. Dalton Trans.* 2423 (1992). A. J. Arce, Y. De Sanctis, A. Karam and A. J. Deeming: *Angew. Chem. Int. Ed. Engl.* **33**, 1381 (1994).
53.   K. Matsubara, R. Okamura, M. Tanaka and H. Suzuki: *J. Am. Chem. Soc.* **120**, 1108 (1998).
54.   R. D. Adams, O. -S. Kwon and J. L. Perrin: *J. Organomet. Chem.* **584**, 223 (1999) and references therein.
55.   P. M. Boorman, X. Gao, J. F. Fait, and M. Parvez: *Inorg. Chem.* **30**, 3886 (1991).
56.   W. D. Jones and R. M. Chin: *J. Am. Chem. Soc.* **116**, 198 (1994).
57.   D. A. Vicic and W. D. Jones: *Organometallics* **18**, 134 (1999).
58.   D. A. Vicic and W. D. Jones: *Organometallics* **16**, 1912 (1997).
59.   M. A. Reynolds, I. A. Guzei and R. J. Angelici: *J. Chem. Soc. Chem. Commun.* 513 (2000).
60.   S. Luo, A. E. Ogilvy, T. B. Rauchfuss, A. Rheingold and S. R. Wilson: *Organometallics* **10**, 1002 (1991).
61.   C. Bianchini, M. V. Jiménez, A. Meli, S. Moneti, V. Patinec and F. Vizza: *Organometallics* **16**, 5696 (1997).
62.   C. A. Dullaghan, X. Zhang, D. Walther, G. B. Carpenter, D. A. Sweigart and Q. Meng: *Organometallics* **16**, 5604 (1997).
63.   C. A. Dullaghan, G. B. Carpenter, D. A. Sweigart, D. S. Choi, S. S. Lee and Y. K. Chung: *Organometallics* **16**, 5688 (1997).
64.   X. Zhang, C. A. Dullaghan, G. B. Carpenter, D. A. Sweigart and Q. Meng: *J. Chem. Soc. Chem. Commun.* 93 (1998).
65.   X. Zhang, C. A. Dullaghan, E. J. Watson, G. B. Carpenter and D. A. Sweigart: *Organometallics* **17**, 2067 (1998). C. A. Dullaghan, X. Zhang, D. L. Green, G. B. Carpenter, D. A. Sweigart, C. Camiletti and E. Rajaseelan: *Organometallics* **17**, 3316 (1998).
66.   K. Yu, H. Li, E. J. Watson, K. L. Virkaitis, G. B. Carpenter and D. A. Sweigart: *Organometallics* **20**, 3550 (2001).
67.   J. Chen and R. J. Angelici: *Organometallics* **18**, 5721 (1999).
68.   U. Riaz, O. J. Curnow and M. D. Curtis: *J. Am. Chem. Soc.* **116**, 4357 (1994).
69.   M. D. Curtis and S. H. Drucker: *J. Am. Chem. Soc.* **119**, 1027 (1997).
70.   E. I. Stiefel and K. Matsumoto, eds., *Transition Metal Sulfur Chemistry: Biological and Industrial Significance*, ACS Symp. Ser. Vol. 653, American Chemical Society, Washington D. C., 1996.
71.   L. S. Byskov, B. Hammer, J. K. Norskov, B. S. Clausen and H. Topsøe: *Catal. Letters* **47**, 177 (1997).

72. C. Bianchini, J. A. Casares, A. Meli, V. Sernau, F. Vizza and R. A. Sánchez-Delgado: *Polyhedron* **16**, 3099 (1997).
73. C. Bianchini, V. Herrera, M. V. Jiménez, A. Meli, R. A. Sánchez-Delgado and F. Vizza: *J. Am. Chem. Soc.* **117**, 8567 (1995).
74. C. Bianchini, A. Meli, V. Patinec, V. Sernau and F. Vizza: *J. Am. Chem. Soc.* **119**, 4945 (1997).
75. C. Bianchini, D. Fabbri, S. Gladiali, A. Meli, W. Pohl and F. Vizza: *Organometallics* **15**, 4604 (1996).
76. C. Bianchini, D. G. Burnaby, J. Evans, P. Frediani, A. Meli, W. Oberhauser, R. Psaro, L. Sordelli and F. Vizza: *J. Am. Chem. Soc.* **121**, 5961 (1999).
77. C. Bianchini, A. Meli, S. Moneti and F. Vizza: *Organometallics* **17**, 2636 (1998).
78. C. Bianchini, A. Meli, S. Moneti, W. Oberhauser, F. Vizza, V. Herrera and R. A. Sánchez-Delgado: *J. Am. Chem. Soc.* **121**, 7071 (1999).
79. D. D. Whitehurst, T. Isoda and I. Mochida: *Adv. Catal.* **42**, 345 (1998).
80. S. Hatanaka, M. Yamada and O. Sadakane: *Ind Eng. Chem. Res.* **36**, 1519 (1997).
81. R. C. Mills, K. A. Abboud and J. M. Boncella: *J. Chem. Soc. Chem. Commun.* 1506 (2001).
82. C. Bianchini, M. Frediani and F. Vizza: *J. Chem. Soc. Chem. Commun.*, 479 (2001).
83. M. Hernández, G. Miralrio, A. Arévalo, S. Bernès, J. J. Gracía, C. López, P. M. Maitlis and F. del Río: *Organometallics* **20**, 4061 (2001).

CHAPTER 5
ACTIVATION OF HYDROGEN
ON METAL COMPLEXES WITH SULFIDE LIGANDS
AND RELATED COORDINATION CHEMISTRY OF H$_2$S

5.1 Introduction

A key feature in HDS catalysis is the ability of metal sulfides to efficiently activate molecular hydrogen. The available experimental and theoretical evidence indicates that on Co-Mo-S and related catalyst formulations, hydrogen activation may occur directly on the metal or on any one of the various forms of surface sulfur atoms (M=S, M-S-M, M(S$_2$), M$_2$(μ-S$_2$), *etc.*); it is often claimed that activation on sulfur actually predominates. Hydrogen atoms activated on a metal-sulfide surface can be considered either as protons (*e.g.* in acid -SH sites) or as hydrides (on the metal centers, M-H); also, bridging H atoms -as in M-H-S, M-H-M, or S-H-S groups- are to be expected to display intermediate acidity. Moreover, adsorbed hydrogen atoms are expected to be highly mobile, being able to migrate throughout the entire catalyst surface via low energy pathways, notably redox processes that easily interconvert H$^+$ and H$^-$ units, as well as M-S, S-S, and S-H bond breaking and forming. These interesting features are related to the dynamic nature of the active sites and to the different reactivities already described in Chapters 1 and 4. Also, the adsorption and reactions of H$_2$S itself at the surface of a catalyst are of great importance in connection with the formation and deactivation of active sites, and with the actual desulfurization processes. In this respect, the chemistry of metal complexes containing sulfur ligands can provide interesting models for the interaction of hydrogen with metal sulfide surfaces. Parkin has extensively reviewed several aspects of the synthesis, structures and reactions of metal complexes containing M=X bonds (X = S, Se, Te) [1, 2]. Peruzzini *et al.* have recently provided a detailed account of the coordination chemistry of metal complexes containing –XH and H$_2$X ligands (X = S, Se, Te) [3].

As mentioned in Section 1.2.5, the question of hydrogen activation in relation to HDS catalysis has been little addressed in organometallic modeling studies (in comparison with *e.g.* C-S bond activation), despite the fact that hydrogen splitting in homogeneous transition metal catalyzed reactions is an extremely well understood and documented phenomenon. The most commonly encountered mechanisms for dihydrogen activation on transition metal complexes are those represented in Fig. 5.1: (a) Oxidative addition of H$_2$ to a low valent metal center [L$_n$M]$^{x+}$ -frequently although not necessarily preceded by the formation of a non-classical metal-η^2(H$_2$) intermediate- yields the corresponding metal dihydrides [L$_m$M(H)$_2$]$^{(x+2)+}$; this reaction is limited to metals having two stable oxidation states separated by two units, and in this case the addition of the

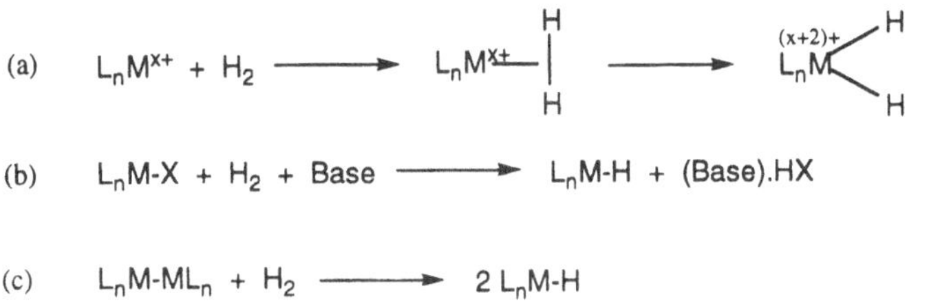

(a) $L_nM^{x+} + H_2 \longrightarrow L_nM^{x\pm}\!-\!\overset{\displaystyle H}{\underset{\displaystyle H}{|}} \longrightarrow \overset{(x+2)+}{L_nM}\!\!\overset{H}{\underset{H}{\diagdown}}$

(b) $L_nM\text{-}X + H_2 + \text{Base} \longrightarrow L_nM\text{-}H + (\text{Base}).HX$

(c) $L_nM\text{-}ML_n + H_2 \longrightarrow 2\,L_nM\text{-}H$

Fig. 5.1. Main mechanisms of hydrogen activation on metal complexes.

two hydrides is concerted and thus it takes place stereoselectively to yield products with mutually *cis* hydrides . (b) Heterolytic splitting of dihydrogen (into H$^+$ + H$^-$) by *e.g.* [L$_n$M-X], commonly assisted by an added base such as Et$_3$N (to remove the proton as HX) produces the corresponding metal monohydride [L$_n$M-H], with no net change in the formal oxidation state of the metal. (c) Homolytic H$_2$ cleavage on binuclear complexes [L$_n$M-ML$_n$] leads also to the corresponding mononuclear metal monohydrides [L$_n$M-H]. Further information on these reactions is easily available in specialized monographs and in major textbooks on organometallic chemistry and homogeneous catalysis [4-7].

An interesting reaction in connection with the activation of hydrogen on HDS surfaces would be the heterolytic activation of hydrogen, in cases where the base required to remove the proton is actually a ligand or part of a ligand already attached to the metal. This type of activation has been little studied, particularly in cases where the acting base is a sulfur atom. Hydrogen splitting on coordinated S^{2-} and S$_2^{2-}$ units is known but further information on these reactions would be welcome, as they represent particularly relevant molecular analogues for mechanistic proposals in which H$_2$ activation during HDS catalysis involves the surface sulfur atoms (see Figs. 1.3 and 1.4). The fact that adequate well-defined metal complexes containing sulfide or disulfide ligands are actually rather scarce [2] may have precluded further developments in this area of organometallic modeling of HDS-related reactions. Nevertheless, a limited number of very interesting examples of reactions of molecular hydrogen with metal complexes containing simple sulfur ligands -related to some of the most abundant forms of surface sulfur in metal sulfide catalysts- have recently become available. They can be considered as interesting models to support some of the proposals in the heterogeneous literature, as described in the following Sections.

Also, some aspects of the chemistry of metal complexes containing H$_2$S and –SH ligands that have been prepared by methods other than hydrogen activation on sulfides, provide interesting information related to HDS catalysis. This type of compounds is actually rather uncommon, mainly due to the ease with which they are decomposed through oxidative processes. Their stabilization is possible by use of highly electron-rich metal centers, coupled with bulky ligands providing a rigid and protective framework; some relevant selected examples of such metal-sulfur derivatives and their reactions will also be described in this Chapter.

5.2 Hydrogen activation on dinuclear complexes containing sulfido or disulfido ligands

The group of Rakowski Du Bois has published extensive studies on the reactions of hydrogen with a series of dinuclear cyclopentadienyl-Mo and -Re complexes containing bridging sulfido and disulfido ligands [8-12]. In a clear parallel to what is thought to happen on MoS_2 surfaces, the complexes $[(Cp'Mo)_2(\mu-S_2)(\mu-S)_2]$ (Cp' = methylated cyclopentadienyls) take up hydrogen –even at room temperature and atmospheric pressure- to yield the corresponding bis(hydrosulfido) derivatives $[(Cp'Mo)_2(\mu-SH)_2(\mu-S)_2]$ [12] (Eq. 5.1), one of which (Cp' = Me-C_5H_4) was characterized by X-ray diffraction by Parkin [2, 13]. This homogeneous splitting process corresponds to an often-invoked surface hydrogen activation pathway that has also been shown by theoretical arguments to be favored thermodynamically and kinetically [14].

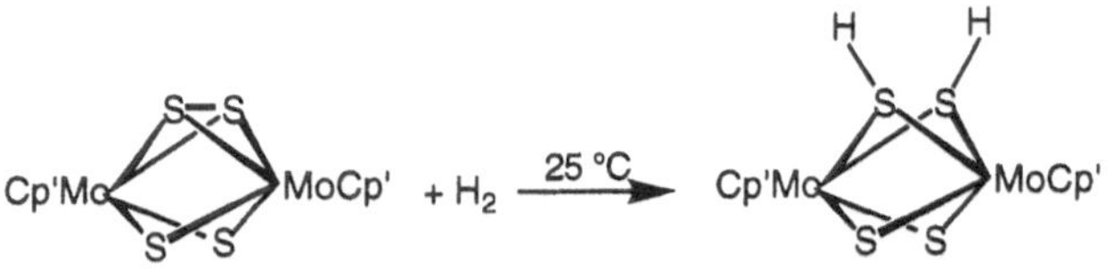

$$\text{(5.1)}$$

This *homolytic* hydrogen addition takes place exclusively at the S_2^{2-} site; it appears not to be reversible, nor does it proceed further to yield H_2S and a "vacancy", as is believed to be the case on solid catalysts. Interestingly, the resulting bis (hydrosulfido) complex does function as a catalyst for H/D exchange under 1 atm H_2/D_2 at room temperature, and also for the homogeneous hydrogenation of elemental sulfur to H_2S under very mild reaction conditions (75° C and 1-3 atm H_2). Furthermore, upon interaction with ethylene, also under moderate conditions, both the bridging sulfido and disulfido moieties become involved, and as a consequence, a hydrogen molecule is evolved and a product containing two ethanethiolate bridges is formed [12]. Similarly, alkynes add to the bis (hydrosulfido) dimer to produce the corresponding alkenedithiolate-bridged derivatives (see Eq. 5.2):

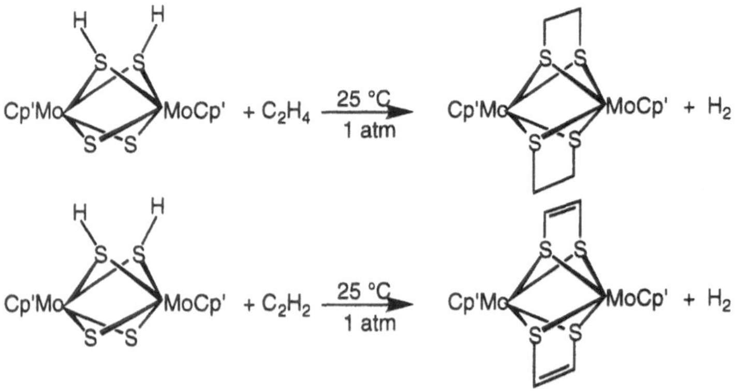

$$\text{(5.2)}$$

Alkyl or aryl thiols (RSH) exchange readily with the -SH ligands of the Mo dimer to liberate H_2S and form the corresponding alkyl or aryl thiolato derivatives (Eq. 5.3), characterized by spectroscopic and crystallographic methods. By use of isotopic labeling (^{34}S) it was possible to demonstrate that the reaction actually proceeds via a Mo-S bond breaking pathway. This is a very nice molecular analogue of a reaction commonly thought to intervene in HDS of thiols (or thiophenes) over solid sulfide catalysts.

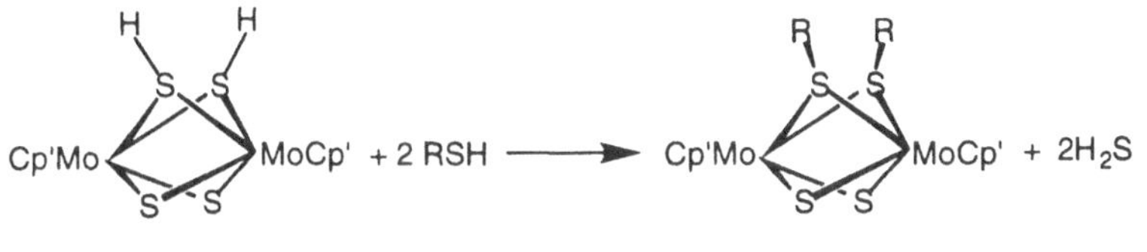

$$\text{Cp'Mo} \quad \text{MoCp'} + 2\,\text{RSH} \longrightarrow \text{Cp'Mo} \quad \text{MoCp'} + 2\text{H}_2\text{S} \quad (5.3)$$

The bis-hydrosulfido Mo dimer was further shown to function as a homogeneous hydrogenation catalyst for a variety of other substrates, demonstrating that the protons bound to sulfur are reactive enough to be transferred to simple unsaturated organic molecules [15]. It is also interesting to note that only the bridging *disulfido* ligand reacts with hydrogen in this case, whereas the bridging sulfides are inert under the same reaction conditions; this coincides with some of the heterogeneous mechanisms that invoke the predominant or exclusive dissociation of hydrogen on surface disulfido (S_2^{2-}) groups.

Nevertheless, in the related complex $[(CpMo)_2(\mu\text{-SR})(\mu\text{-S})(\mu\text{-S}_2CH_2)]$, a bridging *sulfido* ($S^{2-}$) group is actually capable of activating dihydrogen through a process assisted by an external base, as shown in Eq. 5.4 [16]. It is not clear whether the initial interaction of H_2 is with the μ-S and the μ-SR groups, or with the μ-S and the Mo center; what does seem certain is that this H_2 activation on a sulfide group is *heterolytic* in nature, in contrast with the homolytic dissociation on bridging disulfido ligands mentioned above.

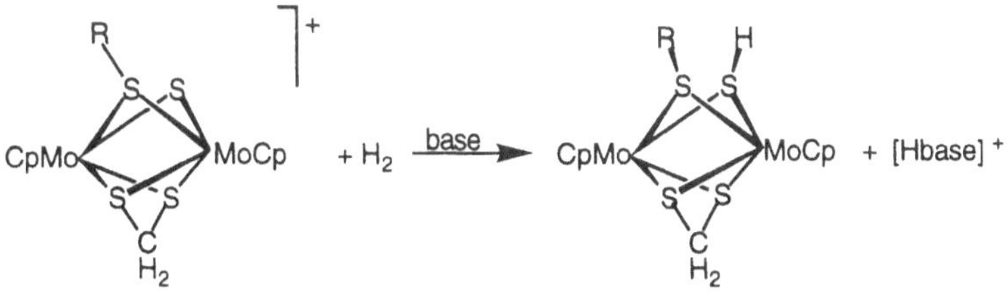

$$\text{CpMo} \quad \text{MoCp} + \text{H}_2 \xrightarrow{\text{base}} \text{CpMo} \quad \text{MoCp} + [\text{Hbase}]^+ \quad (5.4)$$

Some of the molybdenum dimers prepared by this group are also capable of promoting C-S bond breaking reactions that may be related to surface HDS chemistry. These include, for instance, the thermal elimination of olefins from bridging alkanedithiolate ligands (*i.e.* the reverse of Eq. 5.2), or, more interestingly, the desulfurization of heterocycles through thioether exchange, as exemplified for 2,5-dihydrothiophene in Eq. 5.5 [14].

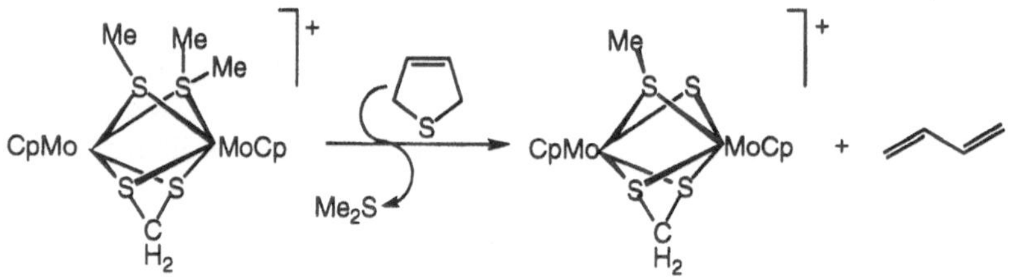

$$\text{(5.5)}$$

More recently, a related rhenium complex [Cp'Re(μ-S$_2$)$_2$]$_2$Cl$_2$ containing two bridging S$_2^{2-}$ units was also synthesized by Rakowski Du Bois and coworkers; this interesting complex proved to react readily with hydrogen under exceedingly mild conditions (25° C, 1 atm) to yield H$_2$S and a trinuclear rhenium cluster tentatively formulated as [(Cp'Re)$_3$S$_4$H$_x$]$^{n+}$ [9, 17]. This unusual hydrogenolysis of bridging sulfides to produce H$_2$S represents an important mechanistic model for the well accepted pathway for the formation of anionic vacancies in metal sulfides; in fact, the difference between the Mo and Re dimers seems to follow the trend in metal-sulfur bond energies observed for the sulfides of those two metals (see Chapter 1). Unfortunately, no intermediate hydrosulfide species could be detected in the reaction of the Re dimer with hydrogen, but an interesting exchange reaction between the Mo-bis(hydrosulfido) complex and [Cp'Re(μ-S$_2$)$_2$]$_2$Cl$_2$ did give a detectable Re-hydrosulfido derivative (Eq. 5.6):

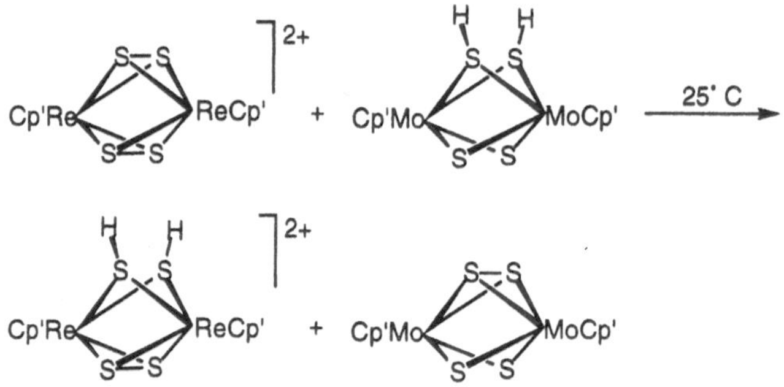

$$\text{(5.6)}$$

Bianchini has provided one clear example of hydrogen activation on the sulfido-bridged Rh dimer [(triphos)Rh(μ-S)$_2$Rh(triphos)]$^{2+}$ (triphos = MeC(CH$_2$PPh$_2$)$_3$) involving a heterolytic splitting mediated by both the metal and the sulfido ligand, to yield the corresponding hydrido(hydrosulfido) derivative [(triphos)Rh(H)(μ-SH)$_2$Rh(H)(triphos)]$^{2+}$ (Eq. 5.7) [18]. In this case, the bridging sulfide is the acting base required to capture the proton.

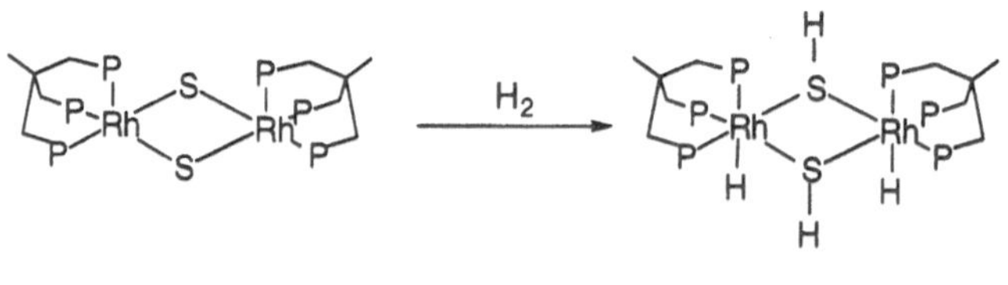

$$\text{(5.7)}$$

This reaction is reminiscent of a proposal frequently encountered in the heterogeneous literature, which has been supported by solid state NMR studies of hydrogenated RuS$_2$ (see Chapter 1, p. 15) [19].

In conclusion, the interactions of these dimeric complexes containing μ-S^{2-} and μ-S$_2^{2-}$ ligands with hydrogen and other related reactions summarized above represent excellent molecular analogues of some of the key steps known or believed to happen on the surface of e.g. MoS$_2$ catalysts. Both homolytic and heterolytic activation pathways have been authenticated in these complexes and it is easy to extrapolate such mechanisms to what may actually happen on the active sites on solid catalysts. The easy interconversion between μ-S$_2^{2-}$ and μ-SH groups, the transfer of hydrogen from hydrosulfides to unsaturated molecules, the exchange of -SH groups by thiols through C-S bond breaking processes, and the desulfurization of coordinated heterocycles constitute a remarkable set of reactions on Mo complexes mimicking a number of important steps of heterogeneous HDS. The reaction of the rhenium dimer containing μ-S$_2^{2-}$ units with hydrogen, leading to the liberation of H$_2$S is a very good model for the mechanism of generation of "anionic vacancies" in metal sulfides. Hopefully, this chemistry will continue to be developed thus providing further insight into the role of surface sulfur species in HDS mechanisms.

5.3 Hydrogen activation on mononuclear Ti=S and Ti(η^2S$_2$) complexes.

Besides the possibility of activating hydrogen on surface S$_2^{2-}$ groups bridging two metal atoms, which have been modeled very elegantly by use of CpMoS and CpReS dimers as described in the preceding section, some other mechanistic proposals for HDS on metal sulfides contemplate either the homolytic splitting of H$_2$ on a monometallic M(S$_2^{2-}$) site to yield M(SH)$_2$, or the heterolytic activation by a terminal M=S group leading to M(H)(SH) species (see Section 1.2.5). The first clear case of these reactions on model complexes in solution was recently reported by Sweeney et al. who achieved the synthesis of two well defined and reasonably stable complexes containing sulfido or disulfido ligands on single metal centers. The Ti(II) complex Cp*$_2$Ti(η^2-C$_2$H$_4$) reacts with elemental sulfur to yield the sulfido derivative Cp*$_2$(py)Ti=S quantitatively, which further reacts with sulfur or with ethylene sulfide to produce the disulfido complex Cp*$_2$Ti(η^2-S$_2$) [20, 21] (see Fig. 5.2). The sulfido derivative reversibly reacts with hydrogen gas at 1 atm to produce the corresponding hydrido-hydrosulfido species Cp*$_2$Ti(H)(SH) in high yields, clearly through a heterolytic splitting in which the sulfur atom acts as the base, possibly via a dihydrogen intermediate that was not observed. In turn, the bis(hydrosulfido) complex Cp*$_2$Ti(SH)$_2$ is gradually produced upon interaction of the disulfido complex Cp*$_2$Ti(S$_2$) with 1 atm of dihydrogen at 70° C, presumably via initial addition of H$_2$ accross a Ti-S bond as shown in Eq. 5.8 although a concerted homolytic splitting could also be conceived. Both products could be well characterized by NMR spectroscopy, but Cp*$_2$Ti(H)(SH) was unstable under those reaction conditions, readily eliminating H$_2$ in solution and reverting back to Cp*$_2$(py)Ti=S. An alternative preparative route for Cp*$_2$Ti(H)(SH) could be developed by treatment of the

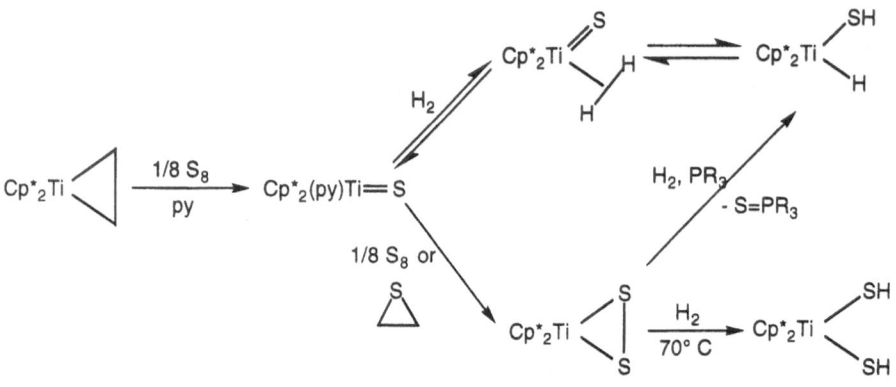

Fig. 5.2. Activation of hydrogen on Ti=S and Ti(S$_2$) complexes.

disulfide Cp*$_2$Ti(η^2-S$_2$) with hydrogen in the presence of a phosphine (PPh$_3$ or PMe$_3$) acting as a sulfur scavenger.

$$\text{(5.8)}$$

This remarkable set of reactions represents another excellent model for hydrogen activation on metal sulfides, and it is to date the only example available of the direct observation of a heterolytic activation of hydrogen assisted by a terminal sulfide ligand. Like Bianchini's Rh complex mentioned above (Eq. 5.7), this is in line with solid state NMR evidence for this latter type of hydrogen splitting on RuS$_2$ which has been provided by Lacroix *et al.*, indicating the formation of Ru-H and -SH species on the surface [19]; it is possible that other metal sulfides are also capable of activating H$_2$ by similar routes.

The number of well-defined metal complexes containing terminal sulfido or disulfido ligands is rather limited [1, 2], their syntheses are seldom simple and the compounds may be highly unstable in solution. Although these features may have precluded reactivity studies, and thus further attempts to model HDS-related reactions, it is very likely that other examples will become available in the near future; particularly welcome would be some advances in the reactions of hydrogen with *e.g.* Mo or Co complexes containing such sulfur ligands, as well as studies of the reactions of M(H)(SH) and M(SH)$_2$ complexes with thiophenes and other related molecules.

5.4. Reactions of H₂S with metal complexes

In a different approach to this problem, several groups have studied the reactions of H_2S with a variety of metal complexes. Only a limited number of complexes containing intact coordinated H_2S have been prepared, as can be seen in the data contained in Table 5.1. Some of them have been shown to exist only in solution by *in situ* spectroscopic measurements or by indirect methods, and very few have been structurally characterized [3, 22-37].

The synthetic methods that have been successful for the preparation of metal complexes of hydrogen sulfide involve one of the following reaction types: (1) Addition to coordinatively unsaturated compounds (Eq. 5.9); (2) displacement of labile ligands (Eq. 5.10), and (iii) protonation of hydrosulfido ligands (Eq. 5.11). Some interesting examples of complexes of molybdenum and tungsten exist, and they can be taken as adequate models of adsorbed H_2S on MoS_2 or WS_2 surfaces. However, it must be pointed out that in the three complexes known (see Table 5.1) the metal oxidation state (0 or I) is lower than it is in the pure metal sulfides. Unfortunately no Co or Ni complexes of H_2S are available as models of adsorption onto the active sites of promoted HDS catalysts. Ruthenium derivatives are by large the most numerous and stable. Besides the terminal binding mode observed or assumed in all the compounds listed, H_2S is, in principle, capable of bridging two metal centers, but this type of coordination has not been reported to date.

TABLE 5.1. Transition metal complexes of H_2S[a]

Complex	Ref.
$(CO)_5Cr(SH_2)$	23
$Cp(CO)_3Mo(SH_2)$	24
$(CO)_5W(SH_2)$	23, 25
$Cp(CO)_3W(SH_2)$	24
$Cp(CO)_2Mn(SH_2)$	26
$(PPh_3)(CO)_4Mn(SH_2)$	27
$[(CO)_5Re(SH_2)]BF_4$	28
$[(triphos)(CO)_2Re(SH_2)]OTf$	3
$[(NH_3)_5Ru(SH_2)](BF_4)_2$	29
$[(NH_3)_4(ISN)Ru(SH_2)](BF_4)_2$	29
$[(LS_4)(PPh_3)Ru(SH_2)]$	30[b]
$[(P-N)(PR_3)RuX_2(SH_2)]$	31[b], 32[b]
(R = Ph, p-tol; X = Cl, Br)	
$[Cp'(PPh_3)_2Ru(SH_2)]OTf$	33
(Cp' = Cp, Thi-Cp)	
$[Ru_2(CO)_5(\mu\text{-etipdp})_2(SH_2)](SbF_6)_2$	34
$[Ru(\eta^3,\eta^5\text{-}C_{10}H_{16})Cl_2(SH_2)]$	35
$[Ir(H)_2(SH_2)_2(PPh_3)_2]BF_4$	36
$[PPh_3)_2Pt(SH_2)]$	37

[a]triphos = $HC(CH_2PPh_2)_3$
ISN = isonicotinamide
$LS_4 = \text{'}SC_6H_4S(CH_2CH_2SC_6H_4S\text{'}$
P-N = $Ph_2PC_6H_4NMe_2$; ThiCp = $C_5H_3S\text{-}CH_2\text{-}C_5H_4\text{'}$
etipdp = N-ethyl(tetraisopropoxy)diphosphazane
[b]X-ray structure available.

$$L_nM\!-\!\bullet \quad \xrightarrow{\text{H}_2\text{S}} \quad L_nM\!-\!S\!\!\begin{array}{c}H\\ \diagdown H\end{array} \tag{5.9}$$

$$L_nM\!-\!L' \quad \xrightarrow[\text{-L'}]{\text{H}_2\text{S}} \quad L_nM\!-\!S\!\!\begin{array}{c}H\\ \diagdown H\end{array} \tag{5.10}$$

$$L_nM\!-\!SH \quad \xrightarrow{\text{HX}} \quad \left[L_nM\!-\!S\!\!\begin{array}{c}H\\ \diagdown H\end{array}\right]^{+} X^{-} \tag{5.11}$$

The main reason for the scarcity of H_2S complexes seems to be mainly the ease with which they can be oxidized, and indeed, their stabilization is usually accomplished by use of electron-rich metal centers in combination with bulky ligands that provide a rigid protective environment. Still, the binding of H_2S to metal ions is in general rather weak, and its displacement by other Lewis bases is very favored. Nevertheless, once it is coordinated H_2S is, as expected, activated toward a number of reactions. For instance, deprotonation can take place by the action of strong bases (Eq. 5.12), and several examples of the oxidative addition of H_2S to give hydrido-hydrosulfido species are known; some of them probably proceed through undetected M-SH$_2$ intermediates, as indicated in Eq. 5.13.

$$L_nM\!-\!S\!\!\begin{array}{c}H\\ \diagdown H\end{array} \quad \xrightarrow{\text{B}^-} \quad \left[L_nM\!-\!S\!\!\begin{array}{c}\\ \diagdown H\end{array}\right]^{-} + HB \tag{5.12}$$

$$L_nM \quad \xrightarrow{\text{H}_2\text{S}} \quad \left[L_nM\!-\!SH_2\right] \quad \longrightarrow \quad L_nM\!\!\begin{array}{c}H\\ \diagdown SH\end{array} \tag{5.13}$$

In some cases the oxidative addition reaction has been observed to proceed further with the elimination of dihydrogen and the consequent formation of a stable sulfido complex, in a way the reverse reaction of the activation of hydrogen by Cp*$_2$Ti=S described in the preceding section. For instance, the Zr(II) complex Cp*$_2$Zr(CO)$_2$ reacts with excess H_2S in the presence of pyridine to yield the corresponding bis(hydrosulfido) derivative Cp*Zr(SH)$_2$. This reaction takes place via an initial oxidative addition of H_2S to yield Cp*Zr(H)(SH), followed by hydrogen elimination in the presence of pyridine to form a terminal sulfido complex Cp*$_2$Zr(S)(py); the latter undergoes a second oxidative addition of H_2S to finally produce the bis(hydrosulfido) complex, as shown in Eq. 5.14 [2].

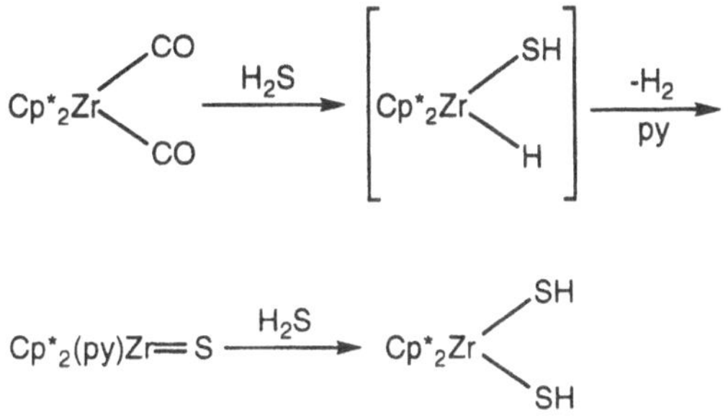

$$Cp^*_2Zr\begin{smallmatrix}CO\\CO\end{smallmatrix} \xrightarrow{H_2S} \left[Cp^*_2Zr\begin{smallmatrix}SH\\H\end{smallmatrix}\right] \xrightarrow[py]{-H_2}$$

$$Cp^*_2(py)Zr{=}S \xrightarrow{H_2S} Cp^*_2Zr\begin{smallmatrix}SH\\SH\end{smallmatrix} \qquad (5.14)$$

An interesting case in Ta chemistry involves the reactions of the trihydride $Cp^*_2TaH_3$ with excess sulfur to yield the hydrido-sulfido species $[Cp^*_2Ta(H)(\eta^2\text{-}S_2)]$ [38], or with H_2S to produce the complex $[Cp^*_2Ta(H)(S)]$, containing both a terminal sulfide and a hydride ligand, as represented in Eq. 5.15 [39].

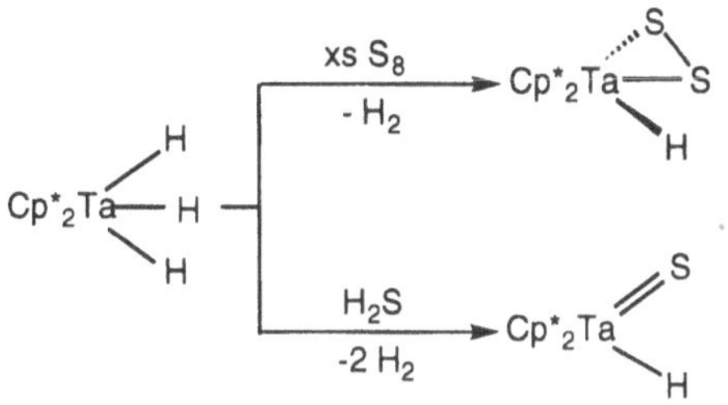

$$Cp^*_2Ta\begin{smallmatrix}H\\H\end{smallmatrix}\begin{cases} \xrightarrow[-H_2]{xs\ S_8} Cp^*_2Ta\overset{S}{\underset{H}{<}}S \\[2ex] \xrightarrow[-2H_2]{H_2S} Cp^*_2Ta\overset{S}{\underset{H}{<}} \end{cases} \qquad (5.15)$$

Molybdenum and tungsten complexes of formula $M(PMe_3)_4(\eta^2\text{-}CH_2PMe_2)H$ (M = Mo, W) readily react with H_2S to yield the bis(sulfido) derivatives $trans\text{-}M(PMe_3)_4(S)_2$. In the case of tungsten, a interesting stable bis(hydrido)bis(hydrosulfido) intermediate can be isolated, which readily dehydrogenates in solution to yield $W(PMe_3)_4(S)_2$, as depicted in Eq. 5.16 [40, 41]:

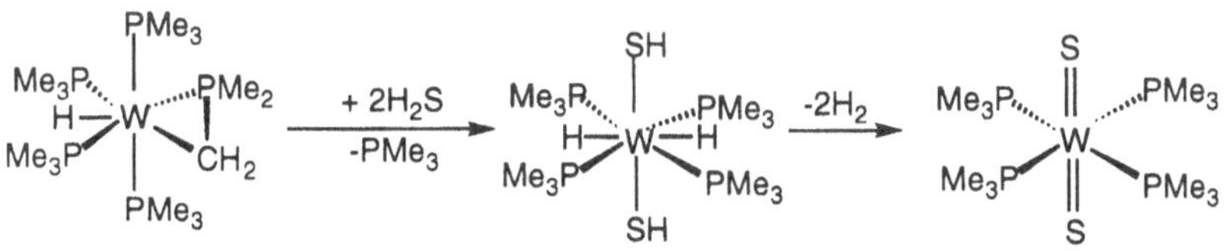

$$\xrightarrow[-PMe_3]{+2H_2S} \qquad \xrightarrow{-2H_2} \qquad (5.16)$$

The most numerous and most thoroughly studied metal hydrosulfide complexes are those of molybdenum, and they have often been used as models for HDS-related reactions and intermediates (for a comprehensive account, see ref. 3). Their syntheses, however, do not involve direct reaction with H_2S very often. For instance, an interesting complex containing a hydrosulfido ligand on a Mo(IV) center in a "sulfur-only" environment is the tris(dithiocarbamate) derivative (a), which resembles some of

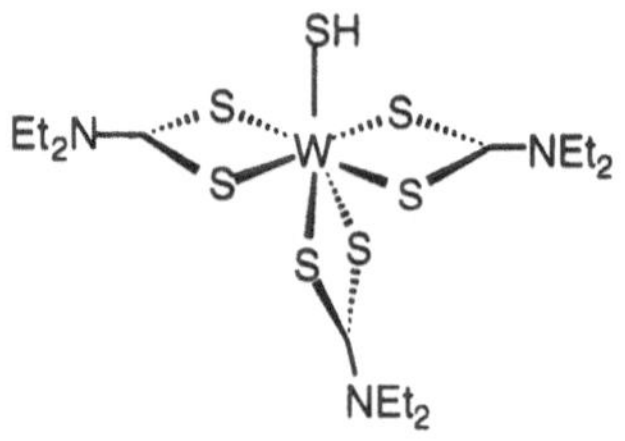

(a)

the proposed geometries for the active sites in MoS_2 [42]. This compound was prepared by sulfur atom transfer from the iron-sulfido cluster $[Cp_4Fe_4S_6]$ to the dicarbonyl $[(CO)_2Mo(S_2CNEt_2)_2]$.

As illustrated in Eq. 5.17, the hydrosulfido complex $[(PP_3)Fe(SH)]BPh_4$, where $PP_3 = P(CH_2CH_2PPh_2)_3$, was obtained by Sacconi and coworkers by briefly bubbling H_2S through an ethanolic solution of $[Fe(OH_2)_6][BF_4]_2$ in the presence of the PP_3 ligand dissolved in acetone [43]. Alternatively, this compound can be prepared from $[(PP_3)Fe(H)(H_2)](BPh_4)$ or $[(PP_3)Fe(H)(N_2)](BPh_4)$ by interaction with H_2S; the latter reaction most likely proceeds via prior elimination of the labile H_2 or N_2 ligand to yield an undetected transient hydrogen sulfide derivative $[(PP_3)Fe(H)(SH_2)](BPh_4)$ from which H_2 is eliminated. Consistent with this idea, the unsaturated $[(PP_3)Fe(H)](BPh_4)$ also readily adds H_2S directly to produce the same hydrosulfido complex [3].

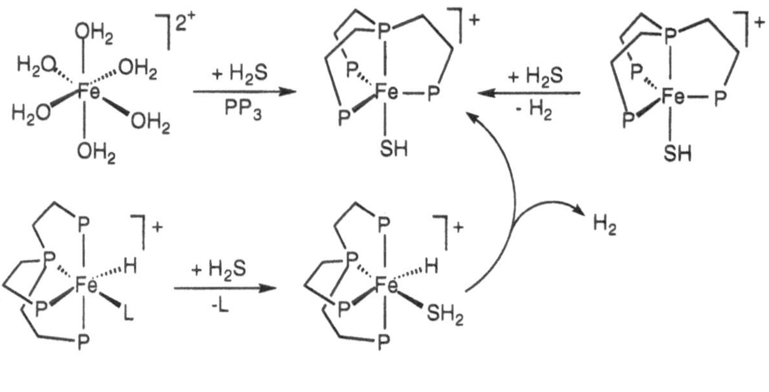

(5.17)

As can be seen in Table 5.1, ruthenium is the only metal for which a number of complexes containing an intact coordinated H_2S molecule has been well characterized. These compounds have been prepared either by displacement of a labile ligand from an appropriate starting material like $[(NH_3)_5Ru(OH_2)]^{2+}$ [29], by a bridge splitting reaction of $[(LS_4)Ru(PPh_3)]_x$ [30], or by direct addition of H_2S to unsaturated pentacoordinated Ru(II) complexes such as $[(P-N)RuX_2(PR_3)]$ [31, 32] under mild and carefully controlled reaction conditions.

Several examples of oxidative addition reactions of H_2S with Ru complexes leading to hydrosulfido derivatives are known. They include, for instance, the formation of $Ru(SH)_2(CO)_2(PPh_3)_2$ from $Ru(CO)_2(PPh_3)_3$ via an unstable intermediate $RuH(SH)(CO)_2(PPh_3)_2$, as shown in Eq. 5.18; also the transformation of cis/trans-$Ru(H)_2(dpm)_2$ (dpm $=$ $Ph_2PCH_2PPh_2$) into the hydrido-hydrosulfido trans.-$RuH(SH)(dpm)_2$ as well as the bis(hydrosulfido) cis/trans-$Ru(SH)_2(dpm)_2$ by reaction with H_2S has been reported by James [22].

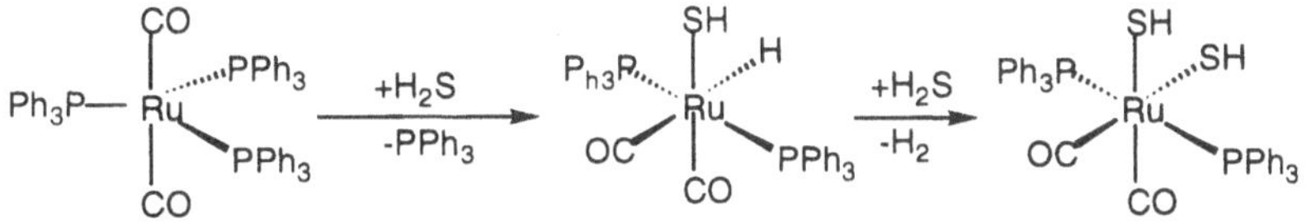

$$(5.18)$$

The set of reactions collected in Eq. 5.19 have been described by Rauchfuss and coworkers [33]; this system nicely models some of the elementary reactions involved in catalytic HDS. The use of a thiophene moiety attached to a cyclopentadienyl ligand through a methylene group, has allowed the synthesis of one of the few stable $\eta^1 S$-bonded thiophene complexes (for details, see Chapter 2, p. 39). Under a H_2S atmosphere, this complex is in equilibrium with a complex containing coordinated H_2S and free thiophene ($K_{eq} = 3.8$), nicely illustrating how these two molecules can compete for a site in a catalytic surface. Indeed, it is clear that in this case, binding of H_2S is preferred over that of thiophene, which is in line with the poisoning effects of H_2S on thiophene HDS frequently observed (see Chapter 1). Moreover, the H_2S complex is also in equilibrium with a hydrosulfido derivative through a deprotonation-protonation process, also very similar to what is thought to happen on metal sulfides.

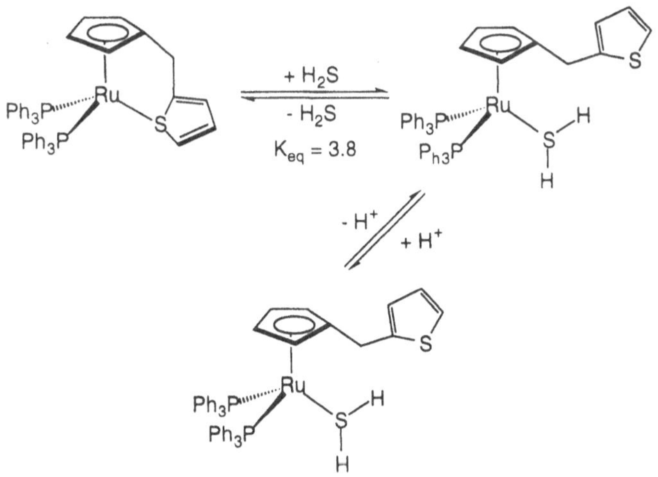

$$(5.19)$$

Although these and other related complexes seem to be very promising as molecular analogues of HDS sites where hydrogen has already been activated, unfortunately, their reactivity has not yet been studied in great detail. In particular, no other reactions related with hydrodesulfurization, like for instance, the interactions of hydrosulfide complexes with thiophenes which would be very interesting to explore, have been reported to date.

James and coworkers have provided a thorough study of another important set of reactions that lead to the decomposition of H_2S into H_2 + coordinated S on dimeric palladium complexes of the type $[(X)Pd(\mu\text{-dppm})_2Pd(X)]$ (X = Cl, Br; dppm = bis(diphenylphosphino)methane); this process is the reverse of the activation of hydrogen by metal sulfide units and the formation of vacancies through complete hydrogenation of surface S^{2-} into H_2S. Extensive kinetic and spectroscopic data led to the mechanism shown in Eq. 5.20; according to this proposal, H_2S adds reversibly across the Pd-Pd bond to yield a complex $[(X)(H)Pd(\mu\text{-dppm})_2Pd(X)(SH)]$ containing a hydride on one palladium atom, and a hydrosulfido on the other. This intermediate then undergoes elimination of dihydrogen via a concerted deprotonation of the -SH group and protonation of the Pd-H bond to produce the bridging sulfido derivative $[(X)Pd(\mu\text{-S})(\mu\text{-dppm})_2Pd(X)]$ [22, 44].

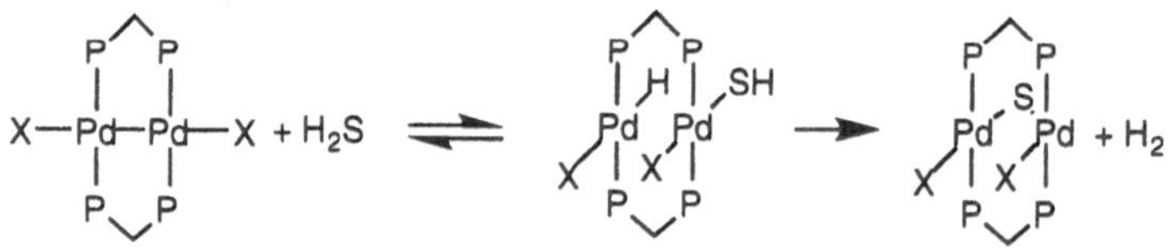

$$\text{(5.20)}$$

Interestingly, this reaction can be effectively coupled with the removal of the bridging S-atom in the final product $[(X)Pd(\mu\text{-S})(\mu\text{-dppm})_2Pd(X)]$, through oxidation by H_2O_2 or m-chloroperbenzoic acid; this leads to an unstable $\mu\text{-SO}_2$ derivative $[(X)Pd(\mu\text{-SO}_2)(\mu\text{-dppm})_2Pd(X)]$which is capable of reversibly losing SO_2. In this way, the starting dimeric complex $[(X)Pd(\mu\text{-dppm})_2Pd(X)]$ is regenerated (see Eq. 5.21); therefore the process can be rendered catalytic, and the overall reaction thus represents the conversion of H_2S into $H_2 + SO_2$. In a variation of this H_2S decomposition catalysis, an excess of the diphosphine dppm can be used instead of the oxidant to abstract the bridging sulfide as dppm(S) [22].

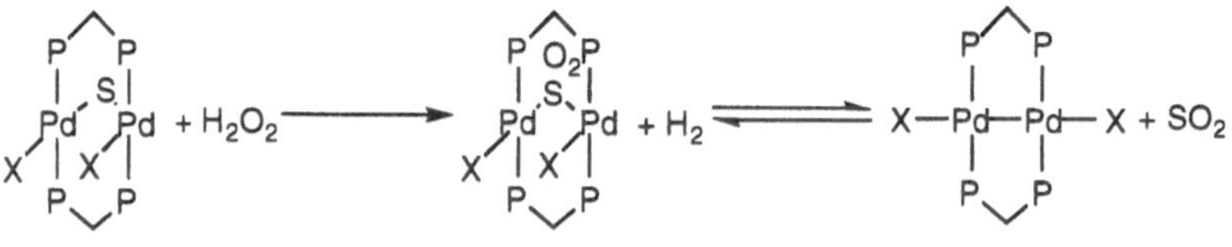

$$\text{(5.21)}$$

5.5 Conclusions and remarks

From the examples presented throughout this Chapter it is clear that transition metal complexes have provided and hopefully will continue to provide good homogeneous models that can be used in trying to define reasonable pathways for the reactions between dihydrogen and activated sulfur, as well as between H$_2$S and metal centers in HDS catalysts. Also, other reactions of metal-sulfido and metal-hydrosulfido complexes illustrate pathways that may be operative also on metal surfaces. Still, this area of coordination chemistry is relatively undeveloped and further models of this type, particularly with heterobimetallic complexes containing terminal and bridging sulfido and disulfido ligands would be most welcome, since their reactions with hydrogen and hydrogen sulfide may help to understand the type of hydrogen activation that is now thought to be dominant on metal sulfide surfaces, since the mechanistic details of such processes are not yet clearly understood.

    Also the study of the reductive elimination of H$_2$S from complexes containing –H and –SH ligands is very worthy of further studies since it could provide important clues to the issue of H$_2$S removal from catalytic surfaces and the regeneration of "vacancies".

References

6.   T. M. Trnka and G. Parkin: *Polyhedron* **16**, 1031 (1997).
6.   G. Parkin: *Prog. Inorg. Chem.* **47**, 1 (1998).
6.   M. Peruzzini, I. De los Rios and A. Romerosa: *Prog. Inorg. Chem.* **49**, 169 (2001).
6.   B. R. James: *Homogeneous Hydrogenation*, Wiley, New York (1973).
6.   J. P. Collman, L. S. Hegedus, J. R. Norton and R. G. Finke: *Principles and Applications of Organo-Transition Metal Chemistry*, University Science Books, Ca. (1987).
6.   B. R. James, in G. Wilkinson, F. G. A. Stone and E. W Abel, eds.: *Comprehensive Organometallic Chemistry*, Pergamon, Oxford (1982). Vol. 8, Ch. 51.
7.   P. A. Chaloner, M. A. Esteruelas, F. Jóo and L. A. Oro: *Homogeneous Hydrogenation*, Kluwer Academic Publishers: Dordrecht, Netherlands, 1993.
8.   M. Rakowski DuBois: *Chem. Rev.* **89**, 1 (1989).
9.   M. Rakowski DuBois, B. Jagirdar, B. Noll and S. Dietz, in E. I. Stiefel and K. Matsumoto, eds.: *Transition Metal Sulfur Chemistry: Biological and Industrial Significance*, American Chemical Society, Washington, D. C., 1996, p. 269.
10.   M. Rakowski DuBois: *Polyhedron* **16**, 3089 (1997).
11.   M. Rakowski DuBois, in R. D. Adams and F. A. Cotton, eds.: *Catalysis by Di- and Polynuclear Metal Cluster Complexes*, VCH Publishers, New York, 1998, p. 127.
12.   M. Rakowski DuBois, M. C. VanDerveer, D. L. DuBois, R. C. Haltiwanger and W. K. Miller: *J. Am. Chem. Soc.* **102**, 7456 (1980).
13.   J. H. Shin and G. Parkin; *Polyhedron* **13**, 1489 (1994).
14.   A. Sierraalta and F. Ruette: *J. Mol. Catal. A* **109**, 227 (1996).
15.   L. L. López, P. Bernatis, J. Birnbaum, R. C. Hatiwanger and M. Rakowski DuBois: *Organometallics* **11**, 2424 (1992).
16.   J. Gabay, S. Dietz, P. Bernatis and M. Rakowski DuBois; *Organometallics* **12**, 3630 (1993).
17.   M. Rakowski DuBois, B. R. Jagirdar, S. Dietz and B. C. Noll: *Organometallics* **16**, 294 (1997).
18.   C. Bianchini, C. Mealli, A. Meli and G. Scapacci: *Organometallics* **2**, 141 (1983).
19.   M. Lacroix, S. Yuan, M. Breysse, C. Dorémieux-Morin and J, Fraissard: *J. Catal.* **138**, 409 (1992).
20.   Z. K. Sweeney, J. L. Polse, R. G. Bergman and R. A. Andersen: *J. Am. Chem. Soc.* **119**, 4543 (1997).
21.   Z. K. Sweeney, J. L. Polse, R. A. Andersen and R. G. Bergman: *Organometallics* **18**, 5502 (1999).
22.   B. R. James: *Pure Appl. Chem.* **69**, 2213 (1997).
23.   M. Herberhold and G. Süss: *Angew. Chem. Int. Ed. Engl.* **15**, 366 (1976).
24.   G. Urban, K. H. Sünkel and W. Beck: *J. Organomet. Chem.* **290**, 329 (1985).
25.   M. Herberhold and G. Süss: *J. Chem. Res. (S)* 246 (1977).
26.   W. Strohmeier and J. F. Guttenberger: *Chem. Ber.* **97**, 1871 (1964).

27. P. J. Harris, S. A. R. Knox, R. J. McKinney and F. G. A. Stone: *J. Chem. Soc. Dalton Trans.* 1009 (1978).
28. K. Raab and W. Beck: *Chem. Ber.* **118**, 3830 (1985).
29. C. G. Kuehn and H. Taube: *J. Am. Chem. Soc.*: **98**, 689 (1976).
30. D. Sellman, P. Lechner, F. Knoch and M. Moll: *J. Am. Chem. Soc.* **114**, 922 (1992).
31. D. C. Mudalige, E. S. Ma, S. J. Rettig, B. R. James and W. R. Cullen: *Inorg. Chem.* **36**, 1809 (1997).
32. E. S. Ma, S. J. Rettig and B. R. James: *J. Chem. Soc. Chem. Commun.* 2463 (1999).
33. J. Amarasekera and T. B. Rauchfuss: *Inorg. Chem.* **28**, 3875 (1989).
34. J. S. Field, R. J. Haines, J. Sundermeyer and S. F. Woolam: *J. Chem. Soc. Dalton Trans.* 2735 (1993).
35. G. Belchem, J. W. Steed and D. A. Tacher: *J. Chem. Soc. Dalton Trans.* 1949 (1997).
36. R. H. Crabtree, M. W. Davis, M. F. Mellea and J. M. Mihelcic: *Inorg. Chim. Acta* **72**, 223 (1983).
37. R. Ugo, G. La Monica, S. Cenini, A. Segre and F. Conti: *J. Chem. Soc. A*, 522 (1971).
38. H.-J. Bach, H. Brunner, J. Wachter, M. Tsunoda, J.-C. Leblanc and C. Moïse: *Inorg. Chem.* **29**, 584 (1973).
39. J. E. Nelson, G. Parkin and J. E. Bercaw: *Organometallics* **11**, 2181 (1992).
40. V. J. Murphy and G. Parkin: *J. Am. Chem. Soc.* **117**, 3522 (1995).
41. D. Rabinovich and G. Parkin: *Inorg. Chem.* **34**, 6341 (1996).
42. N.Dupré, H. M. J. Hendricks and J. Jordanov: *J. Chem. Soc. Dalton Trans.* 1463 (1984).
43. M. Di Vaira, S. Midollini and L. Sacconi: *Inorg. Chem.* **16**, 1518 (1977).
44. A. F. Barnabas, D. Sallin and B. R. James: *Can. J. Chem.* **67**, 2009 (1989).

CHAPTER 6
MODELING HYDRODENITROGENATION.

6.1 Introduction

As mentioned in Chapter 1, hydrodenitrogenation model studies using coordination or organometallic complexes in solution are not as numerous or as detailed as those described for HDS in Chapters 2 and 4. The group of Bianchini very recently published the only review available so far, dealing with the general subject of homogeneous modeling of the HDN reaction [1]; Wigley and coworkers provided in 1997 [2] an overview of HDN model reactions using pyridine-like substrates.

As shown in Fig. 1.13 (p. 28), prehydrogenation of at least the heterocyclic ring of the nitrogen-containing molecules that are abundant in crude oil and refinery cuts is known to take place in heterogeneous HDN reactions, before any C-N bond is actually broken. This is an important feature of this reaction, in which it differs from HDS, where direct C-S bond breaking may indeed take place without any need for prehydrogenation. Thus, hydrogenation reactions of N-heteroaromatics constitute a crucial step in denitrogenation, and they are certainly the most thoroughly studied homogeneous HDN-related reactions. The reduction of N-heterocycles is efficiently catalyzed by a variety of metal complexes in solution under moderate reaction conditions, and the corresponding reaction mechanisms are well understood; this was described in Chapter 3, together with the catalytic hydrogenation of other important organic substrates.

Other important aspects of hydrodenitrogenation modeling have received little attention in the coordination and organometallic chemistry literature. Among them, the most relevant points are possibly: (i) The description of the different binding modes of organonitrogen compounds to metal-containing fragments; (ii) the relations of each binding mode with specific reactivity patterns; and (iii) the actual scission of C-N bonds in metal complexes. The attempts to model such species and interactions that have appeared in the literature conform the subject of this Chapter. As in the case of HDS, the general strategy in HDN modeling has consisted mainly of the synthesis of metal complexes containing N-heterocycles that are representative of the compounds present in real feeds. Once these complexes become available, the study of their structures and reactions by standard methods like NMR spectroscopy and X-ray diffraction can lead to detailed pictures of some structural patterns and reaction mechanisms connected to HDN catalysis.

The substrates most commonly used in HDN modeling studies may be divided in two main classes: (1) The less basic pyrrole (Pyr, pKa 0.4) and indole (In, pKa -3.6), in which the nitrogen lone pair is already involved in bonding, and thus it is not available for interaction with electrophiles. And (2) the strongly basic pyridine (Py, pKa

5.2) and quinoline (Q, pKa 4.9), in which the nitrogen atom is susceptible to be involved in bonding to metal-containing fragments. In both cases it is also important to consider their higher homologues, their alkyl-substituted derivatives, and their partially or completely hydrogenated products.

6.2 Binding Modes of N-Heterocycles in Transition Metal Complexes.

Both the basicity of the nitrogen atom and the presence of carbocyclic rings in addition to the N-containing ring are important in determining the binding modes of N-heterocycles to single or multiple metal sites, and this overall situation ultimately controls the reactivity of the resulting complex. Also, the steric as well as the electronic effects induced by other substituents in the vicinity of the nitrogen atom play a key role in the stability and the chemistry of each type of complex. This interesting combination of effects also makes some of the issues in question more difficult to interpret and to understand.

6.2.1 COMPLEXES WITH PYRROLE, INDOLE AND CARBAZOLE LIGANDS

A number of transition metal complexes containing weakly basic (5-member ring) HDN-related ligands are known. The authenticated bonding modes of pyrrole (Pyr) and pyrrolyl ions (Pyl) -or their alkylated analogues- in mononuclear metal complexes are summarized in Fig. 6.1. Pyrrole is a 5-member aromatic heterocycle in which the lone pair is delocalized over the π system of the ring, and it is therefore an electron rich molecule that reacts readily with electrophiles but is not susceptible to nucleophilic attack.

Table 6.1 contains a list of known compounds of this type [3-25]. No examples of complexes containing η^1(N)-Pyr ligands are known, in accord with the low basicity of the N-atom in this molecule. On the other hand, metal derivatives containing olefin-like ligands, in either one of the tautomeric forms, η^2(C=C)-1H-Pyr or η^2(C=C)-2H-Pyr, have been characterized for osmium(II) complexes [3-5].

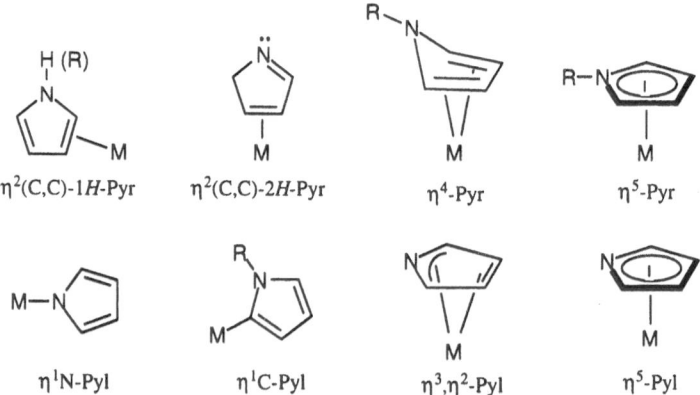

Fig. 6.1. Known bonding modes of pyrrole and pyrrolyl ions in metal complexes.

TABLE 6.1. Metal complexes of pyrrole and pyrrolyl ligands.

Complex	Refs.
Pyrrole complexes	
$[Os(NH_3)5(\eta^2(C=C)\text{-Pyr'})][OTf]_2$	3-5
Pyr' = Pyr, 1-MePyr, 2,5-Me$_2$Pyr, 1,2,5-	
Me$_3$Pyr, 5-EtPyr, 1,4-Me$_2$Pyr	
Cp*Ir[η^4-1^{t}Bu-2,3,4,5(CO$_2$Me)$_4$Pyr]	6
Cr(CO)$_3$(η^5-1-MePyr)	8, 9
[Mn(CO)$_3$(η^5-Pyr)]$^+$	10
[CpFe(η^5-1-MePyr')]$^+$	7, 10
Pyr' = 1,2,5-Me$_3$Pyr, 2,3,4,5-Me$_4$Pyr,	
Me$_5$Pyr	
[Fe(η^5-1-MePyr)$_2$]$^{2+}$	11
[CpCo(η^5-Me$_5$Pyr')]$^{2+}$	12
[(*p*-cymene)Ru(η^5-Pyr')][OTf]$_2$	13
Pyr' = 2,3,4,5-Me$_4$Pyr, Me$_5$Pyr	
[(*p*-cymene)Ru(η^5-Me$_5$Pyr)][OTf]$_2$	13
(PPh$_3$)$_2$Re(H)$_2$(η^5-Me$_5$Pyr)	15
[Cp*Rh(η^5-Me$_5$Pyr')]$^{2+}$	14
(PPh$_3$)$_2$Re(H)$_2$(η^3-Me$_5$Pyr)	15
[Cp*Ir(η^5-2,3,4,5-Me$_4$Pyr')]$^{2+}$	16
Pyrrolyl complexes	
CpFe(η^1-Pyl)L$_2$	17
L = CO, RNC, R$_2$NPF$_2$	
Fe(depe)$_2$H(η^1-Pyl)	18
dmpe = 1,2-bis(diethylphosphino)ethane	
Ru(dmpe)$_2$H(η^1-Pyl)	21
dmpe = 1,2-bis(dimethylphosphino)ethane	
Ni(R)(PMe$_3$)$_2$(η^1-Pyl')	20
R = alkyl, Pyl' = Pyl, 2,5-Me$_2$Pyl	
IrHCl(PMe$_3$)$_3$(η^1-Pyl)	22
Mn(CO)$_3$(η^5-Pyl)	10
[CpFe(η^5-Pyl')]$^+$	7, 10, 17
Pyl' = Pyl, 2,3,4,5-Me$_4$Pyl, 2,5-^{t}Bu$_2$Pyl	
(Carborane)Co(η^5-Pyl')	19
Pyl' = 2,5-Me$_2$Pyl, 2,3,4,5-Me$_4$Pyl	
Cp*Ru(η^5-Pyl)	23
Ru(η^5-Pyl)$_2$	23
(PR$_3$)$_2$RuCl$_2$(η^5-Pyl)	24
[(*p*-cymene)Ru(η^5-2,3,4,5-Me$_4$Pyl)][OTf]	13
[(*p*-cymene)Os(η^5-2,3,4,5-Me$_4$Pyr)][OTf]	13
Cp*Rh(H)(Pyl)	25

In only one case reported by Bergman [6] has the η^4 mode been observed, namely in Cp*Ir[η^4-1^{t}Bu-2,3,4,5(CO$_2$Me)$_4$Pyr], which was synthesized by coupling of two acetylene dimethylcarboxylate molecules onto Cp*Ir(NBu^t). The molecular structure of this complex, determined by X-ray diffraction, displays the pyrrole ligand clearly bonded through the two C=C bonds, while the nitrogen atom is bent away from the ring to a non-bonding distance from the iridium atom, as represented in Fig. 6.1. This is analogous to the structures of η^4-thiophene complexes described in Chapter 2. The most numerous metal complexes of pyrroles are no doubt the η^5-Pyr type, which are known for Cr, Mn, Fe, Co, Ru, Rh, Re, and Ir [7- 16].

Metal complexes containing the anionic pyrrolyl ligands are much more frequent throughout the periodic table, and the η^1N and η^5 binding modes are dominant in the coordination chemistry of such anions [1, 13]. η^1N-Pyrrolyl complexes may be synthesized by *in situ* N-H activation of Pyr by a low-valent metal fragment as in Eq. 6.1 [see *e.g.* 18, 22, 25].

$$[ML_n]^{x+} + \quad \begin{array}{c} H \\ N \end{array} \quad \longrightarrow \quad N{-}[ML_n]^{(x+2)+} \qquad (6.1)$$

Alternatively, $NC_4H_4{}^-M^+$ reagents (M = Na, K) -produced via N-H deprotonation with a base- are commonly reacted with electrophilic metal fragments (Eq. 6.2) to yield the desired compounds [see *e.g.* 17, 19, 20, 23].

$$[X{-}ML_n] + \quad \begin{array}{c} M^{\cdot+} \\ N^- \end{array} \quad \longrightarrow \quad N{-}[ML_n] + M'X \qquad (6.2)$$

Depending on the degree of electronic unsaturation of the reacting metal-containing fragment, the incoming pyrrolyl ligand binds through the N-atom only, or through the entire π-ring. Also, η^5-bonded pyrrolyl complexes may be transformed into η^1-bonded species by addition of extra ligands, as in the case of $CpFe(\eta^5$-Pyl) which reacts with Lewis bases such as CO or RNC to yield $CpFe(\eta^1$-Pyl)L_2[17].

Oxidative addition of C-H bonds leading to η^1(C)pyrrolyl(hydride) metal complexes (Eq. 6.3) has also been described [25].

$$[ML_n]^{x+} + \quad \begin{array}{c} H \\ N \end{array} \quad \longrightarrow \quad [ML_n]^{(x+2)+} \qquad (6.3)$$

The interaction of polynuclear metal complexes with pyrrole-type molecules has also attracted some attention, and it is interesting in connection with possible interactions with surfaces, where vicinal metal atoms may simultaneously interact with the pyrrole molecule. Cluster compounds containing intact coordinated pyrrole ligands are not known, since this reaction invariably involve N-H and/or C-H bond activation to yield derivatives most frequently containing bridging pyrrolyl ligands, as exemplified in Fig. 6.2.

An interesting example provided by Rakowski DuBois and coworkers [26] is the nucleophilic attack by the electron-rich Pyr molecule at the disulfide bond of the tetranuclear Mo complex $[Cp'Mo(\mu\text{-}S)_2(S_2CH_2)]_2{}^{2+}$ (Cp' = C_5H_5, C_5H_4Me) to yield a pyrrolyl-containing product $[(Cp'Mo)_2(S_2CH_2)(\mu\text{-}S)(\mu\text{-}S\text{-pyrrolyl})]^+$, according to Eq. 6.4.

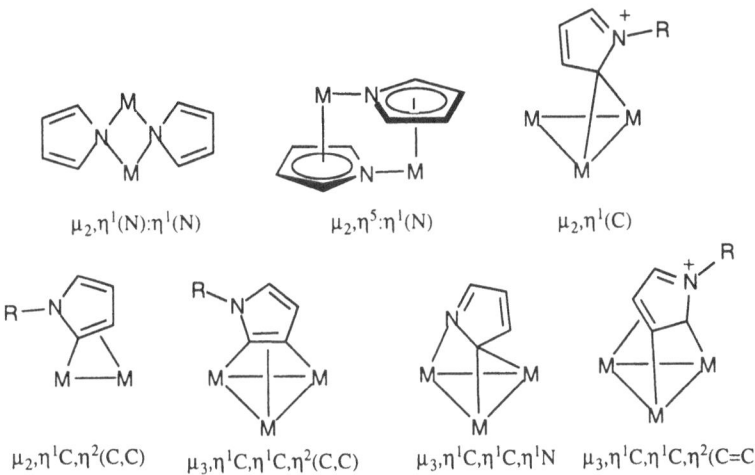

Fig. 6.2. Typical bonding modes of pyrrole-derived ligands in polynuclear metal complexes.

A mechanism has been proposed for this reaction, involving the heterolytic scission of the S-S bond induced by the nucleophilic ring; this is followed by the loss of a proton from the heterocycle to give a 1:1 mixture of μ-pyrrolylthiolate and μ-SH dimers. Alkyl-substituted pyrroles such as 1-MePyr or 1,2,5-Me$_3$Pyr behave similarly although the regiochemistry of electrophilic addition may vary depending on the number and on the position of the substituent. The interest of this reaction lies in the fact that it shows the interaction of pyrrole with a framework containing four Mo atoms in a sulfur-rich environment, resembling a catalytic MoS$_2$ surface.

$$(6.4)$$

Another interesting case is provided by the dimer [Cp*Ir(H$_2$)(μ-H)]$_2$ (Cp* = C$_5$Me$_5$), which has been shown in HDS modeling studies by Jones to cleave the C-S bonds of both thiophene and 2-methylthiophene (see Eq. 4.14, p. 118). Under comparable experimental conditions, this compound promotes the selective C-H bond cleavage of N-methylpyrrole shown in Eq. 6.5, leading to the formation of Cp*Ir(H){μ$_2$,η^{1}C,η^2(C,C)Pyr}(μ-H)$_2$IrCp* [27]. This is consistent with the higher energy barrier to C-N insertion as compared to C-S insertion reported in heterogeneous HDN/HDS.

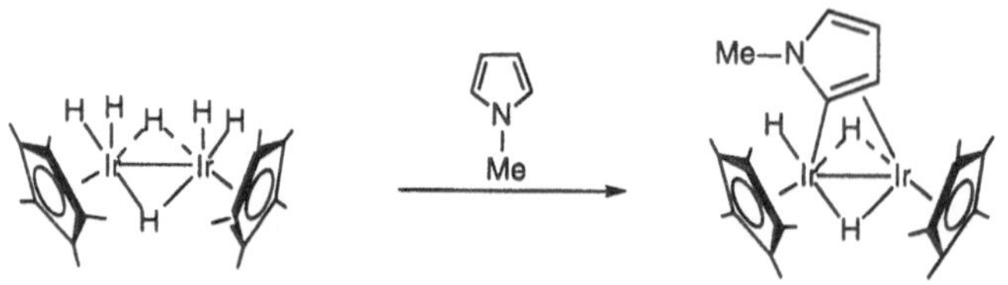

(6.5)

Other examples of metal clusters containing pyrrole-derived ligands, provided by the groups of Arce and Deeming, include the zwitterionic $Ru_3(\mu\text{-}H)(\mu^3,\eta^3\text{-}C_4H_3NMe)(CO)_9$ (**A**) [28], $Ru_3(\mu\text{-}H)_2(\mu^3,\eta^3\text{-}C_4H_2NMe)(CO)_9$ (**B**) [28], $Os_3(\mu\text{-}H)_2(\mu^3,\eta^3\text{-}C_4H_2NMe)(CO)_9$ (**C**) [29], $Os_3(\mu\text{-}H)(\mu,\eta^1\text{-}C_4H_2NMe)(CO)_9$ (**D**) [30].

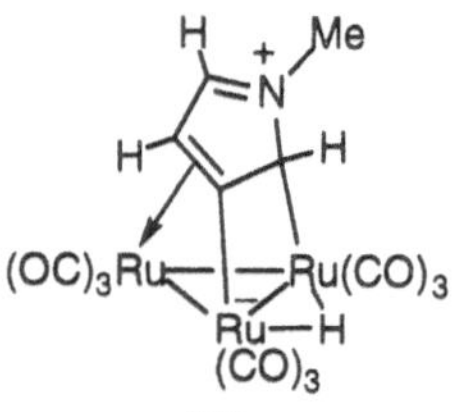

(A)

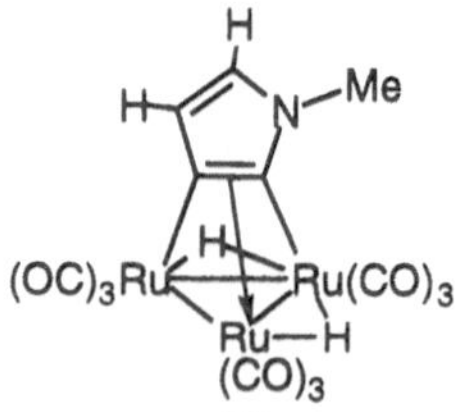

(B)

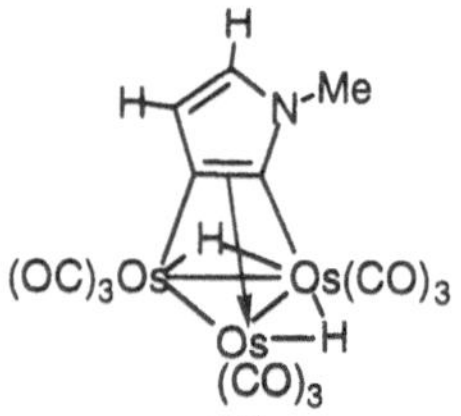

(C)

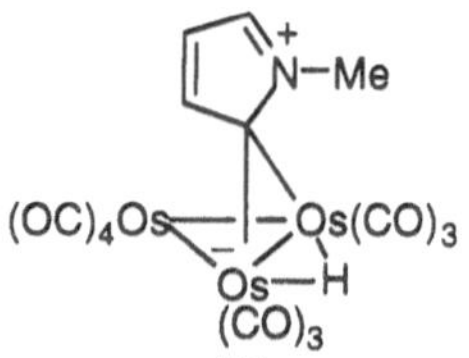

(D)

Indole (In) and indolyl (Inyl) differ from Pyr and Pyl in that they both contain a condensed arene ring. This makes a great difference in the interaction of these compounds and its derivatives with metal centers, since it produces a larger variety of possible coordination modes, as depicted in Fig. 6.3, and summarized in the examples contained in Table 6.2 [31-43].

Some examples of η^1N-bonded indole species have been reported, but they are better described as η^1N-3H-indolenine species. Yamauchi and coworkers have synthesized the N-bonded palladium (II) complexes $PdCl_2[\eta^1(N)\text{-}3H\text{-}In']_2$ (In' = 2-MeIn and 2,5-Me$_2$In) [31], while the group of Rakowski DuBois obtained the complex $Ir(CO)(PPh_3)_2[\eta^1(N)\text{-}3H\text{-}Me_2In]$ [32], and Gladysz reported a series of rhenium compounds of this type $CpRe(NO)(PPh_3)[\eta^1(N)\text{-}3H\text{-}In']_2$ [33]. Also, Arce et al. have described an osmium carbonyl cluster containing an orthometallated form of this tautomer of indole [30]. Nevertheless, in the case of indole and its derivatives, the dominant coordination mode is through the entire carbocyclic ring (η^6), an analogous situation to what has been observed in the case of metal complexes of benzothiophene (see Chapter 2).

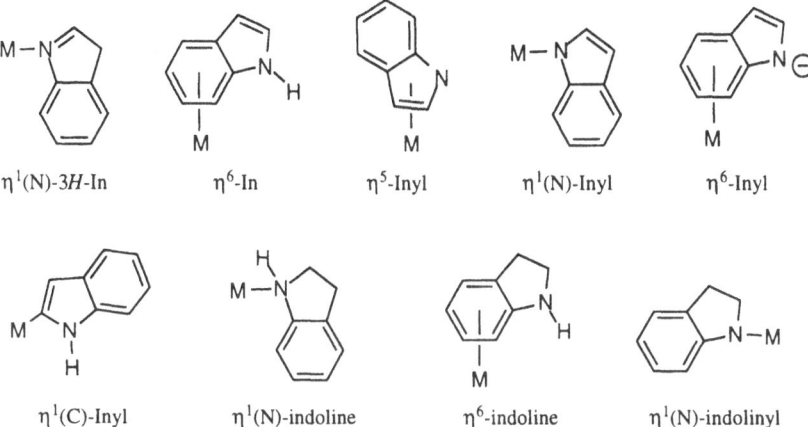

η^1(N)-3H-In η^6-In η^5-Inyl η^1(N)-Inyl η^6-Inyl

η^1(C)-Inyl η^1(N)-indoline η^6-indoline η^1(N)-indolinyl

Fig. 6.3. Known bonding modes of indole and some of its derivatives in metal complexes.

TABLE 6.2. Metal complexes of indole, indolyl and indoline ligands.

Complex	Refs.
Indole complexes	
$PdCl_2[\eta^1$(N)-3H-In']$_2$	31
In' = 2-MeIn and 2,5-Me$_2$In	
$\{Ir(CO)(PPh_3)_2[\eta^1$(N)-3H-Me$_2$In]\}^+$	32
$CpRe(NO)(PPh_3)[\eta^1$(N)-3X-In']$_2$	33
X= H, Me; In' = In, 3-MeIn, 3-EtIn	
$Cr(CO)_3(\eta^6$-In)	34
$[Mn(CO)_3(\eta^6$-In)]$^+$	35
$[Cp'Ru(\eta^6$-X-In]$^+$	36, 37
$[(p\text{-cymene})Ru(\eta^6$-In')][OTf]$_2$	38
In' = In, 1-MeIn, 2MeIn, 2,3-Me$_2$In	
$[Cp^*M(\eta^6$-In)]$^{2+}$	39
M = Rh, Ir	
Indolyl complexes	
$[(p\text{-cymene})Ru(\eta^1$-Inyl)(NCMe)$_2$][OTf]$_2$	40
$CpRe(NO)(PPh_3)(\eta^1$-Inyl')	33
L = CO, RNC, R$_2$NPF$_2$	
$Ir(CO)(PPh_3)_2[\eta^1$(N)-2,3-Me$_2$Inyl]	32
$Mn(CO)_3(\eta^5$-2-MeInyl)	41
$[Cp^*M(\eta^5$-Inyl)]$^+$	39
M = Rh, Ir	
$[(p\text{-cymene})Ru(\eta^6$-In')][OTf]	38
In' = In, 1-MeIn, 2MeIn, 2,3-Me$_2$In	
Indoline Complexes	
$[(\text{cymene})Ru(\eta^1$-Indoline)(NCMe)$_2$][OTf]$_2$	40
$PdCl_2(PPh_3)(\eta^1$-indoline)	43
$Cr(CO)_3(\eta^6$-1-MeIndoline)	42
$[(p\text{-cymene})Ru(\eta^6$-Indoline)][OTf]$_2$	41
In' = In, 1-MeIn, 2MeIn, 2,3-Me$_2$In	

This reflects the well-known very high stability of arene bonding to metal centers, and also conditions the chemistry of such complexes, which is mainly related to the activation of the carbocyclic ring toward attack by nucleophiles. It is not clear whether this type of bonding, so commonly encountered in organometallic chemistry, may be invoked when discussing HDN mechanisms on solid catalysts.

As shown by the data contained in table 6.2, the indolyl anion coordinates mononuclear metal-containing fragments using the nitrogen atom ($\eta^1(N)$ mode), the entire heterocyclic ring (η^5 mode) or the entire carbocyclic ring (η^6 mode), thus making it a very versatile ligand. Some of these species are capable of easily interconverting, as for instance, in the interesting case observed by Maitlis and coworkers [39] of the base-assisted η^6-In $\rightarrow$ η^5-indolyl shift that takes place when the organonitrogen ligand is coordinated to the Cp*M (M = Rh, Ir) fragment (see Eq. 6.6.).

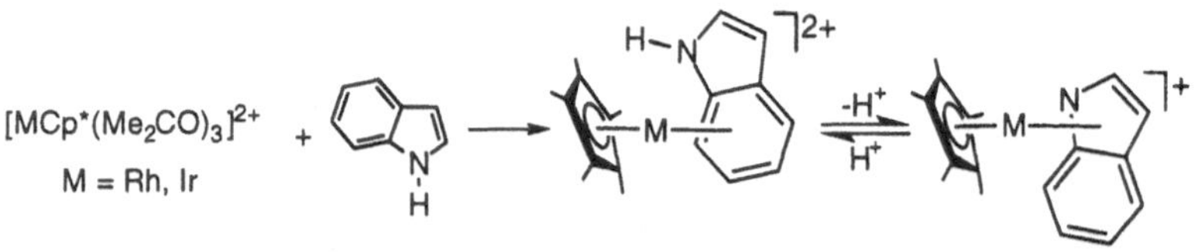

$$[MCp^*(Me_2CO)_3]^{2+} \quad (6.6)$$
$$M = Rh, Ir$$

Also the acid-promoted $\eta^1(N)$-indolyl $\rightarrow$ $\eta^1(N)$-indolenine conversion on CpRh(PPh$_3$)(NO) shown in Eq. 6.7 has been reported by Gladysz [33].

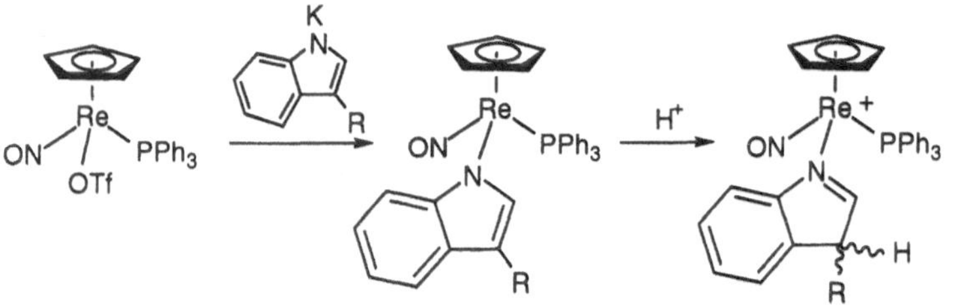

$$(6.7)$$

In an analogous manner to the reaction described above for Pyr [26], 1-Me-In has been found to react with the μ-S$_2$ tetramer [Cp'Mo(μ-S)$_2$(S$_2$CH$_2$)]$_2^{2+}$ to give the electrophilic substitution product (Eq. 6.8).

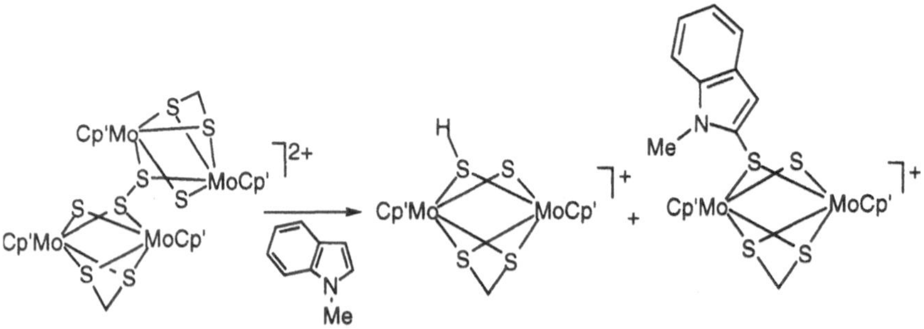

$$(6.8)$$

Electron-rich metal fragments can react with 2-substituted In yielding either the kinetic C-H or the thermodynamic N-H insertion products. C-N bond activation, on the other hand, is not observed, in contrast with C-S bond breaking, which is readily promoted by electron-rich metal centers, as described in Chapter 4.

Despite the fact that carbazole is a very good model compound of the most highly refractory HDN substrates, its coordination chemistry of has not been investigated in any detail, and to our knowledge, there are no examples of metal complexes containing an intact carbazole ligand. Thus, N-H activation leading to metal hydrido derivatives containing the carbazoyl ligand bound through the nitrogen atom has been reported by Jones [25], in the reaction of Cp*Rh(PMe)$_3$(H)(Ph) with carbazole, which yields Cp*Rh(PMe)$_3$(H)(η^1N-carbazoyl), together with the kinetically favored C-H activation isomer (see Eq. 6.9). García *et al.* have also shown that zerovalent Pt(PR$_3$)$_n$ complexes that are very active for C-S bond breaking reactions, are capable of activating the N-H bond of carbazole to produce exclusively the N-bonded derivatives *trans*-Pt(PR$_3$)$_2$(H)(η^1N-carbazoyl) [44].

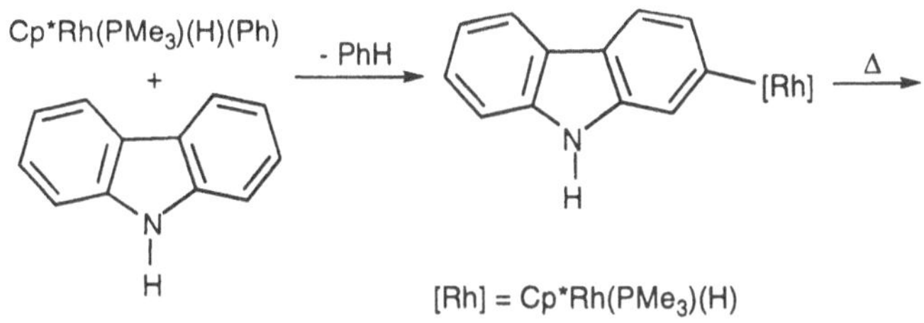

$$[Rh] = Cp*Rh(PMe_3)(H) \tag{6.9}$$

The coordination chemistry of indoline, a compound often obtained by selective hydrogenation of the heterocyclic ring of In, has also been somewhat investigated [43]. Both the η^6 and η^1(N) coordination modes, as well as their interconversion, have been reported [33, 40, 43]. This type of ligand is of particular relevance, since it is known that N-aromatics need to be hydrogenated before nitrogen removal can take place in standard HDN catalysis. Rakowski DuBois has published interesting data on the coordination, acid/base properties and activation of indoline by (*p*-cymene)Ru(II) fragments [40, 43], summarized in Eq. 6.10. The complex [(*p*-cymene)Ru(η^1-indoline)(CH$_3$CN)$_2$)](OTf)$_2$ converts to [(*p*-cymene)Ru(η^6-indoline)](OTf)$_2$ upon gentle heating in CH$_2$Cl$_2$ and can be deprotonated by use of *e.g.* triethylamine to give the corresponding (η^1-indolinyl) complex [40].

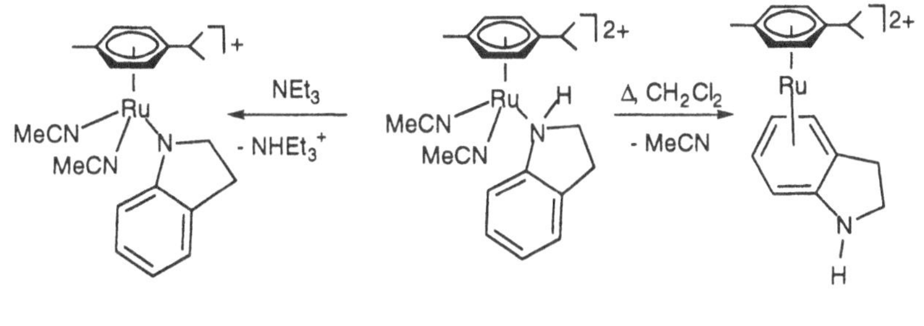

$$\tag{6.10}$$

6.2.2. COMPLEXES WITH PYRIDINE, QUINOLINE, AND RELATED LIGANDS

The more basic pyridines and quinolines are generally better ligands than the 5-member heterocycles, and consequently, the coordination chemistry of such molecules is considerably more developed. The known bonding modes of Py and its alkyl-substituted derivatives to single metal sites are illustrated in Fig. 6.4, and a representative list of relevant metal-pyridine and metal-quinoline complexes is contained in Table 6.3. The most common coordination mode for Py is by far through the nitrogen atom, which uses its lone pair for interaction with Lewis acceptors [45]. Unlike thiophene, which binds to metals in a "bent" fashion (see Chapter 2), the lone pair in the nitrogen atom of Py is located in the ring plane, and thus the M-N vector in η^1(N)-Py complexes is also in the ring plane.

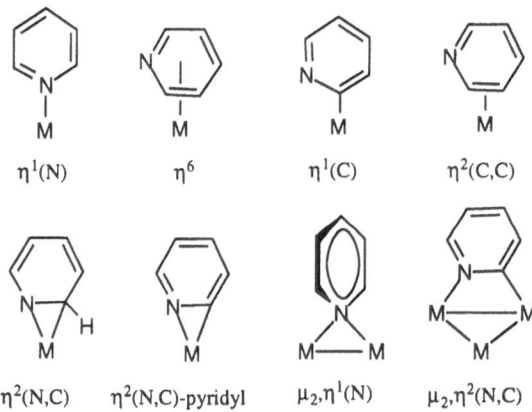

Fig. 6.4. Bonding modes of pyridines in metal complexes.

Indeed, pyridine is one of the most frequently encountered "classical" ligands in coordination chemistry, and there are examples of η^1(N)-Py complexes for virtually every transition metal in more than one oxidation state [45]; therefore no attempt will be made to include such numerous compounds in Table 6.3. The η^1(C)-Py derivatives obtained by replacement of a proton by a metal fragment, usually *via* C-H bond activation by electron-rich metal fragments [46], are also to be considered formally η^1(C)-Py complexes. Another fairly common coordination mode of Py is the η^6 through the entire ring [47-51]. Therefore, many homogeneous studies directed at modeling HDN reactions of Py have centered on the η^1(N) and η^6 modes and the factors controlling their interconversion and/or prevalence.

Fish and coworkers have carried out extensive studies on the chemistry of Ru(II) and Rh(III) complexes of N-heterocycles, in connection with their work on metal complex catalyzed hydrogenation of such substrates (see Chapter 3) [47-5o]. From their results, it is clear that in the absence of any steric constraints, the η^1(N) mode prevails over the η^6 one, particularly with the more electrophilic Rh(III) center. For example,

TABLE 6.3. Examples of unusual metal complexes of pyridine and quinoline.

Complex	Refs.
$\mu^2,\eta^1(N)$-complexes	
$Mo_2O_2[S_2P(OPr^i)_2]_2(\mu\text{-}O)(\mu\text{-}S)[\mu,\eta^1(N)\text{-}Py]$	53
$\eta^2(C,C)$-complexes	
$[Os(NH_3)_5(\eta^2(C,C)\text{-}2,6\text{-}Me_2Py)][OTf]_2$	52
$\eta^2(C,N)$-pyridine and pyridyl-complexes	
$[\eta^2(C,N)\text{-}Py]Nb(silox)_3$	56
$[\eta^2(C,N)\text{-}Py]Ta(silox)_3$	54
$[\eta^2(C,N)\text{-}2,4.6\text{-}tBu_3Py]Ta(OAr)_2Cl$	57, 58
$Ar = 2,6\text{-}^iPrPh$	
$[\eta^2(C,N)\text{-}Q']Ta(OAr)_nCl_m$	2, 63-65
$Q' = Q, 6\text{-}MeQ; Ar = 2,6\text{-}^iPrPh$	
$n = 2, 3, m = 1, 0$	
$Cp^*Lu(\eta^2(N,C)\text{-}NC_5H_4)$	59
$Cp^*Sc(\eta^2(N,C)\text{-}NC_5H_4)$.	60
η^6-complexes	
$[(Cp^*Ru(\eta^6\text{-}Py)][OTf]$	48, 51
$Cp^*Rh(\eta^6\text{-}2,4,6\text{-}Me_3Py)$	49

$[Cp^*Rh(CH_3CN)_3]^{2+}$ reacts with 2-methylpyridine and 2,6-dimethylpyridine yielding compounds of the formula $[Cp^*Rh(\eta^1(N)\text{-}Py')(CH_3CN)_2]^{2+}$, while the trisubstituted 2,4,6-trimethylpyridine prefers to form the η^6 adduct $[Cp^*Rh(\eta^6\text{-}Py')]^{2+}$. In the case of the analogous complex $[CpRu(CH_3CN)_3]^+$, in which the Ru(II) center is more electron-rich than Rh(III), a greater propensity of Ru(II) to stabilize the η^6 mode -*via* π-back bonding from a filled metal orbital to a π^* orbital of Py- is observed. Fish was able to follow by NMR spectroscopy the $\eta^1(N)\rightarrow\eta^6$ interconversion of 2-methylpyridine and 2,4-dimethylpyridine illustrated in Eq. 6.11. The initially formed $\eta^1(N)$ complex slowly loses MeCN and converts to the thermodynamic η^6-Py product. This rearrangement has been proposed to involve an η^4-Py intermediate, but no such complex has ever been actually observed.

$$[CpRu(MeCN)_3]^+ + \underset{R}{\overset{R}{\bigcirc}}N \xrightarrow[\text{5 min}]{} \underset{R}{\overset{R}{\bigcirc}}N\text{-}Ru^+\underset{NCMe}{\overset{NCMe}{|}} \xrightarrow[\text{-2 MeCN}]{\text{20 h, rt}} Ru^+\underset{R}{\overset{R}{\bigcirc}}N\text{-}R \qquad (6.11)$$

With the unsubstituted Py, the stable tris(N-bonded) adduct $[CpRu(\eta^1(N)\text{-}Py)_3]^+$ has been obtained. However, when the permethylated ligand Cp* is employed in order to increase the electron density around the Ru center more, the complex $[Cp^*Ru(\eta^1(N)\text{-}Py)_3]^+$ is a kinetic product that thermally converts to the corresponding η^6 derivative (Eq. 6.12).

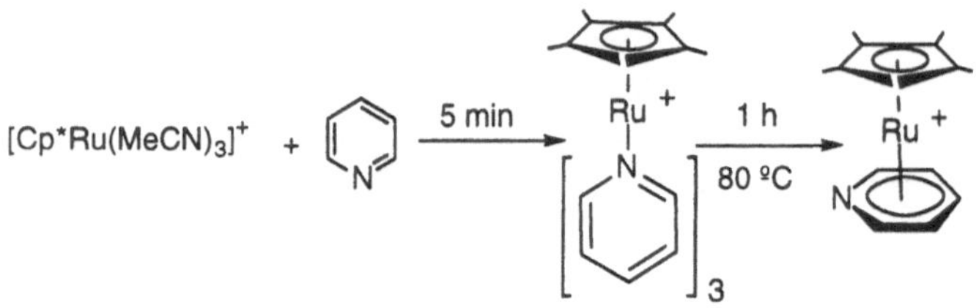

$$[Cp^*Ru(MeCN)_3]^+ \; + \; \text{(pyridine)} \;\; \xrightarrow{\;5\text{ min}\;} \;\; \text{(complex 3)} \;\; \xrightarrow[80\,^\circ C]{\;1\text{ h}\;} \;\; \text{(complex)} \qquad (6.12)$$

A rather rare coordination mode of pyridines is the olefin-like $\eta^2(C=C)$ mode, which is exclusively stabilized by the electron-rich $[Os(NH_3)_5]^{2+}$ fragment, as described by Taube and coworkers for 2,6-dimethylpyridine (lutidine) [52]. $[Os(NH_3)_5]^{2+}$ is also known to bind in this manner to thiophenes, pyrroles and arenes. The olefin-like complex easily rearranges to the $\eta^1(N)$ mode by a one-electron oxidation (see Eq. 6.13).

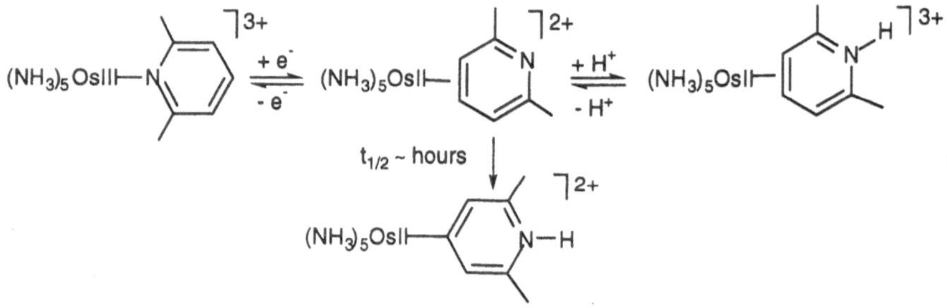

$$(NH_3)_5Os^{III}\text{—N}\;\text{(ring)}\;\;^{3+} \; \underset{-e}{\overset{+e^-}{\rightleftharpoons}} \; (NH_3)_5Os^{II}\text{—(ring)}\;\;^{2+} \; \underset{-H^+}{\overset{+H^+}{\rightleftharpoons}} \; (NH_3)_5Os^{II}\text{—(ring)}\;^{3+}$$

$$t_{1/2} \sim \text{hours} \Big\downarrow$$

$$(NH_3)_5Os^{II}\text{—(ring)}\;N\text{—H}\;\;^{2+} \qquad\qquad (6.13)$$

This $\eta^2(C=C)\rightarrow\eta^1(N)$ rearrangement of lutidine is of great importance, as it relates the bonding mode of the N-heterocycle to the oxidation state of the metal. As discussed in the preceding Chapters, the active metal centers of heterogeneous HDS/HDN catalysts may have different and variable oxidation states due to low energy electron-transfer processes involving both surface and bulk metal atoms. Each oxidation state may be associated with a particular bonding mode and a specific reactivity pattern. Interestingly, the $\eta^2(C=C)$ mode of lutidine rearranges with time to give an Os(II) lutidinium ylide. In contrast, the $\eta^2(C,C)$ mode is maintained when $[Os(NH_3)_5(\eta^2\text{-lutidine})]^{2+}$ is protonated to give a lutidinium derivative.

A further interesting bonding mode of pyridine is the one in which the N atom acts as a bridge between two metal centers [53]. This unique form of coordination has been found in $Mo_2O_2[S_2P(OPr^i)_2]_2(\mu\text{-O})(\mu\text{-S})[\mu,\eta^1(N)\text{-Py}]$ (see Eq. 6.14).

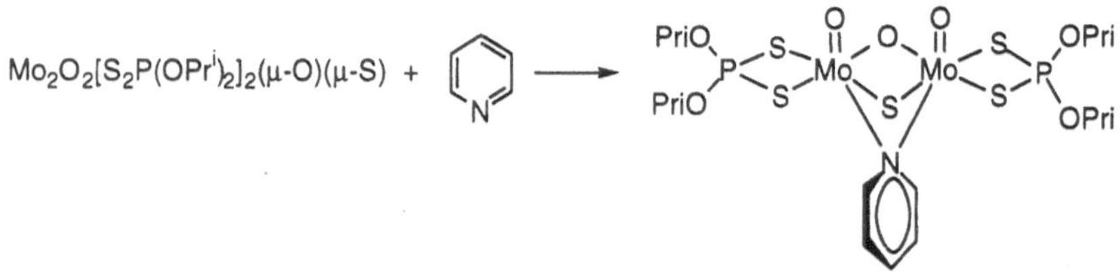

$$Mo_2O_2[S_2P(OPr^i)_2]_2(\mu\text{-O})(\mu\text{-S}) \; + \; \text{(pyridine)} \;\; \longrightarrow \;\; \text{(complex)} \qquad (6.14)$$

This compound is of particular relevance to HDN in the sense that the pyridine molecule is bound to an array of two Mo atoms in a sulfur-rich environment, reminiscent of a MoS_2 surface. In this Mo dimer, the Py ring is almost coplanar with the bridging O and S atoms, while the Mo-N distances are longer than regular Mo-N distances in mononuclear complexes.

Despite the fact that the bonding modes of pyridines hitherto described in this Section represent reasonable models for a variety of forms of adsorption of Py onto HDN catalyst surfaces, none of them has been found to be adequate for C-N bond scission to occur in solution. In contrast, the curious $\eta^2(N,C)$ coordination mode effectively activates the Py ring toward C-N bond activation, as discussed in the following section. The first $\eta^2(N,C)$-Py complex, $(silox)_3Ta(\eta^2(N,C)-NC_5H_5)$ (silox = $^tBu_3SiO^-$), was prepared by Wolczanski and coworkers in 1988 [54] through the direct reaction of $Ta(silox)_3$ with the appropriate pyridine, as illustrated in Eq. 6.15.

$$(silox)_3Ta \quad + \quad \underset{R = H, \ Me}{\overset{}{\boxed{}}} \longrightarrow (silox)_3Ta \quad \quad \quad (6.15)$$

The structure of $(silox)_3Ta(\eta^2(N,C)-NC_5H_4R)$ is best regarded as a Ta(V) metallaaziridine where the aromaticity of Py has been substantially perturbed by strong $Ta(d\pi){\rightarrow}Py(\pi^*)$ back-bonding, which allows the metal to attain its highest oxidation state; also, MO calculations on the model compound $(OH)_3Ta(\eta^2(N,C)-NC_5H_4R)$ have shown that the $\eta^2(N,C)$ coordination mode is preferred over the $\eta^1(N)$ because it avoids a repulsive Ta d_{z^2}-Py $N(\sigma)$ interaction that takes place in the N-bonded isomer [55]. Reaction of $(silox)_3Ta(\eta^2(N,C)-NC_5H_5)$ with 2-picoline and lutidine gives similar $\eta^2(N,C)$ products whose formation has been proposed to occur via a nucleophilic attack by the starting Ta(III) compound at the LUMO of the N-heterocycle (predominantly C=N π^*). A detailed in situ NMR study with the sterically congested lutidine showed that C-H bond activation may kinetically compete with $\eta^2(N,C)$ coordination (Eq. 6.16).

$$(silox)_3Ta \quad \underset{25 \ ^oC}{\overset{20 \ min}{\rightleftharpoons}} \quad (silox)_3Ta \quad + \quad \underset{}{\overset{}{\boxed{}}} \quad \underset{25 \ ^oC}{\overset{> 8 \ h}{\rightleftharpoons}} \quad (silox)_3Ta \quad \quad (6.16)$$

Similar starting materials for other early transition metals, such as $(silox)_3M$ (M = Sc, Ti, V) react with Py in a conventional manner leading to $\eta^1(N)$ derivatives [55]. The absence of four electron repulsion problems in the Sc, Ti and V derivatives has been invoked to account for the selective formation of $\eta^1(N)$ adducts. Reduction of $(silox)_3NbCl_2$ with Na/Hg in Py affords a kinetic $\eta^2(N,C)$-Py product which thermally undergoes C-N insertion (*vide infra*) [56]. A related Ta compound containing an $\eta^2(N,C)$ pyridine ligand is $(DIPP)_2ClTa(\eta^2(N,C)-2,4,6-NC_5H_2{}^tBu_3)$ (DIPP = $2,6-OC_6H_3{}^iPr_2$) obtained by Wigley and coworkers through an indirect route involving insertion of a nitrile into a tantallacyclopentadiene complex [57, 58].

Besides being precursors to C_α-N bond scission, intermediates containing $\eta^2(N,C)$ pyridine ligands are thought to intervene in C-H activation reactions leading to complexes containing $\eta^2(N,C)-NC_5H_4{}^-$ ligands. Watson reported the formation of such a lutetium compound, $Cp^*Lu(\eta^2(N,C)-NC_5H_4)$ [59], while Bercaw achieved the synthesis of the scandium analogue $Cp^*Sc(\eta^2(N,C)-NC_5H_4)$ [60] . Similar Ti derivatives with 2-substituted pyridines have been described by Teuben [61]. A common synthetic procedure to these $\eta^2(N,C)$-pyridyl compounds involves the reaction of the substrate with alkyl or hydride complexes as exemplified for the Sc derivative in Eq. 6.17. The intermediacy of an $\eta^1(N)$-Py adduct prior to C-H insertion has been suggested on the basis of an in situ NMR study [60]. Deeming and Arce have reported the reaction of pyridine with a triosmium cluster, which results in a trinuclear compound $[Os_3(\mu-H) (\mu-NC_5H_4)(CO)_{10}]$, in which Py uses the N atom and the C_α atom for coordination to two metal centers [62].

$$Cp_2{}^*Sc\text{—}R \; + \; \underset{}{\vcenter{\hbox{(pyridine)}}} \quad \longrightarrow \quad Cp_2{}^*Sc\!\!<\!\!\underset{}{\vcenter{\hbox{(pyridyl)}}} \; + \; RH \qquad (6.17)$$

$R = H$, alkyl

The coordination chemistry of quinoline (Q) does not differ significantly from that of Py except for the interactions with metal clusters. In this case, the presence of the carbocyclic ring allows a larger variety of bonding modes, summarized in Fig. 6.5. As for Py, the most common coordination mode of Q in mononuclear complexes is the $\eta^1(N)$ [45, 47-51], and it is often in equilibrium with the η^6-arene form.

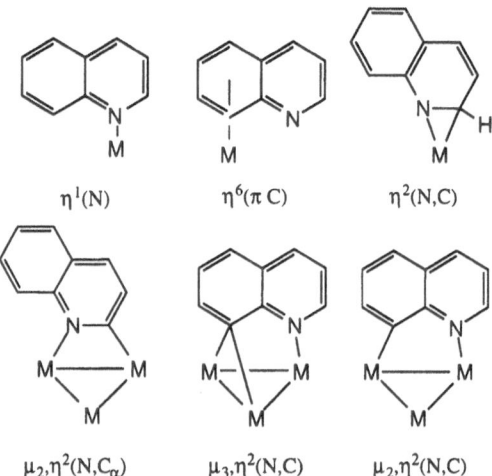

$$\eta^1(N) \qquad \eta^6(\pi\,C) \qquad \eta^2(N,C)$$

$$\mu_2,\eta^2(N,C_\alpha) \qquad \mu_3,\eta^2(N,C) \qquad \mu_2,\eta^2(N,C)$$

Fig. 6.5. Bonding modes of quinolines in metal complexes.

No examples of η^6 coordination involving the heterocyclic ring have been reported. Fish has reported the $\eta^1(N) \rightarrow \eta^6$ rearrangement of quinoline [47, 50] (see Eq. 6.18), analogous to the Py reaction shown in Eq. 6.12. The recursor $[CpRu(CH_3CN)_3]^+$ reacts with Q to yield a kinetic $\eta^1(N)$ product that spontaneously converts to the thermodynamic η^6 derivative through loss of the labile acetonitrile. If Cp* is used instead of Cp the π back-bonding capability of the Ru center is increased and thus the η^6 complex is readily formed while the corresponding $\eta^1(N)$ adduct is not observed [50].

$$[CpRu(MeCN)_3]^+ \; + \quad \xrightarrow{\text{5 min}} \quad \overset{\text{Ru}^+}{|} \quad \xrightarrow[\text{rt}]{\text{20 h}} \quad \overset{\text{Ru}^+}{|} \qquad (6.18)$$

Also similarly to the case of Py, a Ta $\eta^1(N)$-Q complex may be a precursor to the $\eta^2(N,C)$ derivative in the formation of $(DIPP)_3Ta(\eta^2(N,C)-Q)$ (DIPP = 2,6-$OC_6H_3{}^iPr_2$) (Eq. 6.19). This reaction requires an increase of the electron density at the metal center to allow for an efficient $M(d\pi) \rightarrow Q(\pi^*)$ back-bonding interaction [2, 63-65].

$$Ta(OR)_3Cl_2(OEt_2) \quad \longrightarrow \quad \overset{Cl}{\underset{RO}{\diagdown}} \overset{OR}{\underset{Cl}{Ta}} \overset{}{\diagup} \overset{}{\underset{OR}{}} \quad \xrightarrow{Na/Hg} \quad \overset{}{\underset{RO}{}} \overset{H}{\underset{RO}{}} \overset{}{\underset{OR}{Ta}} \qquad (6.19)$$

The activation of Q and its alkylated derivatives on metal clusters such as $Ru_3(CO)_{12}$ [66] and $Os_3(CO)_{10}(CH_3CN)_2$ [67-69] results in the bonding modes summarized in Fig. 6.5. In all cases, N-coordination is accompanied by C-H insertion by an adjacent metal center.

Tetrahydroquinoline (THQ) is another interesting molecule to consider in modeling HDN, since it is known that the reversible hydrogenation of the heterocyclic ring of quinoline is crucial for C-N bond breaking to take place (see Chapter 1). $Ru_3(CO)_{12}$ reacts with THQ to give a complex containing two Q ligands bonded in a $\mu_2,\eta^2(N,X_\alpha)$ mode (Eq. 6.20) [66]. Dehydrogenation of THQ to Q has also been found to occur by reaction with either $Os_3(CO)_{10}(CH_3CN)_2$ [69] or $Os_3(CO)_{12}$ [67].

In summary, Py, Q, and their alkylated derivatives tend to bind through the nitrogen atom -$\eta^1(N)$- to electrophilic metal centers with a single vacant coordination site. If three coordination sites are available and the metal center has filled $d\pi$ orbitals, the preferred bonding mode is the η^6. In the intermediate situation offering two free coordination sites, the electronic nature of the metal ion determines whether the substrate binds to the metal using a C=C bond or the C=N bond.

$$Ru_3(CO)_{12} + \text{(tetrahydroquinoline)} \xrightarrow[130\ ^\circ C]{heptane} \text{(complex A)} + \text{(complex B)} \quad (6.20)$$

$$+ \quad H_4Ru_4(CO)_{12}$$

Electron-rich systems based on late-transition metals prefer the olefinic $\eta^2(C,C)$ mode, while electrophilic early-transition metal systems stabilize the $\eta^2(N,C)$ mode. As will be shown in the next section, the binding mode of the N-heterocycle ultimately controls the reactivity of the activated substrate.

Transition metal complexes of other relevant organonitrogen compounds such as isoquinoline, 5,6-benzoquinoline, 7,8-benzoquinoline, acridine, and phenanthridine are also known, and they contain the N-ligand coordinated in either the $\eta^1(N)$ or the η^6-arene fashion [49, 69, 70]. The triosmium cluster $Os_3(CO)_{10}(CH_3CN)_2$ reacts with polyaromatic N-heterocycles such as 5,6-benzoquinoline and phenanthridine in an analogous manner to Py and Q, yielding $\mu_3,\eta^2(N,C,C)$ complexes [69].

6.3 Reactions of N-Heterocycles in Transition Metal Complexes.

When N-heterocycles are coordinated to a transition metal in a molecular complex their reactivity is usually enhanced with respect to the free molecules. Their reactions, in turn, depend on several factors, notably the electronic and geometric characteristics of the metal-containing fragment, as well as the nature of the organonitrogen substrate itself; however, no general trends can be obviously extracted from the accumulated literature. The most thoroughly studied family of reactions of N-heterocycles on metal complexes are no doubt those involving the homogeneous hydrogenation of the N-containing rings of a variety of substrates, for which very efficient catalysts have been discovered, and the reaction mechanisms have been elucidated in great detail. These homogeneous hydrogenation reactions are discussed at length in Chapter 3.

In this Section examples of other reactions, representative of the some of the elementary steps that are known or believed to occur during HDN *after hydrogenation*, are described. Of particular interest are the rather uncommon reactions leading to C-N bond activation and denitrogenation.

6.3.1. REACTIONS OF COORDINATED PYRROLES, INDOLES, AND RELATED COMPONDS.

As stated in Section 6.2.1 (p.154), no examples of $\eta^1(N)$ pyrrole complexes are known, and the corresponding analogues reported for indole are better described as η^1N-3H-indolenine species. Therefore the "N-only" coordination mode of intact Pyr or In does not represent a very useful model for the purpose of HDN modeling. The η^2(C=C)-bonding mode does produce some activation of the ring, as in the case of $[Os(NH_3)_5(\eta^2\text{-}2,5\text{-}Me_2Pyr)]^{2+}$, which undergoes selective protonation by mild acids to produce a 3H-pyrrolium species, or nucleophilic attack by OH⁻ at the carbon atom adjacent to nitrogen [5]. This type of coordination, however, is very rare –indeed only observed for the Os complex mentioned above- and thus its usefulness for HDN modeling is also limited.

The reactions of pyrrolyl complexes, in turn, are mainly related to protonation-deprotonation processes, as well as rearrangements between the η^1(N) and η^5 forms. For instance, the η^5-Pyl complexes $[(p\text{-cymene})M(\eta^5\text{-}NC_4Me_4)]OTf$ (M = Ru, Os) are readily converted to the corresponding η^5-Pyr by selective alkylation at the nitrogen atom [13]. Interestingly, the resulting η^5-Pyr ligand in $[(p\text{-cymene})M(\eta^5\text{-}MeNC_4Me_4)](OTf)_2$ undergoes regio- and stereoselective hydride addition to an α-carbon atom to produce a curious η^4-MeNC$_4$Me$_4$H derivative. This latter compound further reacts with protic acids leading to the formation of a free iminum salt resulting from the partial hydrogenation of the pyrrole ring by the H⁻/H⁺ couple (Eq. 6.21).

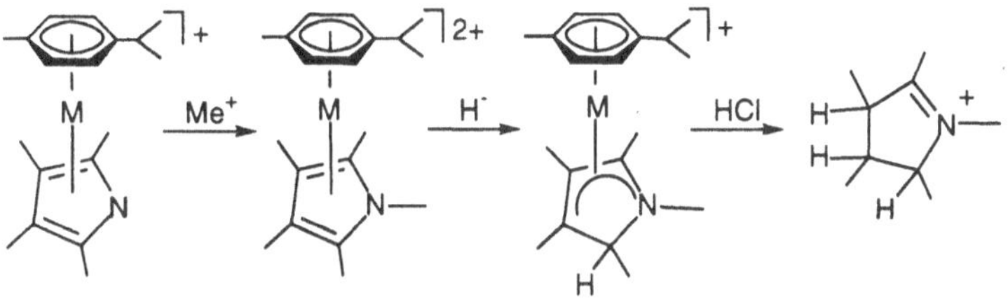

(6.21)

This is an interesting model for the initial hydrogenation steps of HDN, but the compounds involved never experience C-N bond breaking reactions. Similarly, η^6-In complexes often undergo deprotonation of the N-H bond to give η^5-Inyl complexes, which can be protonated back to the corresponding indole derivatives (See Eq. 6.6) [39]. Besides, the η^6-bonded indoles in metal complexes are activated toward nucleophilic substitution *at the carbocyclic ring* [71, 72] as well as to mild hydrogenation of the heterocyclic ring over 5% Rh/C [43]. The latter reaction, disclosed by the group of Rakowski DuBois, is important in that it shows that η^6 coordination through the carbocyclic ring activates the heterocyclic moiety of the molecule toward reduction to the corresponding η^6-indoline complex (Eq. 6.22).

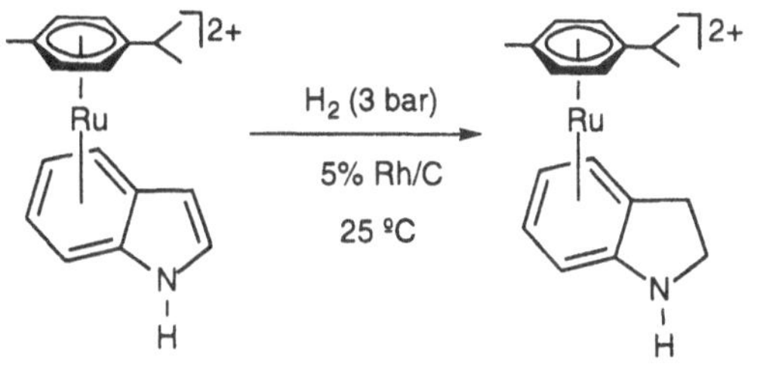

(6.22)

Again, this is an interesting model for a possible way of adsorbing and hydrogenating indoles on solid catalysts at the early stages of HDN. η^1(N)-Indolyl complexes are activated toward protonation at the β-carbon atom, which transforms them into η^1(N)-indolenine derivatives (See *e.g.* Eq. 6.7) [33].

A very interesting observation reported by Sweigart and coworkers is the fact that $Mn(CO)_3(\eta^6$-In) reacts with $Pt(PPh_3)_3$ exclusively by N-H insertion to form the heteronuclear trimer $[(CO)_3Mn(\mu^2,\eta^6,\eta^1$(N)-$C_8H_6N)]_2[Pt(PPh_3)_2]$. The complex fails to react by C-N bond breaking, whereas the analogous $Mn(CO)_3(\eta^6$-BT) (BT = benzothiophene) and $Mn(CO)_3(\eta^6$-BF) (BF = benzofuran) readily undergo C-S or C-O ring opening upon interaction with $Pt(PPh_3)_3$ under similar reaction conditions (see Chapter 4). This is illustrative of the much lower propensity of C-N bonds of heterocycles to react by insertion at the metal atom, as compared to S- and O-heterocycles [1, 73-76]. Metal clusters containing ligands derived from N-H and/or C-H activation of N-heterocycles do not usually react further with H_2, nucleophiles or electrophiles even under harsh experimental conditions, and thus they are not good model compounds for HDN-related reactions.

In a unique case of *homogeneous denitrogenation* of a pyrrole, the dimer $[(tmeda)_2Nb_2Cl_5Li(tmeda)]$ reacts with the lithium salt of 2,5-dimethylpyrrole, 2,5-$Me_2C_4H_2NLi$, through a complicated redox process (Eq. 6.23), to give a mixture of two products characterized by X-ray diffraction (tmeda = N,N,N',N'-tetramethylethylendiamine) [77].

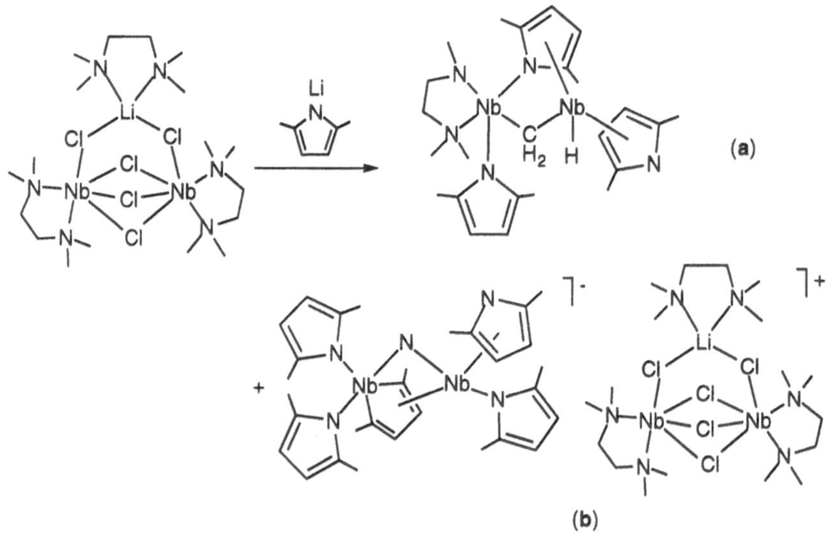

$$(6.23)$$

The neutral product (**a**) contains a Nb(IV) atom coordinated to a terminal hydride ligand and to two η^5-pyrrolyl rings, one of which is also η^1(N)-bonded to the Nb(III) atom. The coordination of the latter is completed by another η^1(N)-pyrrolyl ligand, a tmeda molecule, and a bridging CH_2 group generated through fragmentation of tmeda. More interesting still is the structure of the anion in the second product (**b**)

[{(2,5-Me$_2$C$_4$H$_2$N)$_2$Nb}{(η^5-2,5-Me$_2$C$_4$H$_2$N)(2,5-Me$_2$C$_4$H$_2$N)Nb}(μ-N)(μ,η^1: η^1: η^4-1,4-Me$_2$C$_4$H$_2$)], which displays two Nb(IV) atoms connected by a bridging nitride ligand and by a dienediyl group resulting from *denitrogenation* of a pyrrole ring. This anion was isolated as the [(tmeda)$_2$Nb$_2$Cl$_5$Li(tmeda)]$^+$ salt. This implies that during the reaction, the cleavage of C-N bonds in both the pyrrole ring and the saturated amine takes place. Although the mechanism of the reaction is not known, the formation of the anionic product is thought to involve the cooperative attack of the two Nb(II) centers on one pyrrolyl anion. The four electrons necessary to form the nitride and the dienediyl group are obtained from the five electrons provided by the oxidation of the four Nb(II) centers during the formation of the anion and the cation.

6.3.2. REACTIONS OF COORDINATED PYRIDINES, QUINOLINES, AND RELATED COMPOUNDS

As noted in Section 6.2.2, the η^1(N) mode represents for Py the most common way to coordinate to metal ions in complexes. Nevertheless, this coordination mode does not result in any appreciable activation of the heterocycle, or any particular reactivity of the complexes of relevance to HDN. In the η^2(C=C) mode [52], as in *e.g.* [Os(NH$_3$)$_5$(η^2-lutidine)]$^{2+}$, the pyridine ring can be protonated at nitrogen or intramolecularly deprotonated at carbon as shown in Eq. 6.13 (p. 164). η^6-Py complexes display a reactivity mainly related to the activation produced by coordination toward nucleophilic attack at the carbon atoms, or to some steps connected with the hydrogenation of the ring (see Chapter 3). Although interesting, these transformations cannot be easily related to any reaction leading to the C-N bond scission or the actual removal of nitrogen.

More relevant for HDN modeling is the activation of Py observed when it is coordinated to a metal center in the unusual η^2(N,C) metallaaziridine mode, which has indeed resulted in the subsequent C-N bond cleavage. Wigley and coworkers reported the first example of metal insertion into a C-N bond in 1992. They reacted the complex (ArO)$_2$ClTa(η^2(N,C)-2,4,6-NC$_5$H$_2$^{t}Bu$_3$) (ArO = 2,6-^{i}Pr$_2$PhO) with 1 eq. of the hydride source LiHBEt$_3$ in THF at low temperature, cleanly obtaining the corresponding C-N insertion product (ArO)$_2$Ta[=NC^{t}Bu=CHC^{t}Bu=CHCH^{t}Bu] (Eq. 6.24) [63]. The X-ray structure of the ring-opened product (Fig. 6.6) clearly reveals that the former amido nitrogen has been transformed into a formal imido linkage upon hydride attack, according to Eq. 6.24.

$$(6.24)$$

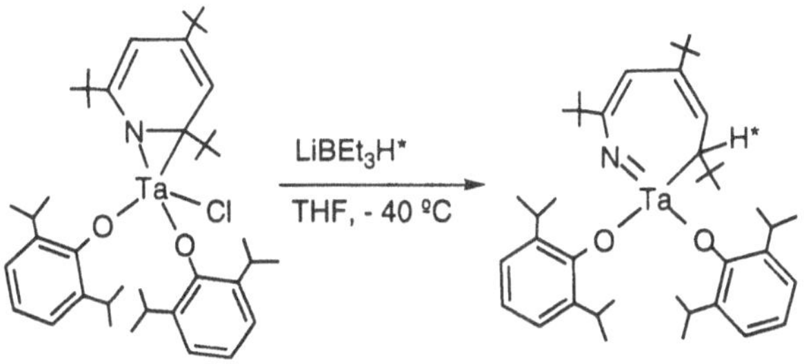

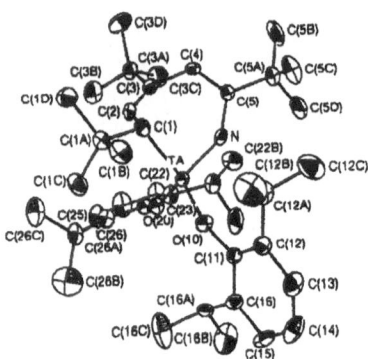

Fig. 6.6. X-ray structure of (ArO)$_2$Ta[=NC^{t}Bu=CHC^{t}Bu=CHCH^{t}Bu]
(Reproduced with permission from ref. 2.)

Also, from a series of additional experiments with other carbon nucleophiles (RMgCl, MeLi) in place of LiHBEt$_3$, it has been confirmed that the reaction occurs *via* initial addition of the hydride to the Ta atom, followed by an intramolecular endo-attack of the metal hydride onto the Py Cα atom (Eq. 6.25).

$$\text{(6.25)}$$

The most interesting conclusion that may be derived from Wigley's study is perhaps the fact that η^2(N,C) bonding of Py produces a perturbation of the formally sp^2 Cα atom toward sp^3 hybridization, rendering this carbon atom susceptible to nucleophilic attack by the hydride. However, it must be pointed out that this is not a general reaction, since the presence of the ^{t}Bu groups in the pyridine ligand in (ArO)$_2$ClTa(η^2(N,C)-2,4,6-NC$_5$H$_2$^{t}Bu$_3$) is crucial for C-N bond cleavage to take place. Indeed, neither the related Ta complex (silox)$_3$Ta(η^2(N,C)-NC$_5$H$_5$) (silox = OSiBu$^t_3$) nor the quinoline derivative (ArO)$_2$TaCl(η^2(N,C)-Q) undergoes C-N insertion upon nucleophilic addition [2]. On the otther hand the η^2(N,C)-bonded Q ligand can be stoichiometrically hydrogenated to THQ under mild conditions, but free Q is not reduced. Within these limitations, Wigley's compound represents one of the very few examples of ring-opening reactions of pyridines that could be envisaged as a model of a transformation that could take place during HDN catalysis over *e. g.* $MoS_2/\gamma-Al_2O_3$ catalysts.

Another interesting case of metal-assisted ring-opening of Py has been reported by Wolczanski using Nb complexes. As shown in Eq. 6.26, the low-valent d^2 $(silox)_3Nb$ fragment coordinates Py in the $\eta^2(N,C)$ mode to yield $(silox)_3Nb(\eta^2(N,C)-NC_5H_5)$, which readily undergoes C-N insertion by thermolysis in benzene at rather moderate temperatures (70 °C) [56]. The stoichiometry of the reaction yields 0.5 eq. of Py and 0.5 eq. of $(silox)_3Nb=CHCH=CHCH=CHN=Nb(silox)_3$ as a thermodynamic mixture of cis,cis-, trans,cis-, trans,trans- and cis,trans-isomers.

$$(silox)_3NbCl_2 \xrightarrow[25\ °C]{Na/Hg,\ Py} (silox)_3Nb \cdots \xrightarrow[- Py]{70\ °C \atop C_6H_6} (silox)_3Nb= \cdots N=Nb(silox)_3 \qquad (6.26)$$

The fact that in these two cases the C-N scission step occurs without prior hydrogenation of the heterocycle is in contrast with the heterogeneous catalysis literature, which points to a dominant mechanism via prehydrogenation. Nevertheless, these models offer clues as to what is needed in a metal center in order to achieve C-N bond breaking. These reactions also present a different perspective worth considering when trying to explain complex reaction schemes. Also these complexes and their reactions suggest a tempting possibility for the design of novel catalysts containing early transition metals that could be able to operate through alternative routes not involving prehydrogenation of the substrates.

6.3.3. HYDROGENOLYSIS OF N-HETEROCYCLES CATALYSED BY TRANSITION METAL COMPLEXES

From the examples described in the preceding Sections and in Chapter 3, Section 3.3.2 (p. 84), it can be concluded that the hydrogenation of N-heterocyles is much more facile than C-N bond breaking reactions when soluble transition metal complexes are employed as models for HDN catalysis; this is in parallel to what has been observed on metal sulfide surfaces. Indeed, while several examples of efficacious homogeneous catalysts for the hydrogenation of N-heterocycles to the corresponding cyclic amines (partially or fully saturated) are available (see Chapter 3), only one case of C-N bond *hydrogenolysis* promoted by a metal-complex in solution has been described so far. Moreover, the real nature of this process is not clearly defined.

Under water-gas shift (wgs) conditions the carbonyl cluster $Rh_6(CO)_{16}$, induces both the hydrogenation of pyridine to piperidine and its hydrogenolysis to various bis(piperidinyl)alkanes (Eq. 6.27) [78].

$$\text{(pyridine)} \xrightarrow[150\ °C]{Rh_2(CO)_{16} \atop CO\ (800\ psi)/H_2O} \text{(piperidine, N-H)} + \text{(piperidinyl)}N-(CH_2)_5-N\text{(piperidinyl)} \qquad (6.27)$$

+ other products

The formation of the latter compounds is thought to involve extrusion of NH_3 from an intermediate 1,4-dihydropyridine to give glutaraldehyde, which in turn reacts with piperidine and hydrogen to form the bis(piperidinyl) products. The absence of both particulate metal in the reactor and the lack of catalytic activity observed when rhodium metal was used instead of $Rh_6(CO)_{16}$ was taken as proof for the homogeneous nature of the process. However, the need for further studies supporting the homogeneity of this HDN reaction, the somewhat harsh experimental conditions (150 °C, 800 psi CO), and the proven capability of heterogeneous catalysts to promote the conversion of Py illustrated in Eq. 6.27 [78] have contributed to raise doubts on the homogeneous nature of the overall process.

6.4   C-N   Bond   Activation   of   Amines,   Imines,   and   Amides   by Transition   Metal   Complexes.

All the common HDN schemes of e.g. pyridine require, after prehydrogenation a first ring-opening C-N bond breaking, followed by the decomposition of n-pentylamine to remove the nitrogen as NH_3. It is also possible to envisage pathways or steps that involve extrusion of nitrogen from related molecules such as imines and amides, and therefore it is of interest to consider at this point some reactions of such substrates induced by transition metal complexes.

Wolczanski [80] has observed that ring-substituted anilines H_2N-Ph-X react with $(silox)_3Ta$ by oxidative addition of the either the N-H or the C-N bond, depending on the substituents present on the ring (Eq. 6.28):

(6.28)

The propensity for C-N vs. N-H activation correlates well with substituent Hammet parameters; groups that increase the basicity of aniline increase the relative rate of N-H activation, suggesting that nucleophilic attack by the amine at an empty d_{xz}/d_{yz} orbital of $Ta(silox)_3$ preceeds   oxidative addition. On the other hand, electron-withdrawing substituents decrease the rate of N-H activation and increase the rate of C-N activation, similarly to the effects observed on electrophilic aromatic substitution. Nucleophilic attack by the filled d_{z^2} orbital of $Ta(silox)_3$ is expected to occur at the arylamine ipso carbon preceding C-N oxidative addition. The carbon-heteroatom cleavages can be accomodated by mechanisms using both electrophilic and nucleophilic sites on the metal center.

Arylamine N-H *vs.* C-N activation is a consequence of energetically similar pathways; electrophilic attack on the nitrogen lone pair is dominant in N-H scission, whereas nucleophilick attack on the arene ring is most important to C-N bond cleavage. This is a very interesting organometallic model of the last step in the heterogeneous HDN of quinoline, which involves the sp^3 C-N bond cleavage of 2-*n*-propylaniline, a reaction pathway available to surfaces of suitable nucleophilicity (see Chapter 1). Considering the composition of classical HDN catalysts (e.g. Ni-Mo-S or Ni-W-S), it is intriguing that this reaction takes place on an early transition metal complex. This is also in contrast with the general N-H activation of primary (alkyl) amines by late transition metal (Pd) complexes to yield the corresponding metal-amido derivatives, reported by Hartwig [81]. Other examples of C-N bond activation by early transition metal complexes are available. Arnold has reported [82] the intramolecular oxidative addition of a C-N bond in the unstable intermediate "[PhC(NSiMe₃)₂]₂Zr" to yield {[PhC(NSiMe₃)₂]Zr(η²-PhCNSiMe₃)(μ-NSiMe₃)}₂. Gambarotta has described C-N bond cleavage of neutral and anionic amides by dinuclear Nb complexes [77, 83].

Late metals, on the other hand, have been used to cleave the C-N bonds of allylamines. Hiraki and coworkers found [85] that ruthenium complexes can cleave the C-N bonds of allylamines [85]. For instance, the complex RuHCl(CO)(PPh₃)₃ reacts with primary and secondary allylamines to yield the corresponding insertion products Ru(CH₂CH₂CH₂NHR)Cl(CO)(PPh₃)₃ (R = H, alkyl). With tertiary amines (*e.g.* N,N-dimethylallylamine), the reaction follows a different course, and instead of inserting the olefinic moiety into the M-H bond, the C-N bond is cleaved and the only metal-containing product is the stable π-allyl complex Ru(η³-C₃H₅)Cl(CO)(PPh₃)₂. The nitrogen atom is eliminated with the metal hydride in the form of dimethylamine [85]; Fig. 6.7 represents possible mechanisms for this reaction, based on deuteration experiments.

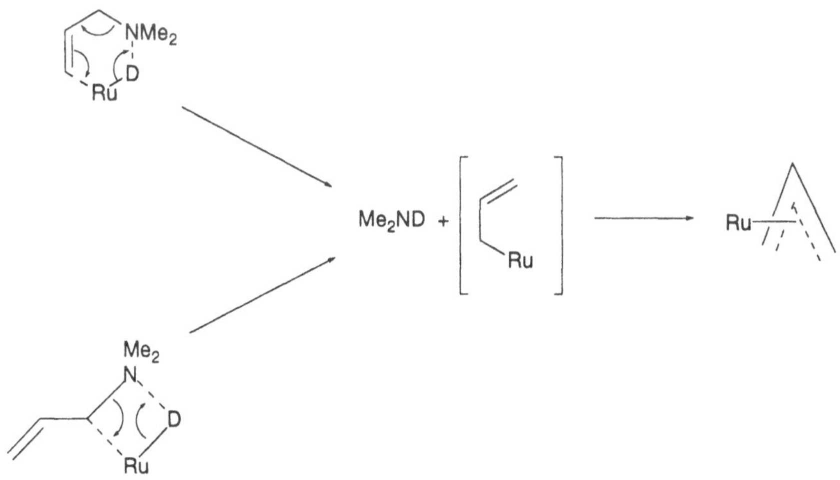

Fig. 6.7. C-N bond activation by Ru complexes
(adapted from ref. 85)

Aresta and coworkers have provided a very interesting example of C-N and N-H bond activation by use of low-valent nickel tricyclohexylphosphine complexes [86]. Alkyl-ammonium or -imminium tetraphenylborate salts readily undergo oxidative addition to $(Cy_3P)_2Ni(\eta^2\text{-}CO_2)$ or $(Cy_3P)_2Ni(\mu\text{-}N_2)Ni(PCy_3)_2$ at or below room temperature, to yield the corresponding Ni(II) derivatives shown in Fig. 6.8.

The corresponding products (a) and (b) contain coordinated NH_3 or imine, besides the hydrocarbon moiety that remains bound to the metal in the form of a η^3-allyl ligand. This model illustrates reaction pathways that may be related to those taking part in HDN processes using nickel-containing heterogeneous catalysts, where the organonitrogen substrates may actually be protonated by surface –OH or –SH groups prior to denitrogenation. The reaction on Ni phosphine compounds, however, does not seem to be general, since changing the substituent on the imminium cation from $\text{-}C_3H_5$ to $\text{-}CH_2Ph$ causes a switch from C-N to N-H activation, as in complex (c), which contains hydrido and amido ligands. Another related example of C-N bond scission of allyl amines involved Pt complexes [87]. Morokuma and coworkers carried out an extensive theoretical study of the mechanism of the Ni-induced C-N bond activation by use of DFT methods [88].

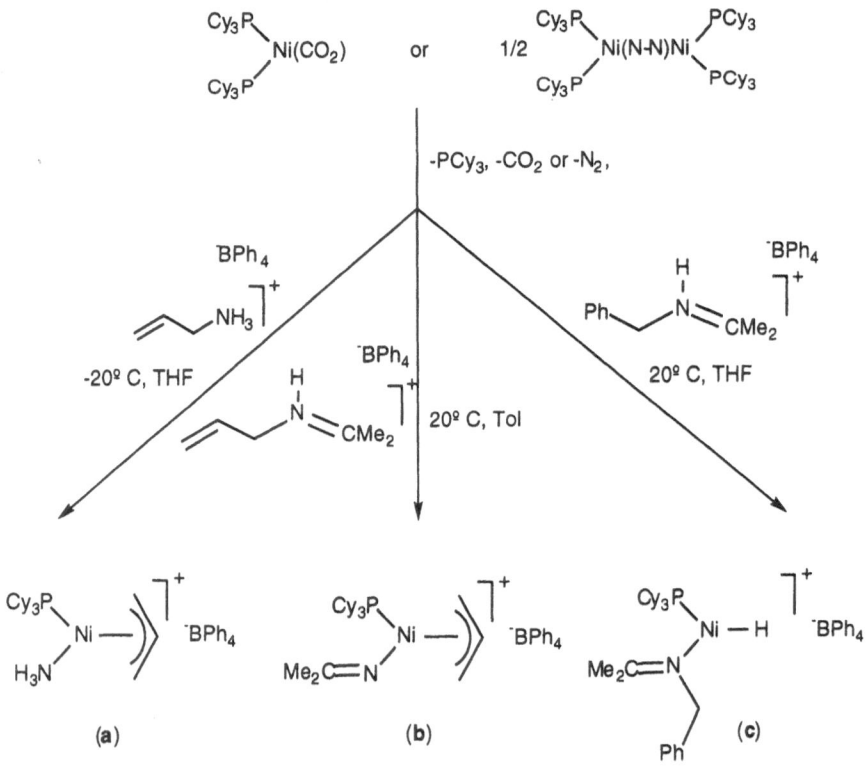

Fig. 6.8. C-N and N-H bond activation by low-valent Ni complexes.

Their main findings were: (1) The actual active catalyst is the (bis)phosphine Ni complex. (2) For allylammonium salts the reaction proceeds by an associative rather than a dissociate mechanism involving coordination of the allylammonium cation to the metal center followed by oxidative addition of the C-N bond to Ni(0). The resulting NH_3 remains coordinated to Ni(II) in a pentacoordinated intermediate, and finally a phosphine is lost to generate the final product. The reaction of the imminium salts, on the other hand, follows a dissociative mechanism.

6.5. Final Remarks.

The results described in this Chapter demonstrate that transition metal complexes can be considered reasonable models for mimicking some of the steps known or believed to occur in heterogeneous HDN. The most relevant points in this organometallic modeling may be summarized as follows:

The $\eta^1(N)$ and η^6 coordination modes of aromatic N-heterocycles are involved in their hydrogenation (see Chapter 3), whereas $\eta^2(N,C)$ binding can be related to C-N bond breaking and hydrogenolysis. In particular, while it is evident that the hydrogenation of the heteroaromatic rings is effectively accomplished by late transition metals, the direct C-N bond cleavage of heteroaromatic rings is best accomplished when the substrate enters into the $\eta^2(N,C)$ form of interaction with an *early* transition metal in relatively low oxidation state; this implies that this type of coordination is favored for rather *electron-poor* metal centers (d^2 Nb(III) or Ta(III), *cf.* Mo(IV) or W(IV)).

This type of reaction may thus not be a very relevant model for "classical" HDN mechanistic schemes, which all require prehydrogenation of the ring –presumably by the promoter Ni active sites- before C-N bond scission. It shows, however, that it is possible to break C-N bonds of aromatic rings, when the right electronic situation is available. Furthermore, it has been clearly established that nucleophilic attack of a metal hydride occurs invariably at the pyridine carbon, rather than nitrogen, to form a metallacyclic imido complex, consistent with Laine's HDN mechanism. It is tempting to speculate that electron-poor, early metal catalysts could be designed so as to promote the denitrogenation of the heteroaromatic rings without previous hydrogenation, a process that would consume much less hydrogen. Still, this interesting chemistry, disclosed mainly by the groups of Wigley and Wolczanski, needs to be further developed before it can be considered a good model for heterogeneous HDN reactions.

The ultimate complete denitrogenation most likely requires the cooperation of two or more metal centers. This has been shown for Nb clusters –although no mechanistic details are available- and it has also been observed for the homogeneous desulfurization of thiophenes (see Chapter 4). A single metal site may not be capable of coordinating an N-heterocycle, inserting into a C-N bond and removing the nitrogen atom.

The fact that Nb complexes are capable of oxidatively adding the C-N bond of anilines takes the organometallic modeling a step closer to the more conventional HDN mechanisms (of *e.g.* quinoline). Some interesting mechanistic considerations have been put forward in this case, but more detailed studies are required in order to better understand this reaction. When other non-aromatic amines have been used as model substrates, electron-rich late metals (Ru, Ni) are better suited for cleaving C-N bonds, more in line with the components of real HDN catalysts, particularly in the case of nickel. The mechanisms reported so far in these cases imply either an intramolecular nucleophilic attack by a metal hydride, or protonation of the N atom, followed by elimination of ammonia or an amine, similarly to the most accepted denitrogenation mechanisms described in Chapter 1. These organometallic examples, however, all refer to allylamines (or ammonium salts), for which the driving force for removal of nitrogen is related to the possibility of forming very stable metal π-allyl complexes. Similar reactions on primary aliphatic amines would be most welcome.

In conclusion, considering the complexity of the HDN process and the many alternative or parallel mechanisms that may be operative on the surface of a typical catalyst, the homogeneous modeling studies summarized in this Chapter must be considered with extreme caution. Nonetheless, many reactions described involving transition metal complexes with N-heterocycles or other N-ligands show some striking analogies with related reactions occurring on the surface of heterogeneous catalysts.

Obviously, organometallic modeling of the HDN reaction is way behind the analogous modeling of HDS, but the research described in this Chapter is certainly enlightening and opens the way for further developments. More work on this largely neglected field of molecular analogues of surface species and interactions related to HDN is very urgently needed. This area should thus be encouraged in every possible way, as a means of aiding in the understanding of the complex issues involved, and in the design of improved or novel catalysts of better performance in practical applications.

References

1.   C. Bianchini, A. Meli and F. Vizza: *Eur. J. Inorg. Chem.* 43 (2001).
2.   K. J. Weller, P. A. Fox, S. D. Gray and D. E. Wigley: *Polyhedron* **16**, 3139 (1997).
3.   R. Cordone, W. D. Harman and H. Taube: *J. Am. Chem. Soc.* **111**, 5959 (1989).
4.   W. H. Myers, M. Sabat and W. D. Harman: *J. Am. Chem. Soc.* **113**, 6682 (1991).
5.   W. H. Myers, J. I. Koontz and W. D. Harman: *J. Am. Chem. Soc.* **114**, 5684 (1992).
6.   D. S. Glueck, F. J. Hollander and R. G. Bergman: *J. Am. Chem. Soc.* **111**, 2719 (1989).
7.   N. Kuhn: *Bull. Soc. Chim. Belg.* **99**, 707 (1990).
8.   K. Öfele and E. Dotzauer: *J. Organomet. Chem.* **30**, 211 (1971).
9.   G. Huttner and O. S. Mills: *Chem. Ber.* **105**, 301 (1972).
10. K. K. Joshi, P. L. Pauson, A. R. Qazi and W. H. Stubbs: *J. Organomet. Chem.* **1**, 471 (1964).
11. N. Kuhn, E. M. Horn, E. Zauder, D. Blaser and R. Boese: *Angew. Chem. Int. Ed. Engl.* **28**, 342 (1989).
12. N. Kuhn, E. M. Horn, E. Zauder, D. Blaser and R. Boese: *Angew. Chem. Int. Ed. Engl.* **27**, 579 (1988).
13. F. Kvietov, V. Allured, V. Carperos and M. Rakowski DuBois *Organometallics* **13**, 60 (1994).
14. R. H. Fish, E. Baralt and H. S. Kim: *Organometallics* **10**, 1965 (1991).
15. H. Felkin and J. Zakrzewski: *J. Am. Chem. Soc.* **107**, 3374 (1985).
16. M. Rakowski DuBois: *Coord. Chem. Rev.* **174**, 191 (1998).
17. A. Efraty. N. Jubran and A. Goldman: *Inorg. Chem.* **21**, 868 (1982).
18. T. Morikita, M. Harano, A. Sasaki and S. Komiya: *Inorg. Chim. Acta* **291**, 341 (1999).
19. K. J. Chase, R. F. Bryan, M. K. Woode and R. N. Grimes: *Organometallics* **10**, 2631 (1991).
20. E. Carmona, J. M. Marín, P. Palma, M. Paneque and M. L. Poveda: *Inorg. Chem.* **28**, 1895 (1989).
21. G. C. Hsu, W. P. Kosar and W. D. Jones: *Organometallics* **13**, 385 (1994).
22. F. T. Ladipo and J. S. Merola: *Inorg. Chem.* **29**, 4172 (1990).
23. W. J. Kelly and W. E. Parthun: *Organometallics* **11**, 4348 (1992).
24. M. Rakowski DuBois. K. G. Parker, C. Ohman and B. C. Noll: *Organometallics* **16**, 2325 (1997).
25. W. D. Jones, L. Dong and A. W. Myers: *Organometallics* **14**, 855 (1995).
26. M. Rakowski DuBois, L. D. Vasquez, R. F. Ciancianelli and B. C. Noll: *Organometallics* **19**, 3507 (2000).
27. D. A. Vicic and W. D. Jones: *Organometallics* **18**, 134 (1999).

28. A. J. Arce, R. Machado, Y. De Sanctis, M. V. Capparelli, R. Atencio, J. manzur and A. J. deeming: *Organometallics* **16**, 1735 (1997).
29. A. J. Deeming, A. J. Arce, Y. De Sanctis, M. W. day and K. I. Hardcastle: *Organometallics* **8**, 1408 (1989).
30. A. J. Arce, J. Manzur, M. Marquez, Y. De Sanctis and A. J. Deeming: *J. Organomet. Chem.* **412**, 177 (1991).
31. O. Yamauchi, M. Takani, K. Toyoda and H. Masuda: *Inorg. Chem.* **29**, 1856 (1990).
32. S. Chen, B. C. Noll, L. Peslherbe and M. Rakowski DuBois: *Organometallics* **16**, 1089 (1997).
33. T. J. Johnson, A. M. Aris and J. A. Gladysz: *Organometallics* **13**, 3182 (1994).
34. N. F. Masters, N. Mathews, G. Nechvatal and D. A. Widdowson: *Tetrahedron*, **45**, 5955 (1989).
35. W. J. Ryan, P. E. Peterson, Y. Cao, P. G. Williard, D. A. Sweigart, C. D. Baer, C. F. Thompson, Y. K. Chung and T. M. Chung: *Inorg. Chim. Acta* **211**, 1 (1993).
36. U. S. Gill, R. Moriarty, Y. Y. Ku and I. R. Butler: *J. Organomet. Chem.* **417**, 313 (1991).
37. S. A. Lomenzo, S. P. Nolan and M. L. Trudell: *Organometallics* **13**, 676 (1994).
38. S. Chen, V. Carperos, B. C. Noll, R. J. Swope and M. Rakowski DuBois: *Organometallics* **14**, 1221 (1995).
39. C. White, S. J. Thompson and P. M. Maitlis: *J. Chem. Soc. Dalton Trans.* 1654 (1977).
40. L. D. Vasquez, B. C. Noll and M. Rakowski DuBois: *Organometallics* **17**, 976 (1998).
41. J. A. D. Jeffreys and C. Metters: *J. Chem. Soc. Dalton Trans.* 1624 (1977).
42. M. F. Semmelhack, W. Wulff and J. L. Garcia: *J. Organomet. Chem.* **240**, C5 (1982).
43. S. Chen, L. Vasquez, B. C. Noll and M. Rakowski DuBois: *Organometallics* **16**, 1757 (1997).
44. J. J. García, A. L. Casado, A. Iretskii, H. Adams and P. M. Maitlis: *J. Organomet. Chem.* **558**, 189 (1998).
45. J. Reedijk in: *Comprehensive Coordination Chemistry* (Eds.: G. Wilkinson, R. D. Gillard, J. McCleverty), Pergamon Press, Oxford, 1987, Vol. 2.
46. H. E. Selnau and J. S. Merola: *Organometallics* **12**, 1583 (1993).
47. R. H. Fish, H-S. Kim and R. H. Fong: *Organometallics* **8**, 1375 (1989).and *Organometallics* **10**, 770 (1991).
48. R. H. Fish, R. H. Fong, A. Than and E. Baralt: *Organometallics* **10**, 1209 (1991).
49. R. H. Fish, E. Baralt and H-S. Kim: *Organometallics* **10**, 1965 (1991).
50. R. H. Fish, H-S. Kim, J. E. Babin and R. D. Adams: *Organometallics* **7**, 2250 (1988).
51. B. Chaudret and F. A. Jalón: *J. Chem. Soc. Chem. Commun.* 711 (1988). B. Chaudret, F. A. Jalón, M. Pérez-Manrique, F. J. Plou and R. A. Sánchez-Delgado: *New J. Chem.* **14**, 331 (1990).
52. R. Cordone, D. W. Harman and H. Taube: *J. Am. Chem. Soc.* **111**, 2896 (1989).
53. M. G. B. Drew, P. J. Baricelli, P. C. H. Mitchell and A. R. Read: *J. Chem. Soc. Dalton Trans.* 649 (1983).
54. D. R. Neithamer, L. Parkanyi, J. F. Mitchell and P. T. Wolczanski: *J. Am. Chem. Soc.* **110**, 4421 (1988).
55. K. J. Covert, D. R. Neithamer, M. C. Zonnevylle, R. E. LaPointe, C. Schaller and P. T. Wolczanski: *Inorg. Chem.* **30**, 2494 (1991).
56. T. S. Kleckley, J. L. Bennett, P. T. Wolczanski and E. B. Lobkowsky: *J. Am. Chem. Soc.* **119**, 247 (1997).
57. J. R. Strickler, M. A. Bruck and D. E. Wigley: *J. Am. Chem. Soc.* **112**, 2814 (1990).
58. D. P. Smith, J. R. Strickler, S. D. Gray, M. A. Bruck, R. S. Holmes and D. E. Wigley: *Organometallics* **11**, 1275 (1992).
59. P. L. Watson: *J. Chem. Soc. Chem. Commun.* 276 (1983).
60. M. E. Thompson, S. M. Baxter, A. R. Bulls, B. J. Burger, M. C. Nolan, B. D. Santarsiero, W. P. Schaefer and J. E. Bercaw: *J. Am. Chem. Soc.* **109**, 203 (1987).
61. E. Klei and J. H. Teuben: *J. Organomet. Chem.* **214**, 53 (1981).
62. A. J. Deeming, M. J. Stchedroff, C. Whitaker, A. J. Arce, Y. De Sanctis and J. W. Steed: *J. Chem. Soc. Dalton Trans.* 3289 (1999).
63. S. D. Gray, D. P. Smith, M. A. Bruck, P. Briggs and D. E. Wigley: *J. Am. Chem. Soc.* **114**, 5462 (1992).
64. S. D. Gray, K. J. Weller, M. A. Bruck, P. Briggs and D. E. Wigley: *J. Am. Chem. Soc.* **117**, 10678 (1995).
65. K. J. Weller, S. D. Gray, P. Briggs and D. E. Wigley: *Organometallics* **14**, 5588 (1995).
66. R. H. Fish, T.-J. Kim, J. L. Stewart, J. H. Bushweller, R. K. Rosen and J. W. Dupon: *Organometallics* **5**, 2193 (1986).
67. R. Eisenstadt, C. M. Giandomenico, M. F. Frederick and R. M. Laine: *Organometallics* **4**, 2033 (1985).
68. E. Arcia, D. S. Kolwaite, E. Rosenberg, K. Hardcastle, J. Ciurash, R. Duque, R. Gobetto, L. Milone, D. Osella, M. Botta, W. Dastrù, A. Viale and I. Fiedler: *Organometallics* **17**, 415 (1998).
69. R. Smith and E. Rosenberg: *Organometallics* **18**, 3519 (1999).
70. B. Chaudret and F. A. Jalón: *J. Chem. Soc. Chem. Commun.* 711 (1988).
71. R. M. Moriarty, U. S. Gill and Y. Y. Ku: *J. Organomet. Chem.* **350**, 157 (1988).
72. L. S. Hegedus: *Angew. Chem. Int. Ed. Engl.* **27**, 1113 (1988).
73. S. Sun, C. A. Dullaghan and D. A. Sweigart: *J. Chem. Soc. Dalton Trans.* 4493 (1996).

74. C. A. Dullaghan, X. Zhang, D. Walther, G. B. Carpenter and D. A. Sweigart: *Organometallics* **16**, 5604 (1997).
75. C. A. Dullaghan, G. B. Carpenter, D. A. Sweigart, D. S. Choi, S. S. Lee and Y. K. Chung: *Organometallics***16**, 5688 (1997).
76. X. Zhang, E. J. Watson, C. A. Dullaghan, S. M. Gorun and D. A. Sweigart: *Angew. Chem. Int. Ed.* **38**, 2206 (1999).
77. M. Tayebani, S. Gambarotta and G. P. A. Yap: *Angew. Chem. Int. Ed. Engl.* **37**, 3002 (1998).
78. R. M. Laine, D. W. Thomas, L. W. Cary: *J. Org. Chem.* **44**, 4964 (1979).
79. K. Kindler, D. Mathies: *Chem. Abstr.* **54**, 19731b (1960).
80. J. B. Bonanno, T. P. Henry, D. R. Neithamer, P. T. Wolczanski and E. B. Lobkovsky: *J. Am. Chem. Soc.* **118**, 5132 (1996).
81. M. S. Driver and J. F. Hartwig: *J. Am. Chem. Soc.* **118**, 4206 (1996).
82. J. R. Hagadorn and J. Arnold: *Organometallics* **13**, 4670 (1994).
83. M. Tayebani, K. Feghali, S. Gambarotta and C. Bensimon: *Organometallics* **16**, 5084 (1997).
84. H. Nagashima, K. Mukai, Y. Shiota, K.-I. Ara, K, Itoh, H. Suzuki, N. Oshima and Y. Moro-oka: *Organometallics* **4**, 1314 (1985).
85. K. Hiraki, T. Matsunaga and H. Kawano: *Organometallics* **13** 1878 (1994).
86. M. Aresta, E. Quaranta, A. Dibenedetto, P. Giannocaro, I. Tommasi, M. Lanfranchi and A. Tiripicchio: *Organometallics* **16**, 834 (1997).
87. H. Kurosawa: *Inorg. Chem.* **15**, 120 (1976).
88. M. Torrent, D. G. Musaev and K. Morokuma: *Organometallics* **19**, 4402 (2000).

EPILOGUE

Despite the fact that hydrodesulfurization and hydrodenitrogenation chemistry has been in continuous development and in industrial application for several decades, the current and future environmental constraints regarding the quality of fuels and the associated commercial perspectives have produced a renaissance of this field. In particular, the search for important breakthroughs is intensifying in the hope of achieving drastic improvements in today's technologies or even of finding entirely new catalytic systems for the production of fuels capable of meeting the present standards.

The complex nature of the HDS and HDN problems requires a broad, transdisciplinary approach in order to try to answer the most varied questions related to these important classes of reactions. The key issues include the practical aspects related to process and product engineering, a precise knowledge of the nature and the composition of petroleum and of refinery fractions, and the thermodynamics and detailed kinetics of the different processes involved. Also, a number of more fundamental solid-state and surface chemistry considerations regarding the preparation, the characterization, and the resulting properties of HDS and HDN catalysts, as well as the complicated reaction mechanisms involved for the various important families of substrates, need to be understood in depth. Even though some very impressive achievements have been disclosed over the last 30-40 years, it seems that some of the major new discoveries desired today may have been held back by the lack of a better understanding of some key issues. Of particular importance are the nature and the structure of HDS-HDN active sites on metal sulfide catalysts, and the intimate details of the elementary reactions implicated in the commonly accepted catalytic schemes.

Organo-transition metal chemistry became a mature discipline essentially during the 1960-1980 interval. In those two decades, vast progress was made in the establishment of sound fundamental principles concerning the structures, bonding, and general reaction patterns of a wide diversity of metal complexes. This was coupled with the mastering of a variety of physical methods specially adapted to the specific synthetic and structural problems being tackled. The development of homogeneous catalysis in turn, seemed to follow the steps of the advances being made in the fundamentals of organometallic chemistry. Some notable exceptions to this are, for instance, the homogeneous Co- or Rh-catalyzed olefin hydroformylation process, and the oxidation of ethylene and other olefins by use of Wacker chemistry, which have been successfully practiced in industry since the 1950's. In the last part of the XX century, however, we have witnessed a major shift in this trend, placing homogeneous catalysis among the common and important alternatives for commercially successful processes. Moreover, in the new millennium it appears that novel developments in both organometallic chemistry and homogeneous catalysis will be driven in good part by the need for new products with special characteristics, rather than by the necessity or the desire for a deeper understanding of the fundamentals alone. This, in turn, will be accompanied by the development of novel more efficient and cleaner technologies.

Also importantly, in recent times many researchers in this field have begun to look in new directions in which the advantages of molecular chemistry may provide the key to help answer some of the important outstanding problems in heterogeneous catalysis. This has been the case for HDS, and to a lesser extent for HDN, which for many years were of the almost exclusive concern of solid state and surface chemists, as well as of chemical and process engineers. This has allowed, to a considerable extent, the unraveling of the complicated science surrounding this problem, together with the practical applications of these important systems. The possibility of preparing metal complexes of important model substrates like the thiophenes or the pyridines, which are stable enough to be amenable to characterization by standard techniques such as NMR spectroscopy and single crystal X-ray diffraction, but at the same time sufficiently reactive to allow the detailed study of their transformations and reaction mechanisms, opened an exceedingly interesting possibility of connecting two fields traditionally considered apart, in an effort to shed new light on a problem of a fascinating fundamental complexity and a tremendous environmental and industrial importance.

Prior to about 1985, less than a handful of metal complexes containing coordinated thiophenes were known, and virtually no mention of them could be found in the HDS literature of that time. Today, not only the number and the variety of such compounds have grown to be very large, but also they have become of obliged reference for workers in the heterogeneous HDS field. In turn, organometallic chemists have become aware of the wealth of information available on the chemistry of thiophenes on solid catalyst surfaces, and of the main standing issues that need to be understood and solved in HDS catalysis.

Some early proposals for the modes of adsorption of thiophenes on metal sulfides have been probed by comparisons with the structures of well-characterized metal complexes; this has allowed the identification of the most reasonable alternatives and of new possibilities not previously considered. Theoretical studies on such complexes at increasing levels of sophistication have also contributed in an important manner to provide a clear and consistent picture of the different possible bonding modes of thiophenes to metal centers. When these theoretical and experimental results from molecular chemistry are combined with the information available from surface techniques and heterogeneous catalysis, the chemisorption of this type of organosulfur compounds on metal sulfides arises as a very well understood phenomenon. This is no doubt one of the most important achievements of the organometallic modeling approach to HDS chemistry.

A fair number of possibilities for explaining HDS reaction schemes have been available for some time in heterogeneous catalysis publications, most of them based on sound experimental evidence and sometimes on extensive calculations. However, because of the intrinsic complexity of the problem, some of the key points of the mechanisms have remained essentially at a controversial or a speculative level over many years. These are mainly related to the nature and the structure of the surface active sites required for each type of reaction involved, and to the details of the various elementary steps of the overall catalytic cycles. By studying the analogous reactions on discrete well-characterized transition metal complexes in solution, a better distinction of the most sensible reaction pathways implicated in HDS from the less likely ones has become possible. A good part of the new knowledge thus obtained can be extrapolated - no doubt with the necessary caution- to analogous surface situations. Perhaps more importantly, some definite patterns have emerged connecting the principal geometric and electronic characteristics of the metal centers, with the occurrence and the preference of the various available types of bonding of thiophenes, as well as with the specific reactions that such activated substrates may reasonably follow in each of these bonding situations.

Similar considerations have been made in the case of HDN-related substrates like pyrroles and pyridines. Nevertheless, the advances in this direction are much more modest and less clear-cut than those made for HDS models, and therefore, an exciting opportunity is open for new research in the organometallic modeling of the adsorption and reactivity of organonitrogen compounds.

Other important subjects related to HDS-HDN mechanisms are also being addressed –albeit to a lesser extent- by means of molecular chemistry, such as the participation of sulfide ligands in the activation of hydrogen and its transfer to the organosulfur or organonitrogen substrate. Further progress is this area is certainly desirable and foreseeable.

The interesting discoveries concerning the mild stoichiometric extrusion of sulfur from highly refractory substrates like 4,6-dimethyldibenzothiophene promoted by Ni or Pt complexes, or the homogeneous *catalytic* systems for the hydrogenation, the hydrogenolysis, and the desulfurization of HDS-related substrates represent impressive achievements in fundamental chemistry. Perhaps more importantly, they open novel promising alternatives for practical applications. This could be of particular use for the purification of refined fuels or refinery cuts, rather than of crudes or residua, especially if the active metal complexes can be immobilized, either in a second inmiscible liquid (*e.g.* water or an ionic liquid), or on a solid inorganic or organic support.

At the beginning of this book we stated our intention to try to relate as much as possible the knowledge gained from coordination and organometallic chemistry with some of the main unsolved questions of HDS and HDN catalysis. This has allowed us to build some interesting conceptual bridges between the fields of molecular and surface chemistry related to HDS-HDN. An attempt has been made to point out these links at various points throughout the preceding chapters, but it seemed useful at this final stage to summarize the most important connections and extrapolations that have been proposed in the text.

In this Epilogue, it is attempted to do that in a systematic and also in a somewhat provocative manner. This will hopefully stimulate new thoughts and further discussions, not only on the general subjects of HDS and HDN, but also on the validity and of the usefulness of the organometallic modeling approach as an additional tool to tackle problems in heterogeneous catalysis. Of course this modeling is not meant to represent a panacea, as we are well aware of the many limitations it contains, which will also be pointed out:

Vacancies. It is clear that any metal center that is required to activate an incoming substrate through bonding must have available orbitals of the appropriate symmetry and energy in order to do so. Also the actual physical space around the active center must be sufficiently large to accommodate the new ligand (adsorbate) without causing strong repulsive interactions with neighboring atoms or groups. Thus, whether in a metal complex, or on a metal sulfide surface, the concept of coordinatively unsaturated sites must always be invoked in relation to catalytic activity, and this has been a constant in both homogeneous and heterogeneous catalysis. The interesting question is if such "vacancies", in the strict sense of electron-avid *empty* areas of space, actually exist for any significant length of time. Alternatively, they may form only if and when needed, as a consequence of an incoming substrate molecule being ready to displace a labile group; the latter possibility seems to be more correct, in view of the accumulated evidence.

In solution, loosely bound ligands may remain coordinated to the metal center until the incoming species displaces it from the coordination sphere. Frequently, solvent molecules perform this function and this is reflected in the rather spectacular solvent effects that are sometimes observed on catalytic rates (see *e.g.* the homogeneous hydrogenation of benzothiophene catalyzed by Rh and Ir complexes described in Chapter 3, p. 81). Also, ligands that are capable of changing their coordination mode through low energy bond breaking and bond forming processes, such as the η^4/η^3 sulfide bridge transformation proposed by Curtis in the mechanism of the homogeneous desulfurization of thiols by CpCoMoS clusters (see Chapter 4, pp. 125-126), can be responsible for liberating a coordination site only when needed. Along this line of thought, it is now more common in organometallic reactions in solution to speak of weakly coordinating -rather than the older version "non-coordinating"- solvents or counterions, and to refer to metal complexes containing labile ligands rather than empty sites.

On a metal sulfide surface, analogous considerations must be taken into account. The active sites, and particularly the vacancies recently observed by use of scanning tunneling microscopy (STM) (Chapter 1, p. 11) correspond to the observation of single layers of MoS_2 and Co-Mo-S nanoclusters carefully grown on Au (111) crystal faces, and studied under ultra high vacuum. This is no doubt one of the most exciting recent developments in metal sulfide chemistry, but the likelihood that vacancies like the ones observed remain empty for any length of time under real catalytic conditions, *i.e.* in the presence of excess H_2, H_2S, substrates and products, is slim. As mentioned several times in the text (see *e.g.* section 1.2.4.2), calculations of metal sulfide surfaces have also indicated that in the absence of incoming substrates physical vacancies, if formed, tend to be filled by migration of sulfur atoms from the bulk to the surface. Therefore, adsorption of the reactive substrate onto the active sites of the catalysts most probably takes place through displacement of labile (weakly adsorbed) groups such as H_2S -formed by hydrogenation of an apical S^{2-} group- or by opening of one or more M-S bridging bonds, as represented in Fig. E1. Both of these situations are thus better described by the term "latent vacancies" introduced by Curtis, and thus we propose that this modified vacancy concept be generalized.

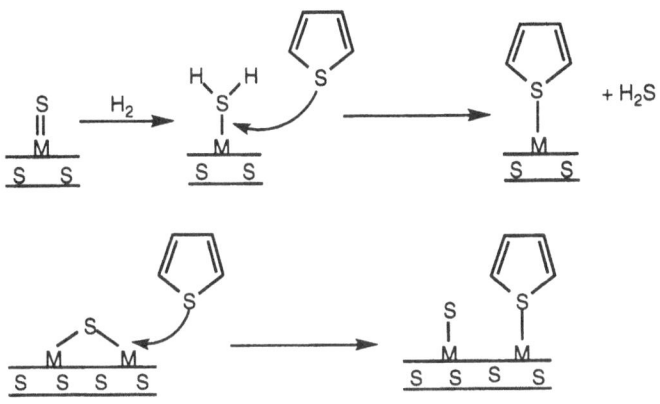

Fig. E1. Latent vacancies in HDS catalysis.

Another important issue related to this point is the possibility of having different sites on a given HDS catalyst, with closely related geometries but defined by different degrees of coordinative unsaturation. Each one of these types of site would be related to specific reactions, or to varying degrees of catalytic activity. This is a very reasonable possibility to consider in metal sulfides, as there is ample experimental and theoretical evidence from surface chemistry studies indicating that sites with 1, 2, or 3 (latent) coordinatively unsaturated sites can be available; their possible implications in the different reactions considered is discussed below. The presence of strongly adsorbing molecules could, in turn, selectively block one or more of these unsaturations, and this would explain many of the poisoning effects that have been reported in HDS and HDN catalysis

The nature and the structure of the active sites. This is perhaps the most important and yet the most difficult question to answer concerning HDS and HDN catalysis. The contribution of organometallic chemistry in this respect has revealed the great relevance of the electronic nature of the site. It is clear now that electron rich metal centers (low oxidation states) are the most capable of easily breaking C-S bonds, whereas more electrophilic sites will be more adapted to promote hydrogenation reactions. Several oxidation states have been found to be present in *e.g.* Co-Mo-S phases, but HDS activity has been related predominantly with Co(II) (d^7) and with Mo(IV) (d^2), which are the lowest oxidation states found in solid sulfides; Co(III) (d^6) and Mo(V) (d^1) can also be present in significant concentrations, and they might be associated with hydrogenation activity.

Also, it is interesting to note that 2-electron redox processes are extremely easy in a metal sulfide. Therefore, the activity of a site could be switched from *e.g.* desulfurization to hydrogenation simply by adding or removing 2 electrons, most likely with the participation of the sulfur atoms which themselves have ample redox capabilities.

As for the geometry of the sites, the Co-Mo-S structures advanced by the Danish groups are no doubt the best models available to date, in good part because they provide a general picture that can be easily adapted and modified for any particular situation. The square-based pyramidal and the tetrahedral models are represented in Fig. E2. For the pentacoordinated structure, the adsorption of thiophene (or of any other HDS or HDN substrate) can take place at the apical position through displacement of H_2S (previously formed, in turn, by hydrogenation of a S^{2-} ligand), or at a site liberated by breaking a Co-S bond in the Co-S-Mo arrangement. In the tetrahedral case, adsorption can occur through simple binding to an empty coordination site, possibly accompanied by a low-energy structural deformation, presumably toward the square base pyramid, without any need for any bond to actually break.

The use of bimetallic and specially heterobimetallic complexes in HDS modeling is still limited, mainly due to the synthetic difficulties encountered in preparing such compounds, and to the increased complexity of their analysis by routine techniques. Hopefully further important advances in this area will be reached in the near future that will allow more elaborate molecular analogues to be compared with surface structures; therefore, all synthetic efforts targeting metal complexes with structures as similar as possible to the ones described in Fig. E2 should be strongly encouraged. Another tempting idea is to anchor well defined metal complexes on MoS_2 surfaces, in an attempt to reach compositions and structures of very high activities in a controlled fashion. Some attempts have been made of decomposing well defined Mo or Co-Mo clusters on alumina, as a means of synthesizing catalysts of well-defined composition, but using the actual sulfides as supports for organometallic complexes, which would introduce an additional functionality to the catalyst, has been little explored and may offer better possibilities.

Hydrogen activation. All the mechanisms known for hydrogen activation in homogeneous solutions (see Ch. 5, p. 139) are in principle available to metal sulfide surfaces. However, the fact that a variety of sulfur groups are present in high concentrations in such solids seems to direct the reactions with dihydrogen toward the mechanisms in which the available reactive sulfur atoms participate. Some of the reaction pathways related to HDS that have been well defined and that are most frequently invoked in reactions of hydrogen with metal complexes involve the homolytic dissociation on S_2^{2-}-bridged dimolybdenum units, and the heterolytic dissociation involving both the metal and terminal S^{2-} or S_2^{2-} ligands.

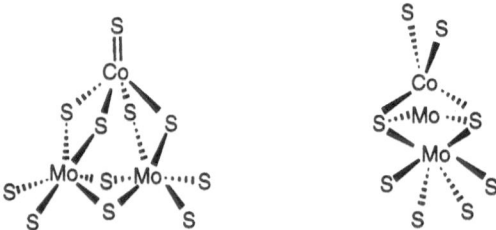

Fig. E2. A representation of the pentacoordinated and the tetracoordinated Co-Mo-S sites.

Analogous mechanisms can be easily adapted to the structures proposed for the active sites, as exemplified in Fig. E3 for the case of the pentacoordinated model. Although the direct experimental evidence available for the heterolytic dissociation of H_2 on solid catalysts is scarce (only one case of direct detection of two types of hydrogen in a hydrogenated RuS_2 catalyst by NMR, see Ch. 5), chemical intuition as well as theoretical arguments clearly point toward a preference for heterolytic splitting on metal sulfide surfaces. Such heterolytic activation results in the initial production of an unstable metal hydride plus a hydrosulfido group. In a metal sulfide surface, the former would be extremely labile and it would thus rapidly migrate onto a sulfur atom, in the form of a proton. Moving the electrons between the metals, the sulfur atoms and the hydrides in an extended array of this type is energetically very facile. The initial splitting may take place on the metal and a terminal sulfide group (as represented in Fig. E3, bottom part), or alternatively, on the metal and a bridging sulfide. In any case, after the rearrangement the end result is the formation of two hydrosulfido groups on the catalyst surface, regardless of whether the activation mechanism is of a homolytic or a heterolytic nature.

In the case of HDN, an additional interesting possibility also consistent with the heterolytic mechanism arises, since substrates like the pyridines -or intermediate alkyl or aryl amines- are sufficiently basic to promote the activation of hydrogen so as to form a metal hydride plus a protonated base (*e.g.* a pyridinium or an alkylammonium cation). Furthermore, some of the most widely accepted amine HDN mechanisms include the initial protonation of the amine nitrogen, followed by elimination of ammonia from the ammonium cation. Therefore, it is very easy to combine the idea of a heterolytic hydrogen activation promoted by, say *n*-pentylamine, with a subsequent degradation by a Hoffmann mechanism, to conform a reasonable HDN catalytic cycle. A simplified representation of this idea is given in Fig. E4.

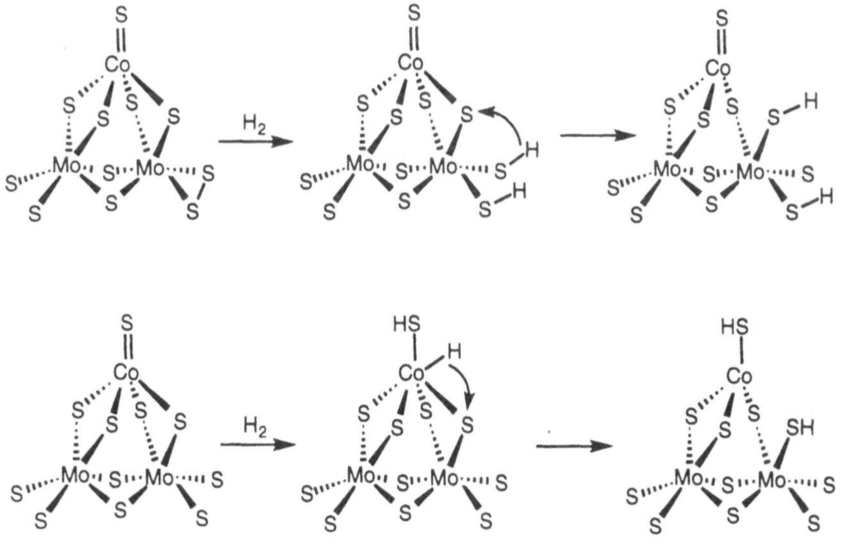

Fig. E3. Some possible pathways for hydrogen activation on a Co-Mo-S site.

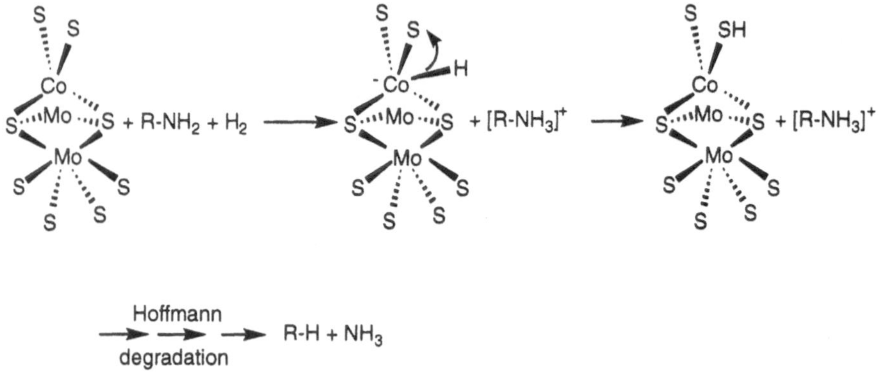

Fig. E4. The involvement of the substrate-promoted heterolytic splitting of H_2 in the HDN of an amine.

Chemisorption modes and reactivity. The possible binding modes of thiophenes to catalytically active surface sites have been the subject of much debate and speculation over the years. However, the many impressive recent developments in the chemistry of metal complexes containing thiophenes as ligands, such as unambiguously defining their detailed molecular structures, and almost simultaneously advancing the theory of their bonding description, provides a sound basis for attempting some correlations betwen coordination types and reactivity patterns. An overwhelming evidence from experimental and theoretical evidence is now available indicating that η^1S-bonding of thiophenes to metal ions is most likely the key form of activation leading to C-S bond rupture. In such coordination mode -which is invariably tilted and not vertical- back bonding from an electron-rich metal center to an antibonding π^* C-S orbital is possible, and oxidative addition to yield the corresponding thiametallacycle relieves the repulsive interaction and leads to a more stable situation. Other bonding modes leading to C-S bond scission have been identified and must be taken into consideration, *e.g.* the η^3(S,C=C) mode observed for (triphos)Ir complexes; nevertheless, because they seem to be rare, their occurrence in metal sulfide surfaces may not be very important. Other types of π-bonded structures (η^4 or η^5) could certainly exist on more highly coordinatively unsaturated sites in sulfide surfaces, but if present, they are most probably peripheral or transient species not directly involved in the HDS process. The olefin-like η^2(C=C) bonding mode is probably common and it is to be associated mainly with hydrogenation reactions of olefins or olefin-like bonds like the C2=C3 bond of benzothiophenes, or of indoles.

   This introduces some interesting considerations: both the η^1S and the η^2(C=C) forms are actually L-type neutral ligands donating 2 electrons. Thus, in terms of coordinative unsaturation of the active site, the requirements for the desulfurization of thiophenes and for the hydrogenation of olefins or olefin-like substrates such as benzothiophenes, could be identical. The difference in reactivity observed is due to the overall electronic nature of the site, beyond the consideration of the degree of unsaturation alone. Highly electrophilic metal centers will prefer to bind to a C=C bond, while electron-rich metal atoms will tend to favor the S-only coordination (see *e.g.* the striking case of (triphos)Ru^x in Fig. 4.16, p. 131). This also provides a reasonable explanation for the poisoning effects of *e.g.* olefins on desulfurization sites that have been sometimes observed, since alkenes are in general better ligands than thiophenes and could effectively compete for the same active sites.

This general situation is exemplified for benzothiophene in Fig. E5. In the higher oxidation state $(n+2)^+$, Co will have a stronger tendency to bind to C=C bonds, and therefore, hydrogenation to 2,3-dihydrothiophene should be favored over desulfurization, while in the lower oxidation state n^+, the S-only bonding mode is preferred, and consequently desulfurization becomes the main reaction. According to this model, a surface site with a well defined and rigid structure could switch its reactivity from desulfurization to hydrogenation by a simple 2-electron redox process – an event easily conceived on an extended metal sulfur array- without undergoing any important geometrical rearrangement. This can be related to Whitehurst's ideas on one single site with several functions (see p. 11); nevertheless, if, for instance, simple alkyl-substituted olefins are present in the mixture, they may indeed coordinate more strongly than benzothiophene, especially –but not exclusively- in the higher oxidation state. Depending on the relative equilibrium constants, this could result in a selective poisoning effect of the olefin on either the hydrogenation or the desulfurization of benzothiophene, or in complete poisoning for both reactions. Similar considerations could be invoked if nitrogen-containing molecules like pyridines or amines –which are known to poison hydrogenation and/or HDS activities- are mixed with the organosulfur substrate.

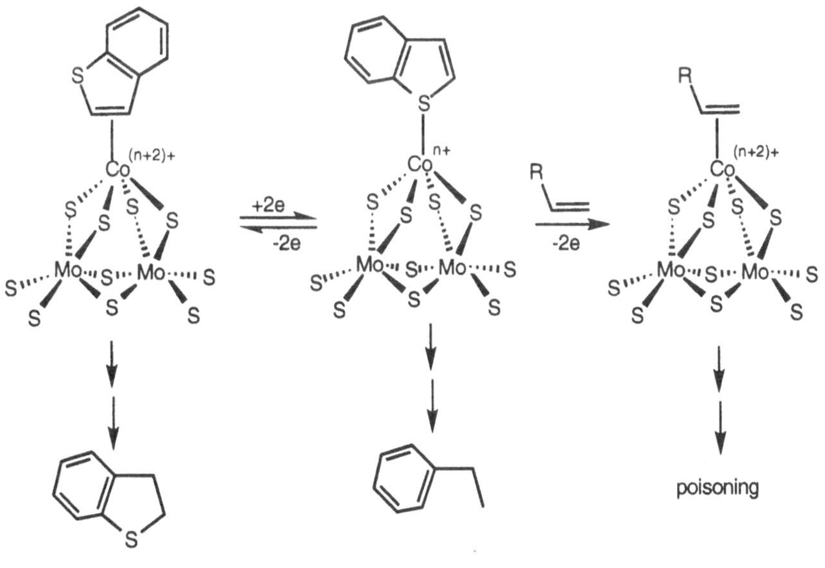

Fig. E5. Hydrogenation, desulfurization and poisoning. A 2e difference?

As for hydrogenation of aromatics, the organometallic modeling approach indicates that the active sites most likely require at least two latent CUS, together with a high degree of electrophilicity and sufficient flexibility to accommodate the substrate in a more sterically demanding disposition (η^4, L_2 type). This has been demonstrated for most of the known mechanisms of arene hydrogenation in solution (see section 3.2.2). If this is so, the sites required for the hydrogenation of aromatic and polyaromatic ring structures are most probably different from the desulfurization sites in both the geometrical structure and the electronic nature. Some aromatics, however, can act as good L_3-, L_2- or even L-type ligands and therefore they are capable of blocking the singly unsaturated and more rigid desulfurization sites through the olefin-like $\eta^2(C=C)$ adsorption. Bonding in this way does not result in hydrogenation of the arene, but it would be responsible for the poisoning effects of aromatics on HDS that have been reported by some authors. A possible sequence depicting such a situation is envisaged in Fig. E6. Again, if strongly coordinating alkyl-substituted olefins are present, they are capable of inhibiting also the hydrogenation of aromatics by blocking at least one of the two unsaturations required for the η^4-binding of the arene.

Sulfur extrusion: We have seen that in solution, ring-opening reactions of thiophenes are easily accounted for on single electron-rich sites with one coordinative unsaturation. However, breaking the second C-S bond of the resulting intermediate thiolates appears to be facilitated when the S-donor atom bridges two metal centers – *cf.* a Co-Mo couple in a Co-Mo-S catalyst. This bridged bonding seems to greatly reduce the C-S bond strength of the adsorbed substrate. According to the mechanism advanced by Curtis for the desulfurization of thiols, the *second* C-S bond scission could actually be taking place through a homolytic pathway with the intervention of both the cobalt and the molybdenum atoms. A possible sequence of events illustrating sulfur removal from thiophene on the proposed pentacoordinated Co-Mo-S site is illustrated in Fig. E7.

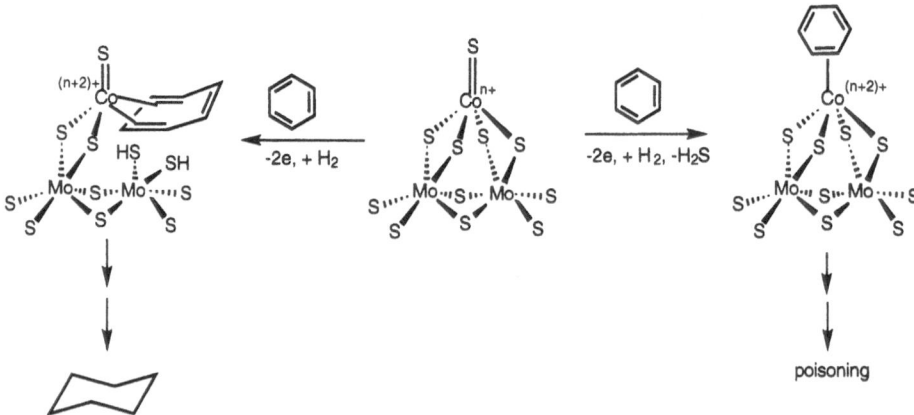

Fig. E6. Aromatic hydrogenation site and poisoning of HDS sites by aromatics.

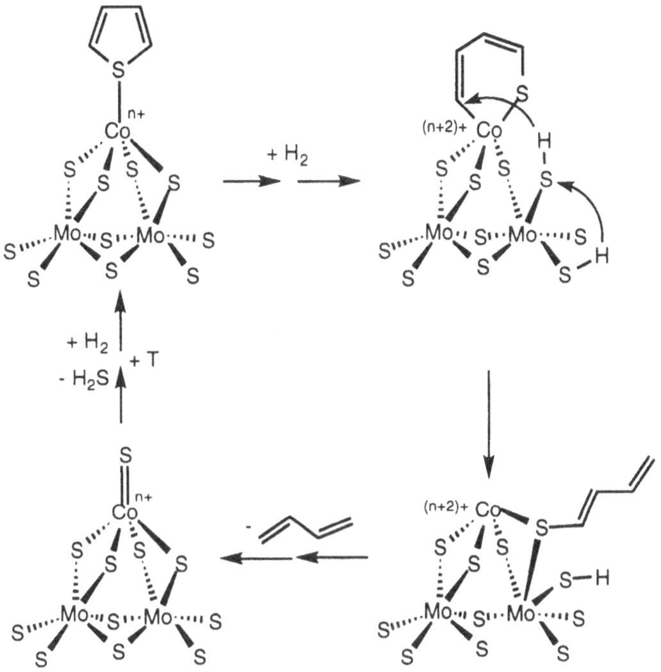

Fig. E7. Cooperative bimetallic effects on sulfur extrusion.

A general mechanism for HDS on Co-Mo-S sites: As a final proposal linked to the general considerations presented above, it is very tempting to attempt a combination of all the individual steps presented above, which emerges as a remarkably coherent picture. On the basis of the extensive HDS-related surface chemistry knowledge available, taken together with the most relevant details carefully extrapolated from the structures and reactions of the organometallic models described throughout this book, a reasonable overall mechanism for the hydrodesulfurization of *e.g.* thiophene on the Co-Mo-S sites proposed above, can be constructed. Fig. E8 is such a representation for the pentacoordinated Co site, and Fig. E9 depicts an analogous series of events for the tetrahedral model. From these reaction schemes the observed catalytic activity for simultaneous competitive reactions may be easily rationalized in terms of any one of the active sites under consideration, since both these models allow the possibility of variable important properties of the metal centers, such as oxidation state, number of vacancies, geometry, and flexibility. It is not difficult to extend these general ideas to the desulfurization of other sulfur-containing substrates, taking into account the specific electronic and steric properties of the particular compounds of interest, like for instance 4,6-dimethyldibenzothiophene. Similarly, the known poisoning effects of olefins, aromatics or nitrogen-containing compounds on HDS reactions can be explained with relative ease by use of the intermediates proposed in Figs. E7 and E8, or close analogues thereof.

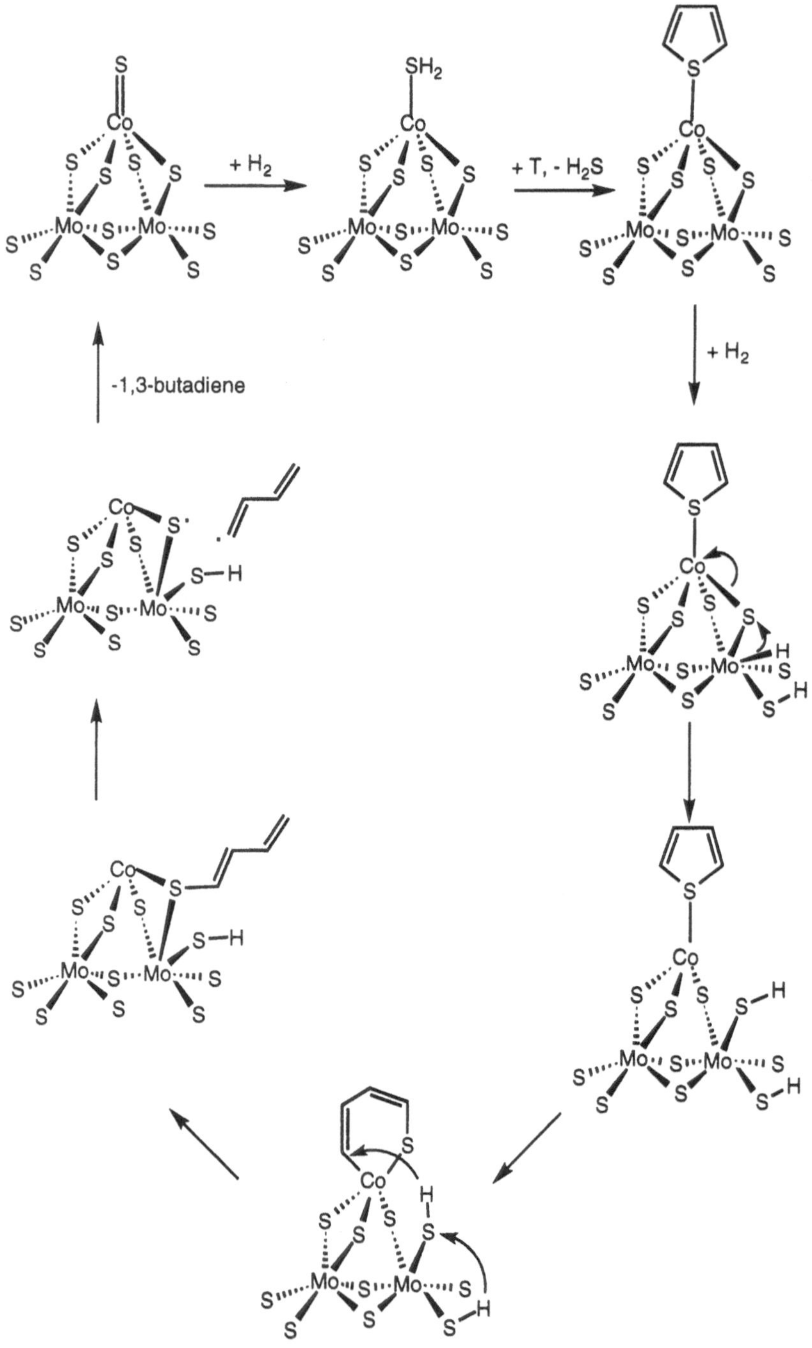

Fig. E8. A possible mechanism for the hydrodesulfurization of thiophene
on a pentacoordinated Co-Mo-S site.

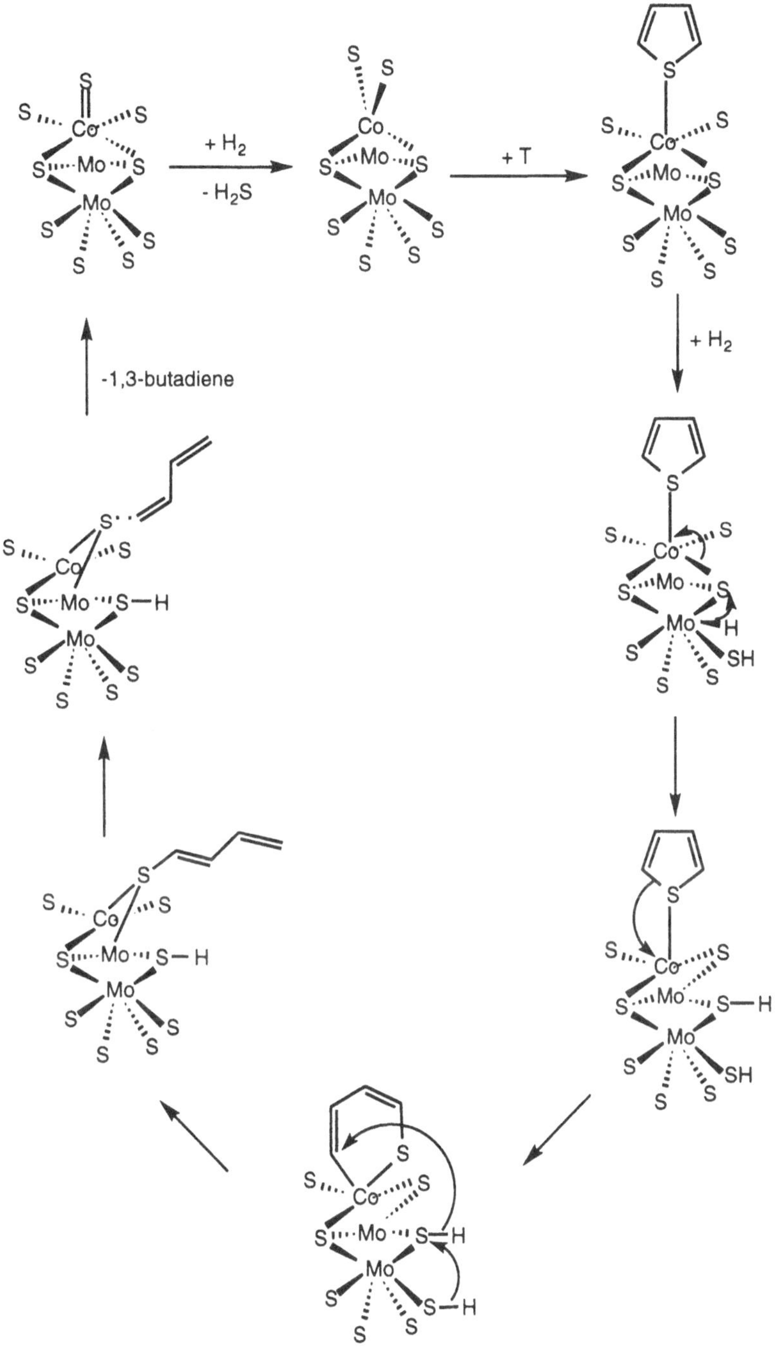

Fig. E9. A possible mechanism for the hydrodesulfurization of thiophene on a tetrahedral Co-Mo-S site.

The activation of H_2 can be associated with either metal or with both, as well as with the sulfur atoms in their various possible forms (terminal or bridged sulfido, disulfido, or hydrosulfido groups). The transfer of hydrogen to the activated substrates is thought to occur predominantly from -SH groups and not from metal hydride units which probably do not persist for any important length of time; instead, they can easily migrate to sulfur to produce another reactive –SH group. Sulfur extrusion is proposed as the result of a homolytic cleavage of the C-S bonds of intermediate thiolates facilitated by simultaneous bonding of the S-donor atom to both Co and Mo.

Hydrodenitrogenation: Many of the concepts and ideas proposed for hydrodesulfurization can be of use in trying to interpret analogous issues related to hydrodenitrogenation. The nature of the active sites in HDN catalysts (*e.g.* Ni-Mo-S compositions) are believed to be similar to those defined for Co-Mo-S phases, but they are certainly understood in much less detail from both the surface chemistry and the organometallic modeling perspectives. Consequently their description relies heavily on the analogy with HDS sites. It is clear that the main mechanistic pathways implicated in HDN include the hydrogenation of at least the nitrogen-containing ring, before C-N bond breaking takes place. The hydrogenation of nitrogen heterocycles is nowadays a well understood reaction, in good part because of the excellent organometallic model studies. It is known that it requires singly unsaturated sites; this is important in that desulfurization sites also require a single unsaturation, and both reactions could even take place on the same sites. Therefore, the presence of nitrogen compounds -generally more strongly coordinating than sulfur donor molecules- can block the desulfurization sites and thus an explanation for poisoning effects is easily found in this model.

As for the actual C-N scission reactions, and the type of site or of binding related to them, the organometallic modeling approach unfortunately still has little to offer. The few examples available of ring-opening reactions of pyridines constitute no doubt a novel and exciting development in this chemistry. However, their possible connection with HDN mechanisms might be limited because the more commonly accepted reaction schemes involve ring opening of *saturated* amine rings, a reaction for which organometallic models are definitely lacking. As for a possible overall HDN mechanism, or details of elementary steps other than hydrogenation, molecular chemistry cannot yet take us beyond the very good proposals already available from solid state and surface chemistry studies, combined with well-established aspects of the organic chemistry of amines. Still, this current limitation of the organometallic approach in HDN can also be considered, on the optimistic side, as a particularly interesting opportunity for future research, since the many challenges that this chemistry offers and the well-recognized importance of this reaction make it a very appealing field of work. Hopefully, some major breakthroughs and interesting developments in HDN-related molecular chemistry will be disclosed before too long.

Concluding remarks: We hope to have demonstrated that a number of exciting conceptual bridges have emerged over the last 15 years or so, connecting coordination and surface chemistry with homogeneous and heterogeneous catalysis, in relation to the very important hydrodesulfurization reaction. As stated earlier in the text, organometallic modeling is simply one more method -and a very powerful one indeed- to study the many aspects of the exceedingly complex HDS mechanisms. As such, it is best used in conjunction with the already available modern analytical arsenal of solid state and surface chemistry.

As with any modeling, the limitations of this approach must always be kept in mind, and all extrapolations must be made with great caution. Solution chemistry of well-characterized complexes allows us to probe deeply into many important questions, but it necessarily ignores, for instance, the consequences of the presence of supports as well as other surface cooperative effects that are known to be of prime importance in heterogeneous catalysis. At the same time, solvent effects that possibly have no importance in the reactions of solid catalysts are introduced in an notable manner when dealing with homogeneous solutions. Some molecular geometries and rearrangements that appear very reasonable in metal complexes, and particularly in clusters, may not be available in more rigid extended solids, and vice versa. Thus, it is of utmost importance to employ the best possible chemical sense when trying to use the results of organometallic models in order to explain surface phenomena.

Still, this is a fascinating example of how two traditionally separated fields can break barriers -such as methodologies, experimental techniques, and even languages- and come together in an effort to solve an outstanding chemical problem that originates in very practical environmental and industrial issues, but requires novel fundamental concepts and approaches. A great deal of information has been derived from relatively simple experiments on well-defined metal complexes; in this way, knowledge at the molecular level has been -and will continue to be- obtained, which is rarely accessible in heterogeneous and surface chemistry studies. This is of paramount importance if one remembers that, after all, any form of catalysis has to deal essentially with making, breaking and rearranging chemical bonds, and it is therefore always a molecular phenomenon.

ABBREVIATIONS

acac	acetylacetonate
bipy	2,2'-bipyridyl
BT	benzothiophene
BNT	benzonaphthothiophene
CNDO	completely neglected differential overlap
COD	1,5-cyclooctadiene
Cp	cyclopentadienyl
Cp*	pentamethylcyclopentadienyl
CUS	coordinatively unsaturated site
DBT	dibenzothiophene
DFT	density functional theory
DHBT	2,3-dihydrobenzothiophene
DHT	dihydrothiophene
dippe	1,2-bis(diisopropylphosphino)ethane
diphos	1,2-bis(diphenylphosphino)ethane
DMDBT	4,6-dimethyldibenzothiophene
DMF	dimethylformamide
DMSO	dimethylsulfoxide
EXAFS	extended X-ray absorption fine structure spectroscopy
HDS	hydrodesulfurization
HDM	hydrodemetalation
HDN	hydrodenitrogenation
In	indole
NBD	norbornadiene
Py	pyridine
Pyr	pyrrole
Q	quinoline
STM	scanning tunneling microscopy
T	thiophene
Tp	hydrotris(pyrazolyl)borate
THQ	1,2,3,4-tetrahydroquinoline
THT	tetrahydrothiophene
TPPMS	*meta*-sulfonatophenyldiphenylphosphine, sodium salt
TPPTS	tris(*meta*-sulphonatophenyl)phosphine, sodium salt
XANES	X-ray absorption near edge spectroscopy
XPS	X-ray photoelectron spectroscopy

LIST OF FIGURES

CHAPTER 1. HYDRODESULFURIZATION AND HYDRODENITROGENATION

Fig. 1.1.   Some major types of organosulfur constituents of petroleum
              And distillates.   4

Fig. 1.2.   Trends in HDS activity for dibenzothiophene (a, d-f)
              and thiophene (b,c).   6

Fig. 1.3.   Schematic representation of Topsøe's Co-Mo-S model
              showing perspective views of: (a) the coordination around
              the Co site and (b) a Mo_2CoS_{11} unit at the edge Co-Mo-S site.   9

Fig. 1.4.   A representation of Whitehurst's model for thiophene,
              -H, and -SH bonded to a unique Co-Mo-S site.   11

Fig. 1.5   A representation of the Co-Mo-S model site from STM
              measurements.   12

Fig. 1.6.   Possible bonding modes for thiophenes on a surface metal site.   17

Fig. 1.7.   Main reaction pathways in HDS of thiophene.   19

Fig. 1.8.   Main reaction pathways in HDS of benzothiophene.   21

Fig. 1.9.   Main reaction pathways in HDS of dibenzothiophene.   23

Fig. 1.10.   Vrinat's mechanism for HDS of dibenzothiophenes.   24

Fig. 1.11.   Main classes of refractory organonitrogen compounds in
              crudes and distillates.   26

Fig. 1.12.   Main adsorpton modes of organonitrogen compounds.   27

Fig. 1.13.   Main reaction pathways in HDN of quinoline   28

Fig. 1.14.   Mechanisms of C-N bond breaking.(a) Hoffmann elimination
              and (b) Nucleophilic substitution   29

Fig. 1.15.   C-N Bond breaking in HDN of 1,2,3,4-tetrahydroquinoline   30

Fig. 1.16.   Laine's mechanism for C-N bond breaking in HDN of piperidine   31

CHAPTER 2. COORDINATION AND ACTIVATION OF THIOPHENES IN METAL
           COMPLEXES

Fig. 2.1.   Bonding modes of thiophenes known in metal complexes.   36

Fig. 2.2.   Thiophene tilt angle (θ) in η^1S complexes.   38

Fig. 2.3.   Energy level diagram for $Cp*(CO)_2Re(\eta^1S$-T).   39

Fig. 2.4.   Structure of $Cp*Ir(\eta^42,5$-$Me_2T)$.   49

Fig. 2.5.   Energy level diagrams for $[Cr(CO)_3(Cp)]^-$ and $Cr(CO)_3(\eta^5$-T).   52

Fig. 2.6.   Structure of $[(\eta^5$-T$)Rh(PPh_3)_2]^+$.   53

Fig. 2.7.   Energy level diagram for $[(\eta^5$-T$)Rh(PH_3)_2]^+$.   54

Fig. 2.8.   Angelici's mechanistic proposal for HDS of thiophene.   57

CHAPTER 3. HYDROGENATION REACTIONS

Fig. 3.1.   Formation, deactivation and degradatio of the active species
              for arenehydrogenation catalyzed by $(\eta^3\text{-}C_3H_5)Co[P(OMe)_3]_3$.   67

Fig. 3.2.   The mechanism of benzene hydrogenation catalyzed by
              $(\eta^3\text{-}C_3H_5)Co[P(OMe)_3]_3$.   68

Fig. 3.3.   Initial steps in the hydrogenation of benzene catalyzed by
              $(\eta^6\text{-}C_6H_6)Ru(\eta^4\text{-}C_6H_6)$.   70

Fig. 3.4.   Hydrogenation of benzene catalyzed by $[(\eta^6\text{-}C_6H_6)_4Ru4H_4]^{2+}$.   71

Fig. 3.5.   The mechanism of thiophene hydrogenation
              by $[Ir(H)_2(\eta^1\text{-}T)_2(PPh_3)_2]\,PF_6$.   76

Fig. 3.6.   The mechanism of benzothiophene hydrogenation
              by $[Cp^*Rh(H)_x]^{2+}$.   79

Fig. 3.7.   The mechanism of benzothiophene hydrogenation
              by $[M(H)_2(BT)_2(PPh_3)_2]PF_6$.   80

Fig. 3.8.   The mechanism of benzothiophene hydrogenation
              by $[(triphos)Ir(H)(NCMe)_2]BF_4$.   81

Fig. 3.9.   A possible two stage process for improved HDS of fuels
              or refinery cuts.   84

Fig. 3.10.   The mechanism of quinoline hydrogenation by using
              $[Cp^*Rh(NCMe)_3]^{2+}$.   87

Fig. 3.11.   The mechanism of quinoline hydrogenation by use of
              $[Cp^*Rh(NCMe)_3]^{2+}$.   89

Fig. 3.12.   The mechanism of quinoline hydrogenation by (a) dihydride
              And (b) monohydride catalysts..   91

CHAPTER 4. RING OPENING, HYDROGENOLYSIS AND DESULFURIZATION
              OF THIOPHENES BY METAL COMPLEXES

Fig. 4.1   Ring-opening reactions leading to η^2C,S thiophene-inserted
              metal complexes.   96

Fig. 4.2.   C-S $vs.$ C-H activation by $Cp^*Rh(PMe_3)$.   98

Fig. 4.3.   Stepwise HDS of thiophene by$[(triphos)Ir(\eta^4\text{-}benzene)]BF_4$.   102

Fig. 4.4.   Reactions of $[(triphos)Ir(\eta^2\text{-}C,S.BT)]^+$ with H^+, H^-, H_2.   106

Fig. 4.5.   Possible mechanism of the hydrogenolysis and hydrogenation
              of $[(triphos)Ir(\eta^2C,S\text{-}BT)]^+$.   106

Fig. 4.6.   Hydrogenolysis of 2,3-dihydrobenzothiophene by (triphos)Ir.   107

Fig. 4.7.   Hydrogenolysis of thiophene by $ReH_7(PPh_3)_2$.   109

Fig. 4.8.   Hydrogenolysis and desulfurization of DBT by Pt complexes.   110

Fig. 4.9.   Mild desulfurization of DBT on nickel.   112

Fig. 4.10.   Desulfurization of BT by $Fe_3(CO)_{12}$.   115

Fig. 4.11.   Desulfurization of BT and DBT by $(Cp^*Ru)_3(\mu\text{-}H)_3(\mu^3\text{-}H)_2$.   116

Fig. 4.12.   Ring opening of THT by nucleophilic attack
              to Cl$_3$W(μ-THT)WCl$_3$.   117
Fig. 4.13.   The structure of Cp*Rh{(η^5-C$_4$Me$_4$Fe(CO)$_3$} and
              Cp*Rh{(η^4:η^1-Me$_4$T)Fe(CO)$_4$}.   120
Fig. 4.14.   The mechanism of HDS of thiols by Cp'$_2$Mo$_2$Co$_2$S$_3$(CO)$_4$.   126
Fig. 4.15.   Hydrogenolysis and hydrogenationof benzothiophene
              by [(triphos)RhH].   130
Fig. 4.16.   Hydrogenation $vs.$ hydrogenolysis of benzothiophene
              by [(triphos)RuH]^x.   131
Fig. 4.17.   Some catalyst precursors used for the homogeneous and
              liquid-biphasic hydrogenolysis of thiophenes.   132
Fig. 4.18.   Homogeneous HDS of DBT by [(triphos)IrH].   133

CHAPTER 5. ACTIVATION OF HYDROGEN ON METAL COMPLEXES WITH
           SULFIDE LIGANDS AND RELATED COORDINATION
           CHEMISTRY OF H$_2$S

Fig. 5.1.   Main mechanisms of hydrogen activation on metal complexes.   139
Fig. 5.2.   Activation of hydrogen on Ti=S and Ti(S$_2$) complexes.   144

CHAPTER 6. MODELING HYDRODENITROGENATION

Fig. 6.1.   Known bonding modes of pyrrole and pyrrolyl ions
              in metal complexes.   154
Fig. 6.2.   Typical bonding modes of pyrrole-derived ligands
              in polynuclear metal complexes.   157
Fig. 6.3.   Known bonding modes of indole and some of its derivatives
              in metal complexes.   159
Fig. 6.4.   Bonding modes of pyridines in metal complexes.   162
Fig. 6.5.   Bonding modes of quinolines in metal complexes.   166
Fig. 6.6.   X-ray structure of (ArO)$_2$Ta[=NC^{t}Bu=CHC^{t}Bu=CHCH^{t}Bu].   172
Fig. 6.7.   C-N bond activation by Ru complexes.   175
Fig. 6.8.   C-N and N-H bond activation by low-valent Ni complexes.   176

EPILOGUE

Fig. E1.   Latent vacancies in HDS catalysis.   185
Fig. E2.   A representation of the pentacoordinated and
              the tetracoordinated Co-Mo-S sites.   186
Fig. E3.   Some possible pathways for hydrogen activation
              on a Co-Mo-S site   187
Fig. E4.   The involvement of the substrate-promoted heterolytic
              splitting of H$_2$ in the HDN of an amine   188
Fig. E5.   Hydrogenation, desulfurization and poisoning. A 2e difference? 189

Fig. E6.   Aromatic hydrogenation site and poisoning
               of HDS sites by aromatics   190
Fig. E7.   Cooperative bimetallic effects on sulfur extrusion.   191
Fig. E8.   A possible mechanism for the hydrodesulfurization of thiophene
               on a pentacoordinated Co-Mo-S site.   192
Fig. E9.   A possible mechanism for the hydrodesulfurization of thiophene
               On a tetrahedral Co-Mo-S site.   193

LIST OF TABLES

CHAPTER 2. COORDINATION AND ACTIVATION OF THIOPHENES IN METAL
      COMPLEXES
Table 2.1.   Known $\eta^1 S$ complexes of the thiophenes.   37
Table 2.2.   Known η^4 complexes of the thiophenes.   48
Table 2.3.   Known η^5 complexes of the thiophenes.   51
Table 2.4.   Known η^6 complexes of the benzo- and dibenzo-thiophenes.   58

CHAPTER 3. HYDROGENATION REACTIONS
Table 3.1.   Homogeneous catalysts for the hydrogenation of aromatic
      Hydrocarbons.   64
Table 3.2.   Homogeneous catalysts for the hydrogenation of benzothiophene.   78
Table 3.3.   Homogeneous catalysts for the hydrogenation of N-heterocycles.   86

CHAPTER 4. RING OPENING, HYDROGENOLYSIS AND DESULFURIZATION
      OF THIOPHENES BY METAL COMPLEXES
Table 4.1   Desulfurization by $Cp_2Mo_2Co_2S_3(CO)_4$.   125
Table 4.2.   Homogeneous catalytic hydrogenolysis of benzothiophene by use of
      $[(triphos)Rh(\eta^3-S(C_6H_4)CH=CH_2]$ 128

CHAPTER 5. ACTIVATION OF HYDROGEN ON METAL COMPLEXES WITH
      SULFIDE LIGANDS AND RELATED COORDINATION
      CHEMISTRY OF H_2S
Table 5.1.   Transition metal complexes of H_2S.   145

CHAPTER 6. MODELING HYDRODENITROGENATION
Table 6.1.   Metal complexes of pyrrole and pyrrolyl ligands.   155
Table 6.2.   Metal complexes of indole, indolyl and indoline ligands.   159
Table 6.3.   Examples of unusual metal complexes of pyridine and quinoline.   163

INDEX

activation
  of C-H bonds by metal complexes, 96
  of C-S bonds and S-bonding, 98, 188
  of C-S bonds and η¹S bonding, 134
  of C-S bonds by Cp metal complexes,
    97
  of C-S bonds by electron rich
    complexes, 97
  of C-S bonds by metal complexes, 95
  of C-S bonds by metal triphos
    complexes, 101
  of C-S bonds by Mo complexes, 100
  of C-S bonds by Tp metal complexes,
    99
  of C-S bonds vs. C-H bonds, 108
  of dibenzothiophenes by Pt complexes,
    114
  of hydrogen by complexes, mechanisms,
    186
  of hydrogen by metal complexes, 15,
    138
  of hydrogen in HDS, 14
  of hydrogen in homogeneous catalysis,
    15
  of hydrogen on sulfide ligands, 183
  of hydrogen, heterolytic, 14, 139
  of hydrogen, heterolytic and homolytic,
    139
  of hydrogen, heterolytic by Ti=S
    complexes, 143
  of hydrogen, homolytic, 14
  of hydrogen, homolytic on metal
    complexes, 140
  of hydrogen, mechanisms of, 14
active sites
  and vacancies, 12
  Co-Mo catalysts, 8
  desulfurization vs. hydrogenation, 107
  in HDS, 185
  in HDS, and vacancies, 16
  structures of, 182
activity
  in HDS and heats of formation, 7
  in HDS and hydrogen activation, 15
  in HDS and steric effects, 17
  in HDS, and Co-Mo-S sites, 9
  in HDS, and M-S bond strentgh, 7
  in HDS, and vacancies, 7
  of metal sulfides in HDS, 5
adsorption
  modes for benzothiophenes, 17
  modes for dibenzothiophenes, 17
  multi point, 17, 35
  of thiophenes in HDS, 16
  one point, 17, 35
aromatics
  hydrogenation and HDS sites, 135
  hydrogenation, active sites, 190
benzothiophene
  2,3-dihydro, hydrogenolysis by Ir
    complexes, 107
  2,3-dihydro, in HDS, 21
  alkyl substituted, in diesel, 4
  biphasic hydrogenolysis of, 132
  desulfurization by Rh-W complex, 120
  HDS of, 21
  homogeneous catalytic hydrogenolysis
    of by Ru complex, 129
  homogeneous hydrogenation of, 78

homogeneous hydrogenation prior to
    HDS, 83
homogeneous hydrogenation,
    mechanisms, 79
hydrogenation, homogeneous catalysts
    for, 78
liquid biphasic hydrogenation in HDS,
    83
ring opening by dinuclear Re
    complexes, 118
stepwise HDS by Ir complexes, 103
$\eta^2(C=C)$ bonding and hydrogenation of,
    45, 83
η^6-bonding, 57
benzothiophenes
    adsorption modes of, 17
bimetallic
    complexes and HDS, 186
biphasic
    catalysis, liquid, 73
    catalysts for hydrogenolysis of BT, 132
bonding modes
    in HDN, 27
breathing
    sulfur, in HDS, 12
bridging
    sulfides and disulfides in hydrogen
        activation, 141
    thiolates and HDS, 135
    thiolates in desulfurization, 117
bridging thiolates
    and HDS, 190
carbazole
    metal complexes of, 161
carbonyl
    clusters in desulfurization, 114
catalysts
    biphasic for hydrogenation of
        benzothiophene, 83, 92
    Co-Mo, 5
    Co-Mo active sites, 8
    for HDN, 5
    for HDS, 5

hydrotreating, 2
    Ni-Mo, 5
    promoted Mo or W, 5
C-H bond
    activation by metal complexes, 96
chemisorption
    modes for thiophenes, 17
    of organonitrogen compounds, 26
    of organosulfur compounds, 16
    of thiophenes, 182
clusters
    complexes in HDS modeling, 114
    complexes in THT activation, 117
    Cp-Co-Mo-S complexes in HDS, 123
    metal carbonyl, in HDS, 114
C-N bond
    activation by metal complexes, 165
    activation of amines, by complexes, 174
    cleavage in complexes, 171
    hydrogenolysis by metal complexes,
        173
    rupture in HDN, 28
Co-Mo-S
    catalysts, HDS mechanism, 191
    model, 9
    phase and HDS activity, 9
    phases, 9
    site structure of, 186
    structure of sites, 10
    Topsøe's model, 9
Co-Mo-S clusters
    in HDS modeling, 123
    mechanism of thiol desulfurization, 124
coordination
    modes of thiophenes and HDS, 188
coordinative unsaturation
    and chemisorption, 16
    in metal complexes, 16
corner sites
    in Co-Mo-S catalysts, 10
Cp*Co complexes
    in C-S bond activation, 99
    in thiophene ring opening, 99

CpMo complexes
  in C-S bond activation, 99
C-S bond
  activation and S-bonding, 188
  activation by Cp*Co complexes, 99
  activation by CpMo complexes, 100
  activation by CpRu clusters, 116
  activation by electron rich complexes,
    97
  activation by electron rich metal
    complexes, 110
  activation by Fe carbonyl clusters, 114
  activation by metal complexes, 95, 96
  activation in THT by metal clusters, 117
  activation vs. C-H activation by
    (triphos)M complexes, 108
  activation vs. C-H activation by CpRh
    complexes, 98
  activation vs. C-H activation by Tp*Rh
    complexes, 99
  activation, of THT, 108
  breaking by heteronuclear complexes,
    120
  breaking by nucleophilic attack, 118
  breaking vs. hydrogenation, 185
  cleavage on dinuclear Mo complexes,
    141
  energies, 96
  homolytic cleavage of, 125
  hydrogenolysis, 18
C-S bond activation
  by W complexes, 100
CUS
  and catalytic activity, 185
  and chemisorption, 16
  and HDS activity, 12
cyclopentadienyl
  complexes in C-S activation, 97
DBT
  4,6-Me$_2$, HDS of by Pt complexes, 113
deep
  desulfurization, 2, 4
  refining, 1

deep desulfurization, 20
dehydrodesulfurization
  of thiophene, 18
denitrogenation
  of pyrrole, homogeneous, 170
desulfurization
  deep, 4, 20
  multimetallic effects for, 120
  of 2,5-dihydrothiophene by Mo
    complexes, 141
  of tetrahydrothiophene by Os clusters,
    117
  of tetrahydrothiophene by W
    complexes, 118
  of thiols by Co-Mo-S clusters,
    mechanism, 124
  of thiophene by dinuclear Ir complexes,
    117, 118
dibenzothiophene
  4,6-dimethyl, HDS of, 17, 20
  4,6-dimethyl, HDS of by Ni complexes,
    113
  4,6-dimethyl, hydrogenation in HDS,
    62, 84
  4,6-dimethyl, ring opening by Pt
    complexes, 111
  4,6-dimethyl, η^6-bonding and HDS, 58
  4,6-Me$_2$, desulfurization by Ni and Pt
    complexes, 183
  4-Me, HDS of by Ni complexes, 113
  HDS, 110
  HDS of, 22
  homogeneous HDS of by Ir complexes,
    132
  hydrogenolysis by Pt complexes, 110
  η^6-bonding, 57
dibenzothiophenes
  adsorption modes of, 17
  HDS of by Ni complexes, 111
  HDS of by Pt complexes, 111
diesel
  sulur content allowed, 3
dihydrothiophene

desulfurization of, by dinuclear Mo
    complexes, 141
disulfido
    ligands and hydrogen activation by
        complexes, 143
edge sites
    in Co-Mo-S catalysts, 10
electron rich
    complexes and C-S activation, 134, 188
    metal complexes in C-S activation, 97
fuels
    quality of, requirements, 181
gasoline
    sulfur content allowed, 3
H₂S
    coordinated, in metal complexes, 145
    decomposition of on Pd complexes, 150
    metal complexes of, 145
    oxidative addition of, 146
    reactions with metal complexes, 145
    Ru complexes of, 148
hapticity
    of thiophenes, 35
HDN, 25
    bonding modes in, 27
    catalysts for, 26
    Hoffman degradation in, 29
    mechanisms of, 28
    nucleophilic substitution in, 29
    organometallic modelling of, 153
    organometallic models of, 31
HDN modeling
    common substrates, 153
HDS
    catalysts for, 5
    definition of, 3
    mechanisms, 182
    of 4,6-dimethyldibenzothiophene, 20
    of 4,6-Me₂DBT by Pt complexes, 114
    of benzothiophene, reaction pathways,
        21
    of dibenzothiophene by Pt complexes,
        110

of dibenzothiophene, reaction pathways,
    22
of dibenzothiophenes, mechanisms of,
    23
of thiophene, concerted mechanism, 18
of thiophene, reaction pathways, 18
of thiophenes, homogeneous, 133
stepwise of thiophene by Ir(triphos)⁺,
    102
heterobimetallic
    complexes in HDS, 121
heterogenized
    catalysts, 132
heterolytic
    activation assisted by sulfide ligand, 144
    activation of hydrogen, 105
    activation of hydrogen, 103
    activation of hydrogen by Ti=S
        complexes, 143
    activation of hydrogen on complexes,
        186
    hydrogen activation, 139
    hydrogen activation by a Ti=S complex,
        143
    splitting of H₂ by Rh-triphos, 142
Hoffmann degradation
    in HDN, 29
Homogeneous
    catalytic HDS, 126, 183
    catalytic HDS by Ir complexes, 132
    catalytic HDS by metal complexes, 127
    catalytic HDS of thiophenes, 133
    catalytic hydrogenolysis, 126
    denitrogenation of pyrrole, 170
    hydrogenolysis of pyridine, 174
homogeneous catalysis
    and hydrogen activation, 15
homolytic
    activation of hydrogen on complexes,
        186
    cleavage of C-S bonds, 125
    hydrogen activation, 139
hydride

Ni complexes in HDS, 113
hydrodenitrogenation, 25
    organometallic modeling of, 153
Hydrodesulfurization
    by metal carbonyl clusters, 114
    definition of, 3
    homogeneous catalytic, 126
    homogeneous catalytic of DBT by Ir
        complexes, 132
    homogeneous catalytic of DBT by
        triphos Ir complexes, 127
    mechanisms of, 7
    of 4,6-dimethyldibenzothiophene by Ni
        complexes, 113
    of benzothiophene by Rh-W complex,
        121
    of dibenzothiophene by Ni complexes,
        111
    of dibenzothiophenes by Pt complexes,
        111
    of thiophene by Co-Mo-S clusters, 123
Hydrogen
    activation by dinuclear Mo and Re
        complexes, 140
    activation by dinuclear Re complexes,
        142
    activation by metal complexes, 15, 138
    activation in homogeneous catalysis, 15
    activation mechanisms, 138, 186
    activation of, 14
    activation on sulfide ligands, 183
    heterolytic activation, 139
    heterolytic activation by dinuclear Mo
        complexes, 141
    heterolytic activation of, 14, 105
    heterolytic activation on RuS$_2$, 144
    homolytic activation of, 14
    homolytic activation on dinuclear Mo
        complexes, 140
    intramolecular heterolytic activation of,
        by Ti=S complex, 143
    oxidative addition of, 15
hydrogenation

    during HDN, 28
    in HDN, 84
    in HDS, 62
    in hydrotreating, 62
    of aromatics and arene bonding, 73
    of aromatics by Cp and arene
        complexes, 69
    of aromatics by Nb and Ta complexes,
        72
    of aromatics, active Co-allyl species, 68
    of aromatics, by metal phosphine
        complexes, 71
    of aromatics, by Ru clusters, 71
    of aromatics, Co-allyl catalysts, 66
    of aromatics, homogeneous, 63
    of aromatics, homogeneous catalysts
        for, 63
    of aromatics, homogeneous vs.
        heterogeneous, 64
    of aromatics, Rh-allyl catalysts, 69
    of benzene by Co-allyl complexes, 68
    of benzothiophene as pretreatment for
        HDS, 83
    of benzothiophene by (triphos)Ru
        complexes, 81
    of benzothiophene by CpRh complexes,
        79
    of benzothiophene in HDS, 21
    of benzothiophene, homogeneous, 78
    of benzothiophene, homogeneous
        catalysts for, 78
    of dibenzothiophene in HDS, 22
    of N-heterocycles, 31, 84
    of N-heterocycles, homogeneous
        catalysts for, 85
    of organosulfur compounds, 18
    of quinoline, mechanisms, 86
    of thiophene by Ir complexes, 75
    of thiophene by Ir complexes,
        mechanism, 76
    of thiophene in HDS, 20
    of thiophenes in HDS, 75
    vs. C-S bond breaking, 185

vs. hydrogenolysis by Ru complexes, 129

vs. hydrogenolysis on Re complexes, 108

hydrogenolysis

catalytic, of thiophenes by Rh complexes, 132

homogeneous catalytic, 126

homogeneous catalytic by triphos Ir and Rh complexes, 127

of 2,3-dihydrobenzothiophene by Ir complexes, 107

of benzothiophene by (triphos)Ru complexes, 82

of benzothiophene catalyzed by Ru complex, 129

of benzothiophene in HDS, 21

of BT, biphasic, 132

of C-S bonds, 18

of pyridine, homogeneous, 173

of sulfides by dinuclear Re complexes, 142

of thiophene by dinuclear Ir complexes, 118

of thiophene by Re complexes, 108

of thiophene C-S bonds, 18

stepwise of benzothiophene by Ir complexes, 103

vs. hydrogenation by Ru complexes, 129

hydrotreating, 3, 7

indole

metal complexes of, 158

indoles

reactions of, in metal complexes, 168

indoline

metal complexes of, 161

indolyl

metal complexes of, 160

Ir

complex as homogeneous HDS catalyst, 132

mechanism

for HDS on Co-Mo-S catalysts, 191

of BT hydrogenolysis, 128

of thiol HDS by Co-Mo-S clusters, 124

mechanisms

HDS, concerted, 14

of HDS, 7

metal complexes

of thiophenes, 182

metal hydrides

and HDS activity, 15

in HDS, 14

metal phosphine complexes

in ring opening of thiophenes, 101

metal sulfides

trends in HDS activities, 5

metal-arene bonding

and hydrogenation of aromatics, 73

multimetallic

effects for HDS, 120

naphtha

prehydrogenation by metal complexes, 83

two stage process for HDS of, 84

N-H bond

activation in complexes, 169

nucleophilic substitution

in HDN, 29

organonitrogen compounds

HDN of, 26

reactions of in metal complexes, 168

organosulfur compounds

chemisorption of, 16

hydrogenation of, 18

in petroleum, 3

oxidation states

and HDS activity, 185

oxidative addition

of C-S bonds to metal centers, 96

Pd

catalysts for decomposition of H_2S, 150

Petroleum

composition of, 3

phosphine

metal complexes in C-S activation, 101
platinum
    complexes in C-S activation, 108
poisoning
    and chemisorption, 17
    of HDS by olefins, 134
    of HDS sites by aromatics, 190
    of HDS sites by olefins, 189
polypyrazolylborate
    metal complexes in C-S activation, 99
promoter
    atoms and adsorption, 16
pyridine
    HDN of, 26
pyridines
    binding modes in complexes, 162
    metal complexes of, 162
    reactions of, in metal complexes, 171
pyrrole
    binding modes in complexes, 154
    HDN of, 26
    ligands on metal clusters, 158
pyrroles
    metal complexes of, 154
    reactions of in metal complexes, 168
pyrrolyl
    binding modes in complexes, 154
    metal complexes of, 154
quinoline
    binding modes in complexes, 166
    homogeneous hydrogenation of, 86
    mechanisms of HDN, 28
quinolines
    binding modes in complexes, 162
    metal complexes of, 162
Rh
    homogeneous catalysts for
        hydrogenolysis of thiophenes, 132
Rh-W complexes
    in desulfurization, 121
ring opening
    of 4,6-Me$_2$DBT by Pt complexes, 113
    of thiophene by CpMo complexes, 99

of thiophenes by bimetallic species, 122
of thiophenes by metal complexes, 96
ring-opening
    of benzothiophene by dinuclear Re
        complexes, 118
    reactions by Pt phosphine complexes,
        108
Ru
    complex in catalytic hydrogenolysis of
        BT, 129
ruthenium
    clusters in desulfurization, 116
S-bonding
    of thiophenes to metals, 17
sites
    coordinatively unsaturated, 184
    coordinatively unsaturated and HDS, 12
sulfide
    ligands in hydrogen activation, 183
sulfido
    ligand, terminal, 112
    ligands and hydrogen activation by
        complexes, 143
sulfonated
    phosphines in liquid biphasic HDS, 132
sulfur
    compounds in fuels, 3
    in petroleum, 3
    ligand, metal complexes, 139
    mobility in desulfurization, 126
    surface atoms, and hydrogen activation,
        139
sulfur extrusion
    in HDS of dibenzothiophenes, 22
surface
    sulfur and hydrogen activation, 139
tetrahydroquinoline
    metal complexes of, 167
tetrahydrothiophene
    desulfurization by Os clusters, 117
    desulfurization by W complexes, 117
thiols

desulfurization by Co-Mo-S clusters, 124

thiophene
bonding capacity, 35
chemisorption modes, 17
coordinated, tilt angle, 38
desulfurization by dinuclear Ir complexes, 118
HDS by Co-Mo-S clusters, 123
HDS of, 18
HDS, concerted mechanism, 18
homogeneous hydrogenation of, 75
homogeneous hydrogenation, mechanism, 76
hydrogenation by Ir complexes, 42
hydrogenolysis and HDS by dinuclear Ir complexes, 118
hydrogenolysis by Re complexes, 108
mechanism of HDS, 56
metal complexes, calculations, 38
ring opening by CpMo complexes, 99
ring opening by CpRh complexes, 97
ring opening by metal complexes, 96
S-bonding and ring opening, 41
S-bonding and ring opening by CpRh complexes, 99
stepwise HDS by Ir complexes, 102
$\eta^2(C=C)$ bonding and C-H activation, 46
η^4-bonded, 47
η^4-bonded, reactions of, 49
η^4-bonded, ring opening, 49
η^5-bonded, 50
η^5-bonded, reactions of, 55
η^5-bonding, calculations, 51

thiophenes
as olefin-like ligands, 43
chemisorption of, 16

coordination modes in metal complexes, 35
desulfurization by $Fe_3(CO)_{12}$, 114
desulfurization by $Ru_3(CO)_{12}$, 115
homogeneous catalytic HDS of, 126
homogeneous hydrogenation of, 75
Mo complexes, 42
ring opening by bimetallic species, 122
ring opening by metal-triphos complexes, 101
ring opening by Pt phosphine complexes, 108
S-bonded, reactions of, 40
η^1S complexes, 37
η^1S-bonded, 37
$\eta^2(C=C)$ bonding, 43
η^5-adsorption and HDS, 57

two site dilemma, 10, 74, 82, 131

vacancies
and HDS activities, 12
and HDS activity, 7, 16
latent, 14, 184
latent in HDS, 126

η^1S-bonding
and C-S activation, 134, 188

η^1S-coordination
and C-S activation, 98

$\eta^2(C,S)$
thiophene ring opening products, 97

$\eta^2(C=C)$ bonding
and hydrogenation, 188

$\eta^2(N,C)$ coordination
of pyridines, and HDN, 165

$\eta^3(S,C=C)$ bonding
and multi point adsorption, 104
in C-S bond activation, 103

Catalysis by Metal Complexes

Series Editors:
R. Ugo, *University of Milan, Milan, Italy*
B.R. James, *University of British Colombia, Vancouver, Canada*

1. F.J. McQuillin: *Homogeneous Hydrogenation in Organic Chemistry.* 1976
   ISBN 90-277-0646-8
2. P.M. Henry: *Palladium Catalyzed Oxidation of Hydrocarbons.* 1980
   ISBN 90-277-0986-6
3. R.A. Sheldon: *Chemicals from Synthesis Gas.* Catalytic Reactions of CO and H_2.
   1983   ISBN 90-277-1489-4
4. W. Keim (ed.): *Catalysis in C_1 Chemistry.* 1983   ISBN 90-277-1527-0
5. A.E. Shilov: *Activation of Saturated Hydrocarbons by Transition Metal Complexes.*
   1984   ISBN 90-277-1628-5
6. F.R. Hartley: *Supported Metal Complexes.* A New Generation of Catalysts. 1985
   ISBN 90-277-1855-5
7. Y. Iwasawa (ed.): *Tailored Metal Catalysts.* 1986   ISBN 90-277-1866-0
8. R.S. Dickson: *Homogeneous Catalysis with Compounds of Rhodium and Iridium.*
   1985   ISBN 90-277-1880-6
9. G. Strukul (ed.): *Catalytic Oxidations with Hydrogen Peroxide as Oxidant.* 1993
   ISBN 0-7923-1771-8
10. A. Mortreux and F. Petit (eds.): *Industrial Applications of Homogeneous Catalaysis.*
    1988   ISBN 90-2772-2520-9
11. N. Farrell: *Transition Metal Complexes as Drugs and Chemotherapeutic Agents.*
    1989   ISBN 90-2772-2828-3
12. A.F. Noels, M. Graziani and A.J. Hubert (eds.): *Metal Promoted Selectivity in Organic
    Synthesis.* 1991   ISBN 0-7923-1184-1
13. L.I. Simándi (ed.): *Catalytic Activation of Dioxygen by Metal Complexes.* 1992
    ISBN 0-7923-1896-X
14. K. Kalyanasundaram and M. Grätzel (eds.): *Photosensitization and Photocatalysis
    Using Inorganic and Organometalic Compounds.* 1993   ISBN 0-7923-2261-4
15. P.A. Chaloner, M.A. Esteruelas, F. Joó and L.A. Oro: *Homogeneous Hydrogenation.*
    1994   ISBN 0-7923-2474-9
16. G. Braca (ed.): *Oxygenates by Homologation or CO Hydrogenation with Metal
    Complexes.* 1994   ISBN 0-7923-2628-8
17. F. Montanari and L. Casella (eds.): *Metalloporphyrins Catalyzed Oxidations.* 1994
    ISBN 0-7923-2657-1

18. P.W.N.M. van Leeuwen, K. Morokuma and J.H. van Lenthe (eds.): *Theoretical Aspects of Homogeneous Catalisis.* Applications of *Ab Initio* Molecular Orbital Theory. 1995 ISBN 0-7923-3107-9

19. T. Funabiki (ed.): *Oxygenases and Model Systems.* 1997 ISBN 0-7923-4240-2

20. S. Cenini and F. Ragaini: *Catalytic Reductive Carbonylation of Organic Nitro Compounds.* 1997 ISBN 0-7923-4307-7

21. A.E. Shilov and G.P. Shul'pin: *Activation and Catalytic Reactions of Saturated Hydrocarbons in the Presence of Metal Complexes.* 2000 ISBN 0-7923-6101-6

22. P.W.N.M. van Leeuwen and C. Claver (eds.): *Rhodium Catalyzed Hydroformylation.* 2000 ISBN 0-7923-6551-8

23. F. Joó: *Aqueous Organometallic Catalysis.* 2001 ISBN 1-4020-0195-9

24. R.A. Sánchez-Delgado: *Organometallic Modeling of the Hydrodesulfurization and Hydrodenitrogenation Reactions.* 2002 ISBN 1-4020-0535-0

KLUWER ACADEMIC PUBLISHERS – BOSTON / DORDRECHT / LONDON

* Volume 1 is previously published under the Series Title:
Homogeneous Catalysis in Organic and iNorganic Chemistry